People and Forest – Policy and Local Reality in Southeast Asia, the Russian Far East, and Japan

Institute for Global Environmental Strategies

VOLUME 3

People and Forest – Policy and Local Reality in Southeast Asia, the Russian Far East, and Japan

Edited by

M. INOUE

The University of Tokyo,
Tokyo, Japan

and

H. ISOZAKI

Iwate University,
Morioka, Japan

Contributing Authors (alphabetical order): T. Fujiwara, K. Harada, A. Hayama, K. Hyakumura, M. Inoue, H. Isozaki, K. Kakizawa, T. Khotpathoum, N. Kitamura, K. Komatsu, S. Matsumoto, S. Nagata, M. Nanang, J. Pulhin, P. Pulhin, D.D. Sam, S. Sasaki, J. Sato, Y. Seki, A. Sheingauz, N. Shiraishi, S. Tachibana, L.Q. Trung, M. Yamane, Y. Yokota

KLUWER ACADEMIC PUBLISHERS
DORDRECHT / BOSTON / LONDON

A C.I.P. Catalogue record for this book is available from the Library of Congress.

ISBN 1-4020-1684-0

Published by Kluwer Academic Publishers,
P.O. Box 17, 3300 AA Dordrecht, The Netherlands.

Sold and distributed in North, Central and South America
by Kluwer Academic Publishers,
101 Philip Drive, Norwell, MA 02061, U.S.A.

In all other countries, sold and distributed
by Kluwer Academic Publishers,
P.O. Box 322, 3300 AH Dordrecht, The Netherlands.

Cover Images:
Photo 1
Rice terraces and a small village in Gunung Halimun National Park, West Java, Indonesia.
Photo Credit: Kazuhiro Harada.
Photo 2
Rice harvesting near Gunung Halimun National Park, West Java, Indonesia.
Photo Credit: Kazuhiro Harada.
Photo 3
Logging equipment clearing a tropical forest in East Kalimantan, Indonesia.
Photo Credit: Makoto Inoue.
Photo 4
A mountain pine forest in the Maya River basin, Khabarovskiy Krai in the Russian Far East.
Photo Credit: Dr. Alexander L. Antonov.

Printed on acid-free paper

Printed in the Netherlands.

TABLE OF CONTENTS

PART 3: LOCAL REALITIES IN FOREST USE AND MANAGEMENT

CONTRIBUTORS

Takashi FUJIWARA, Vice-President (Research Planning & General Affairs), Forestry and Forest Products Research Institute, 1 Matsunosato, Tsukuba City, Ibaraki Prefecture, 305-8687, Japan.
fujiwara@t.nifty.jp

Kazuhiro HARADA, Research Associate, Forest Conservation Project, Institute for Global Environmental Strategies, 2108-11 Kamiyamaguchi, Hayama, Kanagawa, 240-0115, Japan.
harada@iges.or.jp

Atsuko HAYAMA, Division of Southeast Asian Area Studies, Graduate School of Asian and African Area Studies, Kyoto University, 46 Yoshida Shimoadachi, Sakyo, Kyoto, 606-8501, Japan.
hayama@kais.kyoto-u.ac.jp

Kimihiko HYAKUMURA, Research Associate, Forest Conservation Project, Institute for Global Environmental Strategies, 2108-11 Kamiyamaguchi, Hayama-machi, Kanagawa, 240-0115, Japan.
hyakumura@iges.or.jp

Makoto INOUE, Associate Professor, Department of Forest Science, Graduate School of Agricultural and Life Sciences, The University of Tokyo, 1-1-1 Yayoi, Bunkyo-ku, Tokyo, 113-8657, Japan.
minoue@fr.a.u-tokyo.ac.jp

Hiroji ISOZAKI, Professor. Faculty of Humanities and Social Sciences, Iwate University, 3-18-34 Ueda, Morioka, Iwate, 020-0066 Japan.
isozaki@iwate-u.ac.jp

Hiroaki KAKIZAWA, Associate Professor, Department of Forest Science, Graduate School of Agricultural, Hokkaido University, Kita 9 Nishi 9, Kita-ku, Sapporo, 060-8589, Japan.
kaki@for.agr.hokudai.ac.jp

Noriyoshi KITAMURA, Advisor on Forestry Strategy 2020, Department of Forestry, P.O. Box 2932, Vientiane, Lao P.D.R.
nori@laotel.com

Thananh KHOTPATHOUM, Lecturer, Faculty of Forestry, National University of Laos, P.O. Box 7322, Vientiane, Lao P.D.R.
Thanan_ko@hotmail.com

Kiyoshi KOMATSU, Research Associate, Forest Conservation Project, Institute for Global Environmental Strategies, 2108-11, Kamiyamaguchi, Hayama-machi, Miura-gun, Kanagawa Prefecture, 240-0115, Japan.
komatsu@iges.or.jp

Martinus NANANG, Research Fellow, Forest Conservation Project, Institute for Global Environmental Strategies, 2108-11, Kamiyamaguchi, Hayama-machi, Miura-gun, Kanagawa Prefecture, 240-0115, Japan.
nanang@iges.or.jp.

Shin NAGATA, Professor of Forest Policy, Department of Forest Science, Graduate School of Agricultural and Life Sciences, The University of Tokyo, 1-1-1 Yayoi, Bunkyo-ku, Tokyo, 113-8657, Japan.
nagata@fr.a.u-tokyo.ac.jp

Satoru MATSUMOTO, Secretary General, Mekong Watch, 2F Maruko Bldg., 1-20-6 Higashi-Ueno, Taito-ku, Tokyo, 110-0015, Japan.
satoru-m@msi.biglobe.ne.jp

Juan M. PULHIN, Associate Professor and Associate Dean, College of Forestry and Natural Resources, University of the Philippines, Los Baños College, Laguna 4031, Philippines.
jpulhin@laguna.net

Perlyn M. PULHIN, Writer/Research Associate, Ginintuang Masaganang Ani Corn Program-Department of Agriculture, Elliptical Road, Diliman, Quezon City, 1104, Philippines.
perlyn96@yahoo.com

Dr. Do Dinh SAM, Professor, Senior Researcher, Research Center for Forest Ecology and Environment, Forest Science Institute of Vietnam, Dong Ngac, Tu Liem, Hanoi, Vietnam.
ddsam@netnam.vn

Jin SATO, Associate Professor, Institute of Environmental Studies, Graduate School of Frontier Sciences, The University of Tokyo, 7-3-1 Hongo, Bunkyo-ku, Tokyo, 113-0033, Japan.
satoj@k.u-tokyo.ac.jp

Alexander SHEINGAUZ, Deputy Director, Economic Research Institute, Far Eastern Division, Russian Academy of Sciences, 153 Tikhookeanskaya Street, Khabarovsk, 680042, Russia.
sheingauz@ecrin.ru *or* sheingauz@mail.kht.ru

Shiro SASAKI, Professor, Department of Cultural Research, National Museum of Ethnology, 10-1 Senri Expo-Park, Suita, Osaka, 565-8511, Japan.
ssasaki@idc.minpaku.ac.jp

Yoshiki SEKI, Research Assistant, Research Institute of Current Chinese Affairs, Waseda University, Bokusha Bldg., 2F, 1-101 Totsuka, Shinjuku-ku, Tokyo, 169-0071, Japan.
yoshiki@m6.people.or.jp

Norihiko SHIRAISHI, Associate Professor, Department of Forest Science, Graduate School of Agricultural and Life Sciences, The University of Tokyo, 1-1-1 Yayoi, Bunkyo-ku, Tokyo, 113-8657, Japan.
siraishi@fr.a.u-tokyo.ac.jp

Satoshi TACHIBANA, PhD, Senior Researcher, Forestry Socio-Economics Laboratory, Forestry and Forest Products Research Institute (FFPRI), 1 Matsunosato, Tsukuba City, Ibaraki Prefecture, 305-8687, Japan.
binn@ffpri.affrc.go.jp

Le Quang TRUNG, Researcher, Economic Research Division, Forest Science Institute of Vietnam, Dong Ngac, Tu Liem, Ha Noi, Vietnam.
lqtfsiv@yahoo.com

Masanobu YAMANE, Senior Research Fellow, Research Division, Kanagawa Prefectural Natural Environment Conservation Center, 657 Nanasawa, Atsugi, Kanagawa, 243-0121, Japan.
myamane@agri.pref.kanagawa.jp

Yasuhiro YOKOTA, Researcher, Forest Resource Planning Group, Tohoku Research Center, Forestry and Forest Products Research Institute, 92-25 Nabeyashiki, Shimo-Kuriyagawa, Morioka, Iwate, 020-0123, Japan.
yokotaya@ffpri.affrc.go.jp

INTRODUCTION: APPROACHES TO STRATEGIC RESEARCH ON SUSTAINABLE FOREST MANAGEMENT

Forests are important, not just for the production of timber, but also for many social and ecological functions such as the conservation of biodiversity, the supply of water, and the prevention of global warming. The natural characteristics of global forests differ widely, and even for similar forests, appropriate management methods vary from place-to-place because local people have different perceptions and uses of forests. In addition, some countries place more national importance than others on the production and export of timber, making it difficult to engage in scientific and objective discussions of alternative policies and actions based on common global criteria.

Forest conservation was one of the key issues at the 1992 Earth Summit. Although countries adopted Agenda 21, which called for actions to prevent deforestation, and the Forest Principles, the Earth Summit failed to conclude with the creation of a Forest Convention. After the Earth Summit, a number of international initiatives emerged, such as the Intergovernmental Panel on Forests (IPF), the World Commission of Forests and Sustainable Development (WCFSD), and others, in order to find ways to halt worldwide deforestation and the degradation of all types of forestlands. At the Special Session of the General Assembly of the United Nations to Review and Appraise the Implementation of Agenda 21, in June 1997, it was agreed that work should be continued in order to reach an international consensus on forest conservation. The task of shaping concrete actions by the international community was given to the Intergovernmental Forum on Forests (IFF), which was expected to create an international solution for this matter critical to all human societies. But the IFF couldn't achieve its goal and the task was handed over to the United Nations Forum on Forests (UNFF).

The International Tropical Timber Agreement (ITTA), revised in 1994, set out in its basic objectives a goal that the international tropical timber trade should only deal with wood from sustainably-managed forests by the year 2000, and encouraged international cooperation to facilitate the achievement of this goal in developing countries. In 1995, criteria and indicators for the sustainable management of temperate and boreal forests were adopted, but these agreements only set out general obligations and guidelines, and did not specify actual methods for domestic implementation.

In the past, most debates regarding various aspects of forests tended to focus on the forest sector and the direct causes of deforestation and forest degradation, and not on the cross-sectoral aspects or the underlying causes linked to them, such as the connection between forests and societies. After the UN Special Session, non-governmental organizations took the initiative on one of the most pressing agendas and started researching the underlying causes of deforestation and forest degradation.

In recent years, the world's forests have been affected by over-harvesting, over-grazing, pests and diseases, high global temperatures, floods, droughts, storms, air pollution, and forest fires, as well as economic crises in Asia and other regions—all leading to an overall decrease in the world's forest cover. The forests of Asia, in particular, have been strongly impacted. A number of initiatives have suggested forest policy reforms, and the need for the sustainable management of forests has been widely recognized and encouraged. But because implementation of reforms at the local level has been insufficient, it is imperative that local people begin to effectively participate in forest planning and management as well as in protected-area management.

The Forest Conservation Project, launched in April 1998 by the Institute for Global Environmental Strategies (IGES), has carried out research activities on forest strategies, including policy analysis and on-site surveys. This book gives an overview of the project's research activities in its first three-year phase (April 1998–March 2001). Since viable forest strategies work best when based on the involvement of local people, this report is addressed to stakeholders in the communities of the relevant countries, including local people and authorities, community-based organizations, experts, national agencies, and international institutions.

The first phase of the IGES Forest Conservation Project attempted to identify principles and measures for sustainable forest management, based on experiences in the Asia-Pacific region. Extensive studies were carried out within a framework of four interrelated sub-themes by in-house research staff, visiting researchers, and many outside collaborators. Based on the project's diversified research activities, including four international workshops and three local policy dialogues, the research team reached many useful conclusions related to each sub-theme. The structural analysis of forest destruction confirmed that its root causes can be traced to an insufficient base of local participation and community rights, the impact of market forces under an incomplete market system, a predominant forest development paradigm with industrial emphasis, and economic and political instabilities.

Researchers categorized existing participatory forest management into several types, based on the main actors involved and the legal status of forestland and activities, and made policy recommendations based on their examination of internal and external constraints on participation. They conducted time-series economic analyses of the timber trade in the Asian market, and collected data for space-equivalent analysis of the timber trade. They also focused on international legal measures related to sustainable forest management, international processes for policy dialogue on forest issues, and domestic legal and administrative measures relating to participatory forest management. The sub-theme researchers elaborated the principles and measures for sustainable forest management with special reference to local participation in sustainable forest management, in cooperation with other sub-themes. With respect to project implementation, valuable networks were constructed with researchers, non-governmental organizations, local communities, and government officials.

This book is based on the result of the first phase of IGES's Forest Conservation Project and parts of the second phase (April 2001—March 2004). Part 1 provides case analyses of the major causes of forest loss and degradation, and examines necessary countermeasures—especially legal, administrative, economic, and political measures. Based on these, the project reaffirmed the importance and usefulness of participatory measures for sustainable forest management. These can improve the processes for the development of management plans and economic policies, and the implementation of laws and regulations. Chapter 4 recommends a series of supportive measures for the effective application and implementation of forest management, and concludes with a set of specific policy recommendations for local participation, based on field research activities, for several individual countries in the Asian region. Part 2 covers relevant policies and legislation on forest management in Asian countries, and Part 3 illustrates the actual situation of the forests and the local communities around them from the local perspective.

In the IGES Forest Conservation Project's second phase, the project team is working to develop and propose detailed guidelines and measures for both local and national levels, based on the first-phase recommendations, for three targets: Indonesia, Laos, and the Russian Far East.

Hiroji Isozaki

PART 1

FRAMEWORK FOR SUSTAINABLE FOREST MANAGEMENT

Chapter 1

UNDERLYING CAUSES OF FOREST LOSS IN THE ASIA-PACIFIC REGION

Masanobu Yamane
Kanagawa Prefecture Natural Environment Conservation Center

1. INTRODUCTION

During the past decade, especially after the 1992 Earth Summit, forest loss in many regions received considerable attention and led to numerous initiatives by governments and intergovernmental agencies at various levels, from national to global. The Food and Agriculture Organization of the United Nations (FAO) noted that it was a time of strengthened international cooperation (FAO 2001), and it is no exaggeration to say that such active and frequent worldwide dialogue on forest conservation was an entirely new, historic event. Unfortunately, these efforts failed to make substantial progress in preserving forests, especially in tropical countries.

In the 1980s, when tropical forests were rapidly disappearing and the crisis occupied media and public attention, various researchers began to explore the mechanism of tropical forest degradation, and by the middle of the 1990s had produced meaningful results. The initial phases of this research focused on the proximate causes of tropical forest destruction and identified activities, such as shifting cultivation, expansion of pastures and croplands, and export-oriented timber extraction, as primary causes (i.e., Myers 1980; Allen and Barnes 1985; Kummer 1992). Amelung and Deihl (1992) reported that agricultural activities, including shifting cultivation, accounted for more than 90 percent of tropical forest loss, while forestry-related activities, including timber extraction itself, accounted for less than 10 percent.

Once the main culprits, or leading proximate causes, of forest conversion in tropical countries were identified, analysis of the indirect causes of unsustainable forest development—agricultural expansion and timber extraction, for example—became the focus. Various issues were analyzed, such as institutional factors, including land tenure systems and logging concession regimes in tropical forests (i.e., Repetto and Gillis 1988; Southgate 1990), political factors (Kummer 1992), and economic factors (i.e.,

M. Inoue and H. Isozaki (eds.), *People and Forest—Policy and Local Reality in Southeast Asia, the Russian Far East and Japan, 3–31.*

Brown and Pearce 1994; Rundle and Roper 1996). In Japan, which imported a large amount of tropical timber from tropical countries in southeast Asia, research on the mechanisms of tropical deforestation progressed significantly during the 1990s (i.e., Nectoux and Kuroda 1989; Nagata et al. 1994).

The academic research mentioned above contributed greatly to clarifying the complex structure of deforestation in tropical countries. These results also indicate to international society that taking comprehensive, cross-sectoral countermeasures over policies on economic development, trade, agriculture, forests, land development, and human rights are indispensable for overcoming the problems.

In contrast to the progress of research on the mechanisms of tropical deforestation, the pace of deforestation did not decelerate, and thus international society continued calling for stronger measures to be put in place that reflect the local realities. The 1990s was also a decade of worsening forest degradation in Europe, caused by air pollution, and a time of intensified commercial forest extraction of boreal forests in Russia and North America. In addition to the degradation of these non-tropical forests, the role of forests in preventing global warming garnered significant attention, renewing public concern for conserving the world's forests.

Research activities to address the direct and indirect causes of forest loss and generate possible forest conservation measures intensified again in 1995, when the United Nations Commission on Sustainable Development established the Intergovernmental Panel on Forests (IPF) to address a wide range of forest-related issues, including one element, entitled "Underlying Causes of Deforestation and Forest Degradation." In early 1997, the IPF produced a final report containing a set of 135 proposals for action that governments have agreed to implement. At the first meeting of the Intergovernmental Forum on Forests (IFF), held October 1997 in New York, the delegates decided to include analysis of the underlying causes of deforestation and forest degradation in the program of work.

The Forest Conservation Project of the Institute for Global Environmental Strategies (IGES) responded to these circumstances, and then launched a research team in 1998 to address the underlying causes of forest loss (deforestation and degradation) in the Asia-Pacific.

2. IGES'S APPROACH TO ADDRESS UNDERLYING CAUSES OF FOREST LOSS IN THE ASIA-PACIFIC

2.1 Definition of Terms

The term *forest loss* refers to not only the decline of forestland area (through replacement or deforestation) but also the decline of forest quality (modification or degradation of the forest itself), including tree health, ecosystems, and biodiversity. The term *deforestation* describes the complete, long-term removal of tree cover. In a definition taken from the FAO publication, *Forest Resources Assessment 1990: Tropical Countries*, deforestation refers to change of land use with depletion of tree-crown cover to less than 10 percent.

Change within the forest class from closed to open forest that negatively affects the stand or site and, in particular, lowers its production capacity, is termed *forest degradation*. Many activities that modify forests can be accurately described as forest degradation. In a joint publication by the UNECE (United Nations Economic Commission for Europe) and the FAO (UNECE and FAO 1992) the terms *replacement* and *modification* are used in place of *deforestation* and *degradation*. The term *replacement* means replacement of natural forest or other wooded land by another land use. The term *forest modification* refers to modifications that may be regressive (degradation) or progressive (recovery or enhancement). The frequency and intensity of human intervention determines the degree of human impact on forest loss. Continued forest uses, such as repeated over-harvesting of resources, may lead to significant damage. Extreme degradation can, of course, lead to total forest replacement with other uses. Thus we should consider how to use forest resources in a sustainable way to prevent further serious forest degradation.

Forest loss can be attributed to many different causes or factors. Some causes operate directly on the forest itself and are often easily recognizable in the field; these are referred to as "direct causes" (UN Economic and Social Council 1996). Behind these direct causes may lie a whole sequence of causes, each more indirect or remote than the preceding one; these are referred to as "underlying causes" (UN Economic and Social Council 1996).

2.2 Dialogical Approach

In this study we dealt with forest loss caused directly or indirectly by human activities observed in selected countries and areas in the southeast Asian sub-region (Indonesia, the Philippines) the Mekong River basin sub-region (Thailand, Laos, Vietnam, and Cambodia), and in northeast Asia (the southern part of the Russian Far East).

We avoided quantitative analysis of identified causes, since there are not sufficient reliable forest resource statistics available to clarify the causes of forest changes one-by-one and the causes of forest loss are mutually related, leading to difficulties in identifying individual contributions. Instead we employed the so-called dialogical approach to address the structure of forest loss in the Asia-Pacific, as described below.

In the first step, joint researchers familiar with the local reality in each region selected the recent major proximate causes of forest loss, which were identified through literature surveys and field surveys.

In the second step, the selected proximate causes were sorted by identifying the common elements among target regions, as well as the unique elements of each country and sub-region. We then analyzed each proximate cause of the forests changing from primeval to other land uses and examined the underlying causes by specifying the measures, purposes, and main actors involved in forest loss.

In the final step, the structure of forest loss in the study areas, focusing on underlying causes and examined through the procedure mentioned above, was discussed and validated through several policy dialogues with local forest authorities.

3. OVERVIEW OF FOREST LOSS IN THE ASIA-PACIFIC

The major proximate causes of forest loss in the target areas, which were identified by the procedure mentioned above, are listed in Table 1. In the following section, forest loss in the countries studied is briefly outlined prior to a synthetic analysis of forest loss in the region.

Table 1: Recent major proximate causes of forest loss in the target areas addressed.

Area	Recent major proximate causes
Philippines	Export-oriented unsustainable commercial logging, the failure of industrial plantations, frequent forest fires caused by local people, mining operations, forest conversion for agricultural expansion, upland farming, and government projects such as dams, land clearing for the landless.
Indonesia	Export-oriented unsustainable commercial logging, logging for domestic plywood industry, transmigration projects and paddy field development projects, non-traditional shifting cultivation, industrial tree plantations, frequent large-scale forest fires, and oil-palm plantation development.
Thailand	Logging, forest clearance for cash crops, shifting cultivation, rubber plantations, shrimp farming, land zoning, and expansion of protected areas.
Laos	Direct and indirect impacts of the Second Indochina War, land clearance for rice self-sufficiency, shifting cultivation, hydropower development, and commercial logging.
Vietnam	Direct and indirect impacts of the Second Indochina War, land clearance for rice self-sufficiency, in-country migration, coffee plantations, shifting cultivation, financial resources for the military from logging.
Cambodia	Land clearing for crops, logging under the protection of powerful people and the military.
Southern Russian Far East	Export-oriented unsustainable commercial logging and large-scale forest fires.

3.1 Indonesia[1]

Full-scale forest development in Indonesia started in the 1960s, fueled by several factors including a rapid increase in demand for tropical timber in consumer countries, the abundant forest resources then still remaining, the establishment of large-scale

[1] This part was extracted from discussion papers by Inoue (2000) and Okamoto (2000). The author also referred to Araya (1998) for the development of the plywood industry in the country.

logging operations with logging machines, plus an inflow of foreign investment for timber extraction. Towards the end of the 1950s, Japanese demand for hardwood logs as raw material for its wood industry had begun to soar, and production rose rapidly through the 1960s. The group of Shorea species known as *meranti*, abundant in the lowland forests, was the first target of demand. New technologies, especially the one-man chainsaw, first developed in the 1950s, coupled with transport improvements, such as outboard motors, bigger and more powerful trucks, four-wheel drive vehicles, bulldozers, and crawler tractors for hauling logs, were used to exploit these abundant timber resources and exponentially increased log production. When Indonesia's president at the time, Suharto, introduced a policy welcoming foreign investment under the country's first five-year economic development plan (1969–1973), full-scale, intensive forest development was essentially launched. Based on the Basic Forestry Law, the government provided foreign companies with preferential conditions and issued large-scale logging concessions to them. These policies opened up access to logging concessions to foreign companies and joint venture capital, and provided the foundation for export-oriented forest development in Indonesia by the tropical timber importing countries themselves. In consequence, by the end of 1969, there were more than 151 applications for logging concessions and the area slated for production reached more than 18.28 million hectares, equivalent to 43 percent of the country's forests. Under these conditions, all indices of log production, log exports, and export amounts increased sharply.

In 1970, immediately after the logging boom began, the Suharto government issued an order that required all companies with logging concessions (*hak pengusaha hutan* [HPH]) to establish processing plants for forest products, such as plywood, within three to ten years of commencing logging. The HPH companies started building plywood plants in 1973, and at first these were almost all producing solely for the domestic market, but a 1980 change in forest policy obligated the them to supply logs for domestic use. Restrictions on log exports gradually tightened, ending in a ban in 1986. Indonesia's development policy was switched from emphasizing the export of primary products, like oil, natural gas, and logs, to emphasizing exports of industrial products. A resulting effect of this was the rapid development after 1979 of the plywood industry into an export industry, with plywood exports of about 120,000 cubic meters growing thirty-fold in six years to over 3.5 million cubic meters in 1985 and then to 9.6 million cubic meters in 1993. As a result, in 1985, plywood outstripped the traditional export industries, such as coffee, tin, processed rubber products, and shrimp, and came out in third place behind oil and natural gas, with the biggest employment capacity among large and medium-sized industries.

In the 1990s, Indonesia's government energetically promoted industrial tree plantations (*hutan tanaman industri* [HTI]), which signified the planting of production forestland by business concerns holding an industrial tree plantation concession (*hak pengusaha hutan tanaman industri* [HPHTI]). Its purpose was to strengthen the domestic wood industry while promoting environmental conservation by planting trees in deforested areas. HTIs are areas used to grow trees to produce woodchips for pulp as well as for other purposes.

The development of oil palm plantations is a recent key cause of forest loss, not only because it has led to further deforestation but also due to its direct links to forest fires. During 1967 to 1977, the rate of expansion of oil palm plantations remained low. But since 1978 it increased by 21.7 percent for industrial private plantations and 2.9 percent for state-owned plantations. In 1996, the total area planted with oil palm in Indonesia was about 2.2 million hectares. The most recent figures suggest that there are now 2.4 million hectares of oil palm, of which state-run companies possess 443,000 hectares of older, productive plantings, small-holders have 824,000 hectares, and private companies have the rest—primarily new, immature plantations. By 1997 the total area under oil palm cultivation had reached 2.5 million hectares. Conglomerates now dominate the sector, with just eight of them owning land totaling 2.1 million hectares out of the 5.4 million hectares officially allocated for oil palm. When agricultural plantation is conducted, the forest is clearcut and then intentional burning is broadly conducted on the newly cleared land. This method of site preparation sparked out-of-control forest fires, caused by escaped fires, as well as the expansion of forest fires. In reaction to the runaway forest fires of 1997 and 1998, the government restricted the use of fire to clear land, but the number of new plantations is still expected to grow. The restrictions have probably had some impact, but many doubts remain about their real effectiveness.

3.2 The Philippines[2]

At the beginning of the twentieth century, 70 percent of the Philippines were covered by forests. Even just after World War II, an estimated 55 percent of the land was still forested. By 1996, however, forest cover had dropped to approximately 18.3 percent, according to government statistics, although included in this figure are secondary forests growing in logged-over areas already considerably degraded. Virgin forests, consisting mostly of mossy forests in the mountainous upland areas, account for no more than 7.0 percent of the land. The deforestation of natural forest occurred rapidly after the country gained independence. The rate of forest cover dropped sharply as the land was changed into secondary forest and meadow, while the structure of the forests deteriorated at the same time.

Full-scale commercial logging in the Philippines started in the era under American rule, during the 1910s. By 1930, timber exports to Japan exceeded exports to the United States. Upon restoration of trade relations with Japan after World War II, as a result of the San Francisco Peace Treaty, timber exports to Japan expanded again. At the beginning of the 1960s, Philippine log exports were the highest in the world, surpassing Canada and the former Soviet Union. In only half a century, however, logging to meet the overseas demand depleted all the timber resources that could be logged. After 1986, conditions in the Philippines declined to the extent that the country reversed its exporter status and became a net importer of timber from Malaysia and other countries.

Forest conversion to commercial ranching also caused large-scale deforestation. In the Philippines, the development of logged land for agricultural use was not legally

[2] This part was extracted from a discussion paper by Seki (2000).

sanctioned, but nonetheless, logged land was often used for commercial ranching. "Pasture fires" were a common means to open up a forest for commercial ranching. As a result of repeated fires, secondary forests were converted to grasslands covered with herbs, such as cogon (*Imperata cylindrica*), and vast grassland scenery emerged in the country. The fires from pasture burning spread beyond the limits of the commercial pastures and over a wide area of surrounding second-growth forests, with a devastating effect.

The logged-over areas of the Philippines gradually became grasslands because of the fires, which occur almost annually. The outbreak of fire was highly unlikely in the original natural tropical forest, where humidity levels can reach 90 percent or higher. After being logged, however, the forest's humidity drops and the incursion of fires into the forest becomes easier. In this sense, logging is an indirect cause of forest fires. Overall, human activities are, for one reason or another, the direct cause of fires in the upland areas.

The failure of industrial plantations also caused a certain amount of forest loss. The promotion of large-scale plantation projects started in the latter half of the 1980s, and with restrictions on local people's access to commercial plantation areas, land problems occurred anew. The industrial plantation projects, such as the Contract Reforestation Program (CRP) and Industrial Forest Management Agreements (IFMAs) that were promoted actively in the 1980s, have not produced good reforestation results due to resistance and conflicts with local people. In effect, these projects supported the modification of commercial logging sites into grassland.

3.3 Thailand[3]

Thailand has lost at least half of its forest cover since the early 1960s. Officially, about one-quarter of the country is still forested, but it is widely accepted that this is an overestimate.

The loss of forests in Thailand is associated largely with various development processes. Until 1989 logging concessions covered large parts of the forested areas that lay outside national parks and wildlife sanctuaries. Concern over the effects of legal and illegal logging led to a logging ban in 1989, after disastrous floods in southern Thailand led to heavy losses of life. These floods were attributed in part to timber extraction. Other development pressures that have led to widespread forest clearance include clearing land for planting cash crops, such as cassava, kenaf, and sugar cane.

Northeastern Thailand's more open, dry dipterocarp and savanna forests, and land previously forested with dipterocarps, as well as the dry and moist evergreen forests on the eastern and western edges of the Central Plains, have been largely cleared and the land is now used to grow cash crops. Road construction facilitated the clearing and provided for the cheap transport of products, while ambiguous tenure relations in forest reserve areas and market pressures further fueled land clearing, often in the wake of earlier logging operations. In northern Thailand, shifting cultivators have received much

[3] This part was extracted from a discussion paper by Hirsch (2000).

of the blame for the loss of the teak and mixed deciduous forest in upland areas; this has been a hotly contested issue for some time. Crop substitution programs often exacerbated the problem, as the cabbage and other cash crops planted in place of opium require large areas of land, heavy chemical applications, and attract those with capital to invest—raising questions over the extent of actual substitution involved. Southern Thailand has seen its moister, mainly evergreen forests replaced by rubber plantations over wide areas. Additionally, peninsular southern Thailand's mangrove forests have been decimated along the eastern shores of the Gulf of Thailand, while the western Andaman shoreline now faces similar pressures on coastal intertidal forests. Shrimp farming has been the main reason for mangrove clearance, and this has been exacerbated by poor land tenure definition in the mangrove forests.

3.4 Laos[4]

Laos has a much higher proportion of its land still forested than Thailand or Vietnam. Officially, the country has about half its territory under forest cover, down from two-thirds in the early 1960s. Many regard this as a gross overestimate, depending on the level of degradation required for reclassification. While there has been a steady loss of forest since the early 1960s, the causes have most likely changed considerably over time. During the Second Indochina War, large areas were bombed along the Ho Chi Minh Trail, near the border with Vietnam, and in the northern central province of Xieng Khouang. After 1975, the socialist regime maintained a policy of rice self-sufficiency, which encouraged northern provinces in particular to clear upland forests for rice cultivation. Meanwhile, after 1975, when one-quarter of the country's population was displaced by U.S. bombing, a massive internal refugee problem led to clearing of more new land. This was—and continues to be—exacerbated by the problem of unexploded ordnance, placing otherwise fertile lowland areas out of safe cultivation and requiring further clearance. Another powerful pressure for over-exploitation of the forest has been the dependence, until the early 1990s, of provincial budgets on logging revenues, which encouraged unsustainable logging practices.

Shifting cultivators are often blamed by the government for deforestation. A central plank of government policy has been the resettlement of all shifting cultivators by the year 2000, though this is clearly unachievable. While shifting cultivation is practiced in all regions, the Mon-Khmer communities of southern Laos practice a more stable pattern of rotational swidden farming.

The sustainability of logging is an issue in all regions. Most logging in Laos is now carried out under the auspices of one of three military-owned logging companies operating in the northern, central, and southern regions, respectively. The most intensive logging has been carried out in areas earmarked for hydropower development, most notably to clear the designated reservoir areas of vegetation. This has occurred, however, before contracts were in place, so it is highly likely that some areas cleared will not actually be flooded for use as reservoirs. The most intensive logging operation

---

[4] This part was extracted from a discussion paper by Hirsch (2000) and Yamane and Chanthirath (2000).

in recent years has been the clearing of 450 square kilometers of land due to be flooded by the Nam Theun II Dam in central Laos, but the World Bank has yet to make a decision on the loan guarantees for this build-own-operate-transfer project.

The exploitation of Lao cypress is one of the leading causes of the degradation of the country's primary forests. Lao cypress has been logged and exported, mainly to Japan, since the 1980s, without any effective conservation measures. The exploitation of the Lao cypress has undoubtedly been aggravated in recent years, and several Lao cypress habitats are on the verge of depletion.[5]

3.5 Vietnam[6]

Vietnam has been heavily deforested over the past several decades. During the Second Indochina War, the United States Air Force had a deliberate policy of destroying forest cover used by the communist forces in the South. This led to the destruction of at least two million hectares of forest in the central highlands and mangrove forests along the southern coasts. After 1975 the principal forces driving deforestation changed. In northwestern Vietnam, a rapid loss of forests occurred in provinces such as Son La and Lai Chau, in part due to a policy of rice self-sufficiency similar to the one in northern Laos. The central highlands were the destination of several million lowland Vietnamese who moved to new economic zones under land settlement schemes to overcome the crowding of the lowlands, to rebuild lives disrupted by the war over previous decades, as a government policy to maximize resource utilization and, in one analysis, to "colonize" a part of the country mainly inhabited by ethnic minorities.

Shifting cultivators in the central highlands are even more vilified than those in the North, despite the fact that, like their Mon-Khmer kin in southern Laos, they practice quite stable rotational patterns of cultivation. Logging has increased dramatically in the central highlands, as has clearing of land for cultivation. Coffee and other cash crops are grown in plantations in areas that were forested until recently.

Large swathes of mangroves that survived or had recovered from the wartime devastation have now been cleared again. This time, however, the destruction is mainly for shrimp farming. Ca Mau Province in the far south of the country has been particularly affected by mangrove clearance. The World Bank is currently embarking on a large coastal protection project in this area that will concentrate on mangrove forests.

Economic reforms since 1986 encouraged the military to secure its own sources of finance through logging, mainly in the central highlands. Relations with Cambodia and Laos also give the military an advantage in timber exploitation and trade along border areas.

---

[5] According to Mr. Xeme Samoutry, Director General Department of Forestry, Laos, in the Prime Minister's Decree No. 10 on Control of Forest Operation and Timber Business, 4 October 2000, logging of five species, including Lao cypress, is banned, along with other provisions ordering strict enforcement of regulations concerning harvesting and sales of logs in general.

[6] This part was extracted from a discussion paper by Hirsch (2000).

3.6 Cambodia

Cambodia was, until recently, the country in the Mekong region with the largest proportion of its territory under forest cover. It is also the country which has experienced the most dramatic pace of deforestation over the past decade, mainly associated with poorly controlled logging operations. These have taken place both in government-controlled areas and in parts of the country that were until recently under the control of opposition forces, notably the now-defunct Khmer Rouge.

In western Cambodia, logging and deforestation were long associated with the Khmer Rouge and its need to raise funds to purchase weapons for its fight against the Hun Sen regime. Timber interests, however, have survived the decline of the insurgency. Close links with Thai interests have facilitated this continued timber trade, and Thailand remains the main market.

Northeastern Cambodia is the most heavily forested part of the country and, indeed, of the entire Mekong region. Large logging concessions, however, have been granted by the central government to Indonesian logging and plantation interests, threatening the forest and the minorities of Ratanakiri and Mondolkiri provinces. Sensitivity over Thailand's role as a market for Cambodian timber has led to an increase in export trade through southern Laos, but the ultimate destination remains Thailand, with long queues of log-bearing trucks visible daily at Chong Mek on the Lao-Thai border, mainly carrying Cambodian timber.

Forest resources have been exploited intensively, especially since 1993, during the post-UNCTAC (United Nations Transitional Authority in Cambodia) period. With most of the country's forests under concession licenses, including the heavily forested provinces, logging under the protection of powerful people and the military is dominant throughout the country. Log production reached its highest level in 1997, at 4.3 million cubic meters (7 million hectares in area). Illegal timber felling accounted for at least 92 percent of total production.

3.7 Regional Issues in the Mekong River Basin[7]

In addition to the country-specific aspects of deforestation in the Mekong basin, it is also worthy to consider the regional issues that arise in an increasingly interlinked regional resource economy. Thailand's logging ban led to an immediate shift of attention among Thai timber traders to neighboring countries; the pattern of exports from Burma, Laos, and Cambodia reflects this clearly.

A second regional issue is the tendency to lay blame for deforestation on the more vulnerable groups, notably shifting cultivators. There is now a convergence in many areas of policy between countries in the region, and a tendency to focus blame on more proximate causes and actors. Ironically, in Thailand it is the poor occupants of forest reserve land who have been identified as proximate agents in deforestation, yet their settlement of such land has its roots in the Thai state's long-standing policy to diversify away from rice production by promoting dryland export crops.

---

[7] This part was extracted from a discussion paper by Hirsch (2000).

The Greater Mekong Subregional Cooperation Program, promoted by the Asian Development Bank with the support of many bilateral and multilateral agencies, involves large-scale infrastructure development that has implications for regional forest cover. The greater accessibility created by the expanded regional road network can be expected to increase the rate of log extraction and, more generally, encourage settlement and land clearance for cash cropping in hitherto isolated parts of Laos and Cambodia in particular. Hydropower projects promoted under the same program also involve forest clearance and increased access to hitherto remote forest areas. The market development promoted by the program also puts pressure on forest products previously used mainly for local subsistence.

3.8 Southern Part of the Russian Far East

Deforestation in Russia has not occurred in the same way as in tropical countries, but still the degradation of forests has progressed steadily. One can observe broadly changing forest dynamics in the Russian Far East (RFE) in that the mature forests, mainly coniferous, have decreased and young deciduous secondary forests have naturally regenerated on the vacant lots. The rate of such degradation is faster than in other regions in Russia, and the speed is likely increasing due to the present situation.

Logging operations were located mainly in Primorskiy Krai (state) at first. After the 1960s, full-scale logging also started up in Khabarovskiy Krai, and the speed of forest development accelerated. From 1980 to 1990, in both krai the total logging area was around one million hectares, with a volume of about 20 million cubic meters. In those days logging operations in Russia paid little attention to sustainability and used a "cut and move" strategy—moving to new logging frontiers when timber resources were exhausted in one area. It was said that such logging methods originated in legal and institutional arrangements aimed at achieving production quotas under the planned economy, where achieving the volume of harvest assigned by Moscow was recognized as the most critical goal. Thus neither profitability nor efficiency was paid much attention. Timber products were sold at very low prices and the forest industry was sustained by generous federal subsidies. Most timber produced in the region was transported to European Russia and then exported. The profits went elsewhere and were not redistributed for forest production needs, such as forest management, technology modernization, or improvements bringing more efficient resource use.

After the collapse of the Soviet Union, on top of the shortcomings of the legal and institutional base, weakened forestry governance caused by political disorder and economic crisis has since increased extensive logging even more. Log production for export has gone up, especially due to the rapid increase of domestic transportation costs and sharply lowered domestic demand, plus strong timber demand emerging in China due to a logging ban of its own forests.

Meanwhile, many areas in the RFE have suffered fires every year, with middle to large-scale forest fires occurring frequently. The forest fires of 1998 in Khabarovsk, when more than two million hectares of forests burned, were the worst since 1954 and 1976.

Recent official statistics on forest resources reveal the changes caused by extensive forest development and frequent forest fires in the form of degradation of forest resources, with a decrease of mature forests and an increase of young deciduous forests, as well as a decrease in the volume of forest stocks.

4. STRUCTURE OF KEY PROXIMATE CAUSES OF FOREST LOSS IN THE ASIA-PACIFIC

4.1 Process for Forest Loss in the Area

Much of the forest loss in the target areas generally started with the logging of primeval forests. Logging caused further degradation of forests, but not connected directly with the major deforestation. Often, logging is only the first stage of forest loss. Most logging operations target valuable timber for commercial purposes. In the pre-stage of forest conversion in every tropical country, factors such as settlement programs, agricultural farm development (in many tropical countries), hydropower development (in Laos), mining (in Indonesia), settlement (in all tropical countries), and many extensive large-scale logging operations were observed.

After logging, three paths leading to forest loss were observed. The first is the absence of management. When logging sites are not managed, the forestland of both tropical and boreal zones progresses into secondary forest, and finally—usually more than one or two hundred years later (if allowed)—into climax forest with almost the same structure as the original forest. However, cutover land in tropical forests in the target study areas quite frequently experienced an invasion of settlers, who came along with the roads built for the logging operations. Settlers began slash-and-burn agriculture or non-traditional swidden farming (shifting cultivation). Forests where non-traditional swidden farming was conducted often turned into unproductive land, due to the ecological properties of tropical forest soils, the vulnerability of soil to erosion, meteorological factors, and other causes. Without careful fire control, forest fires occurred frequently and, as a consequence, forests occasionally were transformed into grasslands or barren lands. In this sense, the increase of non-traditional swidden farming that originated from natural forest extraction is a key proximate cause of deforestation.

The second path leading to forest loss is the creation of tree plantations. In many cases, where cutover areas were planted with seedlings after logging, the large-scale planting of fast-growing tree species for industrial purposes was promoted. Since forest conversion projects largely degrade the ecological and productive functions of a forest, industrial tree planting is an activity that can lead to further forest degradation.

The third path is conversion to other land uses. One activity common among tropical countries is reclamation of land for paddy fields or croplands. These developments have proceeded and been closely connected with settlement policies.

Another special path was pointed out in the Mekong River basin: physical destruction caused by regional and internal wars. In Laos, during the Second Indochina

War from 1964 to 1973, large forests were bombed along the Ho Chi Minh Trail, and in Vietnam, the United States Air Force had a deliberate policy of destroying forest cover by chemical defoliation.[8]

By sorting the causes of forest loss observed in the target areas according to a mutual relation in changes, a schematic diagram has been drawn (Figure 1). It shows the possible process leading to forest loss in the target areas, but the actual experience in each country or area was different according to their historical background of forest loss. For example, the process of forest destruction in the southern part of the Russian Far East is rather simple; it progressed through a combination of unsustainable commercial logging practices and frequent forest fires. In contrast, tropical forests in countries in the southeast region experienced more complex processes. Thailand and the Philippines, which experienced serious forest losses in the past, have gone through most of the processes shown in the diagram. Indonesia also has passed through almost all the processes for forest loss, excluding destruction by war. As for Laos and Cambodia, the move from logging to planting has not been established in earnest, but all other routes are identified.

4.2 Purposes of Forest Development

The proximate causes of forest development mentioned above often depended on the purpose of forest use, such as commercial logging, conversion to cash crops or plantations, industrial tree plantations, land clearance for rice self-sufficiency, shifting cultivation, land clearing for migration, or hydropower development.

In the target areas, commercial logging of primary forests has been conducted widely in both tropical and boreal areas, because timber resources are a key source of foreign exchange and timber extraction is one easy means to acquire it. In many cases, commercial logging is conducted through the allocation of a logging concession by paying a forest-use charge to the national treasury. But many past timber extractions served as financial sources for powerful people and members of the military with close connections to domestic and/or foreign forest industries, because forests are usually state-owned and managed by government authority. Commercial timber harvest operations have often been undertaken in a large-scale and extensive way in order to gain enormous profits. These operations have favored special groups, without any consultation with local people. In the Philippines, fires resulting from arson in many cases originated in land conflicts between local people and concession holders or logging companies. In Indonesia, commercial logging was conducted to produce materials for the domestic plywood industry after the exportation of raw logs was banned. In the southern part of the Russian Far East, local forest management bodies often harvested timber illegally, in the name of forest management or pest control, to

---

[8] According to Mr. Xeme Samoutry, Director General of the Department of Forestry, Laos, the Indochina War and bombing by U.S. forces resulted in tens of thousands of refugees, and destruction of forests by bombs themselves and by the refugees. Vast areas still remain barren and unexploded ordnances (UXOs) prevent access to rehabilitation of them or management of existing forests.

compensate for serious budget shortfalls. Extensive commercial logging had strong connections directly or indirectly with forest fires due to the drop in forest humidity after logging, the careless handling of fire by logging workers, and the increased number of small fires started by citizens.

Plantation developments, such as rubber (Thailand), coffee (Vietnam), and oil palm (Indonesia), have also brought about large-scale forest conversions. Additionally, the conversion of forestlands to commercial ranches (the Philippines) and shrimp farms (Thailand) is also identified as a major proximate cause. Since land clearing in these activities often employs low-cost intentional burning, escaped fires frequently cause forest fires (the Philippines and Indonesia). As a result, vast grasslands and unproductive lands were created due to repeatedly occurring forest fires. In the target areas, forest conversion was closely connected with commercial timber extraction, and various kinds of loopholes and illegal acts were widely observed. In many cases, since the company's real objective was to extract the wood, the company abandoned the site after logging. The common feature of these forest conversions is that products from the newly developed land are aimed at acquiring foreign currency, supported by strong demand in consumer countries.

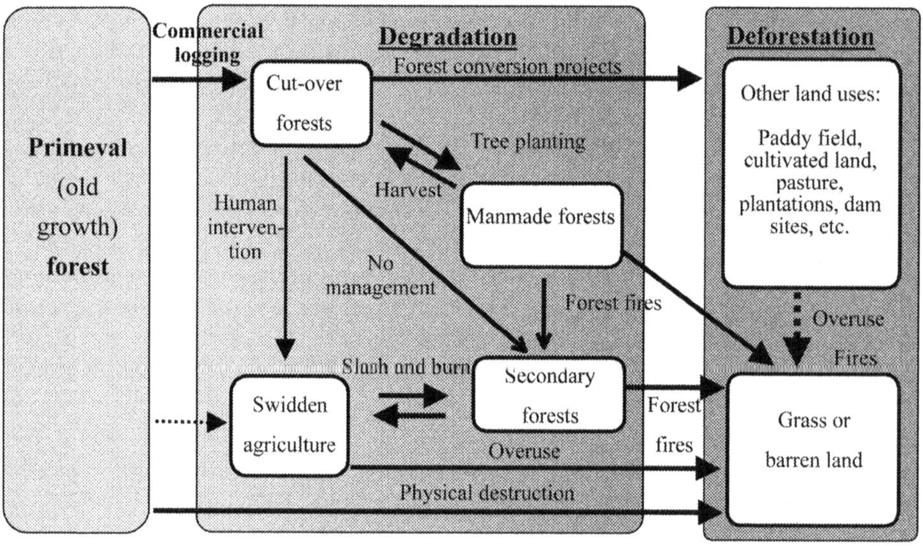

Figure 1: Schematic diagram of forest loss in the Asia-Pacific.

Industrial tree plantations were promoted actively once natural forest resources were depleted (the Philippines and Indonesia). Large companies proceeded forcibly with industrial tree plantations in the Philippines, without consulting with local people and stakeholders. As a result, forest fires started by local people due to land conflicts occurred frequently. For land clearing on project sites in Indonesia, low-cost intentional burning was also broadly employed and frequently caused forest fires, even after an official circular that prohibited burning was distributed. In short, forest conversion projects become a key proximate cause leading to deforestation if the procedures and controls are inappropriate.

Land clearance for rice self-sufficiency has grown in Indonesia, Laos, and Vietnam since 1975, along with the notable population increases of these countries. There is a shared feature, which is that this activity was prompted by national policy in each country.

Shifting cultivation is often pointed out as a leading proximate cause of forest loss in tropical areas. Many studies, however, revealed that shifting cultivation has been a traditional way of forest use in tropical forests and never destroyed entire forests. These activities have only recently been viewed as questionable, for example, since the 1960s in Thailand and after 1975 in Vietnam. In general, increasing population pressure or competition with other land uses, due to social and economic development, has transformed traditional shifting cultivation into being unsustainable, leading to forest loss (e.g., Laos).

Forest loss was prompted, not only by government programs of in-country colonization and settlement, but also by spontaneous settlements due to population pressures or competition for land use. In the Philippines, the clearing and settlement program continued until 1978, where it is estimated that 1.2 million persons moved to Mindanao from other islands from 1948 to 1960. In Indonesia, where the "Transmigration" program started in 1983 and continues today, it is estimated that more than 724,000 hectares of forests were converted from 1994 to 1997 (Inoue 2000). In the Mekong River basin, in-country migration, which was planned as government policy, resulted in forest clearing, which became serious for the first time since the Second Indochina war.

In the Mekong River basin, many large-scale dams were constructed, leading to the destruction of vast forests. In Laos, hydropower development started with construction of the Nam Ngum Dam in the 1960s, while development accelerated in the 1990s. Often, logging concessions were granted to clear the sites planned for dam development. Such permission is said to provide favorable opportunities for extensive timber extraction.

4.3 Actors in Forest Development

There are various kinds of domestic agents involved in activities that have led to forest loss—actors such as central and local governments, domestic logging companies, industry and military authorities, powerful people, and local people. At the same time, foreign agents and actors, such as timber-importing countries, foreign capital from importing countries, and foreign aid institutions have also played crucial roles.

4.3.1 Central and Local Government

The roles of governments vary between target areas, but they can be classified according to their functions as planners and executors of economic development policies, planners of forest policies, and bodies responsible for forest management.

In terms of the design and implementation of economic development policy, all central governments promoted timber cutting as a major means of acquiring foreign exchange and recognized forestland as a resource to be converted for agriculture and other land uses, which then progressed on to forest conversion projects. Such large-scale forest development programs were done with a top-down approach and pushed forward forcibly without consultation with local people and other stakeholders, because, in many cases, the forests are owned by the state. Governments in these countries have strong authority. Besides forcible forest development, governments also promoted a set of policies aimed at industrialization and transition to a market economy.

As forest policy planners, natural resource policies aimed at extracting valuable wood were promoted by governments in every area. In these policies of forest exploitation the logging operations were extensive, yet reforestation programs were not included. The governments wielded enormous power in the allocation of logging concessions. When natural forest resources became depleted, many governments aggressively promoted large-scale industrial tree plantations. These forest policies were frequently implemented forcibly using a top-down approach without environmental impact assessments (EIA), social impact assessments, or consultations with affected parties, in particular the local communities and authorities. As bodies responsible for forest management, governments often failed, because they allowed corruption and the violation of rules. Insufficiencies in the management of resources are common indirect causes of forest loss in the target areas. The concrete effect of the inability of governments to manage their forests is evident in logging concessions (the Philippines, Indonesia, and others), forest conversion projects (Indonesia), and illegal logging (Indonesia, Cambodia, and the southern part of the Russian Far East).

4.3.2 Domestic Logging Companies and Industries

Domestic logging companies and industries, tightly connected with the government, military authorities, and powerful people have played a leading role in forest loss in many places. In many cases, logging companies conducted forest extractions through the allocation of logging concessions by paying forest-use charges to the national treasury. Large forestry concessions to companies were allocated in a very opaque way, while proper monitoring and control by forestry administrative bodies was quite limited. Many logging companies employed a "cut-irresponsibly-and-get-out" strategy, which became the most efficient way to maximize their profits (the Philippines [Seki 2000], Indonesia [Inoue 2000; Siscawati 2000], the southern part of the Russian Far East [Sheingauz 2000a]). Even in forest conversion projects, such as oil palm plantations in Indonesia, in many cases the company's real objective was forest extraction, leaving many cut-over areas abandoned after logging (Okamoto 2000).

4.3.3 Military Authority

In Indonesia, the military authority had strong powers over allocating logging concessions, and they promoted natural forest extraction because of their tight connections with the domestic forest industry, Chinese merchant capital, and foreign capital from timber-importing countries. In Vietnam (Hirsch 2000) and Cambodia (Bottomley 2000), the forest has been both legally and illegally exploited as a financial resource by the military. These activities were identified as the most serious causes of forest loss in the area.

4.3.4 Powerful People

Powerful people are also key domestic actors. In the Philippines in the past, members of Congress established forest policies and forest regulatory systems that enabled logging companies to obtain enormous profits (Seki 2000). Many of the politicians were also concession holders that profited from logging at the same time. In Indonesia, powerful people, such as Chinese merchants who were closely connected to government and military authorities, proceeded with natural forest extraction and established a domestic plywood industry and oil palm plantations, leading to large-scale forest loss from the 1970s to the present (Siscawati 2000).

4.3.5 Local People

Local people have played a significant role in the occurrence of forest fires through not only non-traditional swidden farming but also through local conflicts and the careless handling of small fires. They seem to be passive actors, however, affected by various social and economical underlying causes. In terms of the increase of non-traditional shifting cultivation, various factors involved were identified in our studies, such as poverty and population increases in rural areas, political instability, in-country migration (internal colonization), timber extraction, clearance of community forests, and others. Local people were the major culprits of arson, causing frequent forest fires as a result of land conflicts in the Philippines (Seki 2000). In the southern part of the Russian Far East, citizens' and local peoples' careless handling of small fires (such as bonfires and cigarettes) greatly contributed to recent large-scale fires (Sheingauz 2000a).

4.3.6 Importing Countries

Many studies have indicated that foreign actors, such as importing countries and foreign capital, have played significant roles in forest destruction, because of the strong demand for forest products in consumer countries. The extraction of teak is a typical example, while Lao cypress forest extraction in Laos is a special case (Yamane and Chanthirath 2000).

It is difficult to talk about forest development in the Philippines and Indonesia without mentioning the strong demand for timber in importing countries—Japan, in particular, and the contribution of Japanese investors (Kuroda 2000). It is well known that Japanese investors have extracted valuable timber resources, depleted them, and then shifted production to another country—one after another. When the supply of tropical raw logs decreased greatly or faced import restrictions, the Japanese plywood industry shifted from tropical countries to Russia as a source of timber. The rapid increase of demand in consumer countries has contributed greatly to the recent expansion of forest conversion projects aimed at industrial plantations in the Philippines (Seki 2000) and in Indonesia (Siscawati 2000). In terms of drastic increases in oil palm plantation development in Indonesia, the strong demand in importing countries has been boosted by corporate image strategies claiming that vegetable oils are good for the environment and human health (Okamoto 2000).

4.3.7 Adjoining Countries

Besides these cases, intervention in adjoining countries, such as policy changes that led to decreases in domestic timber production at the same time as strong wood consumption, has accelerated forestland development in timber supplier countries—in China, for example (Yamane 2002). The effects of such intervention were found in the recent increases in border timber trade from Russia to China (Yamane and Lu 2001), and from Laos and Cambodia to China and Thailand (Hirsch 2000). Such regional trade will be more active in the near and middle terms.

4.3.8 Foreign Aid Institutions

Many non-governmental organizations (NGOs) and academic scholars have already pointed out that foreign aid institutions have played a negative role leading to forest loss in developing countries. Our studies confirmed two causes leading to forest loss: structural adjustment programs and individual project support. Both in Indonesia (Okamoto 2000) and in the southern part of the Russian Far East (Sheingauz 2000b), the structural adjustment programs of the International Monetary Fund (IMF) and the World Bank forcibly pushed ahead hasty economic policy reforms, including forest policy reforms. Our study of oil palm plantation developments indicates that the "improved" policies still contain various defects, which have brought about still more forest destruction (Okamoto 2000). In terms of individual project support, the case of international aid for industrial tree plantations in the Philippines is interesting (Seki 2000). After the Aquino government came to power, developed countries provided "environmental aid," and the introduction of participatory forest policy was connected with this funding. In continental southeast Asia, the Greater Mekong Sub-regional Cooperation Program, promoted by the Asian Development Bank along with support from many bilateral and multilateral agencies, also belongs in this category. This project prompted large-scale infrastructure development (Hirsch 2000) and the expansion of regional road networks, increasing the rate of log extraction and, more generally, encouraging settlement and land clearance for cash crops. Hydropower projects prompted under the same program are encouraging further forest clearance. The Market Development Program promoted by the Mekong program also put pressure on forest products previously used mainly for local subsistence purposes. In more than a few cases, foreign aid programs for natural resource extraction, such as mining, functioned indirectly to promote forest destruction.

5. UNDERLYING CAUSES OF KEY PROXIMATE CAUSES OF FOREST LOSS IN THE ASIA-PACIFIC

5.1 Unsustainable Logging

We have many observations from such places as the Philippines, the southern part of the Russian Far East, and the Mekong basin (Thailand, Vietnam, Cambodia, and Laos), where clear-cut logging and wasteful and/or intensive selective logging have been major means of acquiring foreign exchange and financial sources for the military and forest authorities. In these cases, many domestic and foreign actors are mutually related. Logging companies, central government authorities, the military, powerful people, concession owners, and local forest management authorities are the domestic actors. Foreign actors include import and/or consumer countries and development assistance from timber-importing countries. The underlying causes of unsustainable logging that we identified in each target areas are summarized below.

In the Philippines, the intensive market demands of timber importing countries, especially Japan, and market forces, compounded by an insufficient legal and administrative base, were identified as the leading underlying causes of the wide use of the "cut-and-run" strategy. In the southern part of the Russian Far East, various underlying causes of widespread unsustainable logging operations—originating from such ultimate causes as economic and political instability after the collapse of Soviet Union, the economic crises in 1998, and an insufficient legal and administrative base for sustainable forest resource use—are closely and mutually connected. Extensive and wasteful logging activities, including harvesting only high quality logs whenever possible, originated in very few wood-processing facilities for low quality wood, unreasonable pricing of timber, and insufficient controls on the violation of rules. The favorable conditions for exports that emerged after the devaluation of the ruble, plus internal and external market forces such as the demand for foreign currency and the increased timber demand of the Asia-Pacific region, especially from Japan and China, have amplified export-oriented, unsustainable timber extraction. Based on these observations in the Mekong River basin, the power of the military in the political structures of countries due to ongoing war and political rivalry (political instability) is a recent leading underlying cause that emerged. In addition, trade aimed at increasing the amount of foreign exchange—demonstrated by the strong regional demand from Thailand and Vietnam (market forces)—and a lack of funds, institutional capacity, and willpower on the part of central and local-level authorities (insufficient legal and administrative base) has fueled this situation. In Laos the degradation of Lao cypress forests has various underlying causes of both domestic and foreign origin. The strong demand and consumption for Lao cypress in Japan, however, appears to be the ultimate cause, and the insufficient legal and administrative base has allowed unsustainable resource exploitation.

In conclusion, commercial logging in the areas of this study has been caused by internal and external market forces, such as the demand for foreign currency and the increase in demand for timber in consumer countries. Various underlying causes that have encouraged widespread unsustainable logging operations—originating in economic and political instability, insufficient legal and administrative bases, and economic and forest development policy—are closely and mutually connected.

5.2 The Failure of Industrial Plantations with Large-scale Reforestation

The experience in the Philippines provides a very instructive case that illustrates the failure of large-scale reforestation in industrial plantations, where foreign aid institutions played a significant role, along with domestic actors, such as governmental forest authorities and local people. The largest forest-related project, the CRP, was controlled by the central government, based on the principle of state control of plantations. But large-scale restrictions on local people's access to land intensified conflicts over land, and the enormous flow of project funds spread corruption, leading to the non-payment of contract fees to plantation workers. At the same time, the Department of Environment and Natural Resources (DENR) promoted the government-managed CRP and IFMAs, and leased several thousand hectares to one company. When an unbroken area of several thousand hectares is granted to one company, it is almost impossible to avoid conflicts with local people's land use. It is not surprising that, as with the CRP, the IFMAs have not produced very good reforestation results. The failure of industrial plantations stemmed from the defects in forest policy that used a top-down project approach to establish large areas for plantation lands, while disregarding forest use by local people. Moreover, the shortage of good governance is thought to have led to a lack of appropriate control over the projects. In the background, the behavioral patterns of authorities that led to the centralization of power and profit-seeking, conspicuous since the Marcos-era, were still prevalent. In conclusion, an insufficient legal and administrative base, which disregarded the forest uses of local people, led to the failure of industrial plantations in the Philippines.

5.3 Frequent (Large-scale) Forest Fires

One of the more recent significant proximate causes of forest loss in the Asia-Pacific is frequent, large-scale forest fires, as shown in the cases of the Philippines, Indonesia, and the southern part of the Russian Far East. Local people have played a prominent role, along with logging companies and local forest management authorities.

In the Philippines, the underlying causes of this type of forest loss originated in past forest policies which promoted commercial logging, commercial ranching, and industrial plantations. These policies not only disregarded the rights of land possession and forest resource use of public forestland, but also directed the profits of the projects to specific parties, such as logging companies. These forest development projects went against the reality that indigenous people and pioneers from the lowlands had inhabited public forestlands since World War II, and that they had traditionally used various forest resources there.

In the Russian Far East, political and economic instability and an insufficient legal and administrative base led to inadequate forest institutions and regulation. A wasteful resource use policy, a legacy of the Soviet era, is also a significant underlying cause. These underlying causes, mutually connected to an increased number of small fires and logging sites with high fire risks, amplify the problems of the weak fire control system and have caused frequent forests fires on a large scale. Forest management operations, which mainly depend on natural regeneration, may also accelerate forest degradation.

Recent large-scale forest fires in Indonesia were caused mainly by national economic development policies that promoted large-scale forest conversion projects with an inappropriate land allocation policy, supported by strong demand for forest products in consumer countries, insufficient enforcement of fire prevention in the forest development programs, and insufficient forest fire control systems.

The underlying causes in the three areas were different from each other, but unsustainable forest development regime associated with an insufficient fire control system is a common characteristic.

5.4 Forest Conversion for Plantations and/or Cash Crops

Converting forests to plantations and/or cash crops through land-clearance using intentional burning or clear-cutting was widely observed in Indonesia, Thailand, Laos, and Vietnam. Both domestic and foreign actors were involved, such as governments, conglomerates, logging companies, domestic and foreign consumers, and foreign aid institutions.

In Indonesia, the development of palm oil plantations is a recent key proximate cause of forest loss. Most of Indonesia is suited to palm oil cultivation, and there are vast lands available that could be converted to plantations. On the other hand, the area under cultivation in Malaysia, currently the world's top producer, has reached the saturation point. Accordingly, such natural factors have created strong pressure for conversion, along with the rapid growth of demand in consumer countries and favorable domestic policies that do not pay enough attention to the realities of current land use, even after policy reforms recommended by the IMF were instituted.

The introduction of industrial rubber and eucalyptus tree plantations in the so-called degraded forests and shrimp farming in cleared mangrove forests is recognized as one of the major causes of deforestation in the Mekong River basin. In the region, the leading cause identified was that industry ignored the local subsistence use of natural resources, set against the background of strong demand in consumer countries.

In conclusion, domestic economic development policies with industry emphasis, associated with strong demand in consumer countries, have led to increased rates of forest conversion to plantations and cash cropping.

5.5 Direct and Indirect Destruction caused by War and/or Local Conflict

In Laos and Vietnam, forest loss caused directly and indirectly by war and/or local conflict was observed, where the use of heavy firearms caused massive destruction, compounded by the influences of governments, militaries, and refugees. It is apparent that political instability was a key underlying cause of the Second Indochina War. The war also had an indirect impact—the internal resettlement of refugees and repatriates, which was closely connected with national policies.

5.6 Land Clearance for In-country Migration (Internal Colonization)

This key proximate cause was identified in Vietnam and Laos, where the central governments, landless farmers, refugees, and repatriates were key actors. Our research team did not conduct an intensive analysis of the situation, but clearly, the rapid population growth in cities and the incomplete policies of internal colonization under incomplete economic/forest policies were leading underlying causes.

6. ROOT CAUSES OF FOREST LOSS IN THE ASIA-PACIFIC

Based on discussions at three regional workshops in Indonesia, Laos, and the Russian Far East, five root causes of forest loss were addressed, which essentially affected the recent major proximate and underlying causes of forest loss in the region.

The first root cause is ***the lack of recognition of the real value of forests***. Actors related to forest development have, in many observed situations, ignored the real value of forests, including their environmental functions and their function as the base of the local community. Instead, they aimed to maximize economic profits through the development of the forests. In the background, lack of appropriate knowledge and parameters of forest biodiversity, ecosystem management, and traditional forest use on the part of the parties concerned seems to have encouraged narrow-minded resource use policies.

The second root cause is ***the impact of market forces under incomplete market systems***. Our researchers frequently observed that the forest developments in target areas were driven in unsustainable ways that centered on monetary benefits, foreign exchange in particular, and on market forces that originated in strong consumer demand for products, especially in importing countries.

The studies also indicated that ***economic and forest development policies with an industrial emphasis*** were a leading cause of forest loss in the region. So far, forest policies have used the benefits of timber extraction and forest conversion as a means of obtaining foreign exchange or as a financial source for powerful people. Moreover,

insufficient political will and lack of compliance to stop destructive forest development were recognized as key root causes of ongoing forest loss. Many institutional causes can be extracted from these root causes, such as an insufficient legal and administrative base for sustainable forest management, incomplete public forestland policy, and insufficient attention to local people's rights.

Although not examined in detail in our studies, it is certain that many direct or indirect causes of forest loss were strongly affected by ***political disorder and economic difficulties***. This root cause brought about many other causes of forest loss, such as the inability to govern the use of forests, the expansion of rule violations and corruption, increases in poverty, and the consequent rise of non-traditional shifting cultivators. Moreover, various impacts of forest loss themselves often brought on further political, social, and economic instability. In this sense, the progress of forest loss itself is the root cause of forest loss.

7. TOWARDS FOREST CONSERVATION IN THE ASIA-PACIFIC

Based on our findings, in terms of addressing the root underlying causes of forest loss discussed above, in order to overcome forest loss in the Asia-Pacific we propose the following direction under five recommendations.

7.1 Ensure the Multiple Functions of Forests are Protected in Forest Use

Based on our analysis, in order to manage forest resources sustainably for the long term, all stakeholders involved, especially forest users, should follow the basic requirement of ensuring that the multiple functions of forests are taken into account. In the target study areas, most of the recent forest developments employed unsustainable means, with the goal of maximizing monetary benefits, and led to forest loss. The multiple functions of the forest were almost completely ignored. Forest loss has not only had an economic impact, but has also impacted on the environment and people's livelihoods, and created land conflicts and land alienation. In many cases, the impacts of forest loss exceeded the economic benefits from forest development. Thus we should clarify the true value of forests and develop conservation measures to protect and manage these multiple values.

To achieve this, an urgent requirement is to develop a method to evaluate the multiple functions and values of the forest, including the traditional knowledge uses of forest resources, and all social, environmental, and economic costs, and the price of timber products under the initiative of the international community, such as the United

Nations Forum on Forests (UNFF), when considering the benefits of any land or forest development.[9]

7.2 Consumption of Sustainably Produced Forest Products

Unsustainable forest development has accelerated, both in forests and on land converted from forests, because of strong demand for natural resources. In many cases, illegal or uncontrolled timber extraction originated in strong demand from consumer countries, coupled with an inability of supplier countries to govern their forest resources.

In order to better control the consumption of forest products and promote sustainably produced forest products in consumer countries, resource use reduction and the trade of forest products produced sustainably should be promoted effectively, as well as the deployment of environmental education, and the employment of economic mechanisms such as voluntary certification. The central and local governments of timber-exporting countries, in addition to major importers of timber, must take responsibility and exert control over the origin of imported wood products, and refuse transactions of timber of illegal and non-sustainable origin. For effective monitoring and to control illegal trade, the capacity of relevant officials in the institutions of both producing and consuming countries should be enhanced.

7.3 Reform of Economic and Forest Policies Towards Sustainable Forest Management, with Emphasis on Local Participation

Forest loss in the target areas was directly and indirectly promoted through top-down forest development, which ignored the customary rights and uses of forests and provided insufficient governance. To move towards sustainable forest management in the target areas, economic and forest policy reforms should consider the following policy options, with a special emphasis on strong governance using a participatory approach.

As a basis for this change, central governments must continue to show the political will to regulate and monitor forestry with strong measures and balance the interests of the state, business, and local communities with a master plan for appropriate, sustainable, and equitable development.

In addition, central and local governments must make decisions transparently, consult stakeholders on all decisions regarding the forestry sector, and provide for participatory processes that actively engage a wide range of stakeholders, while recognizing the rights and abilities of local communities to manage lands and forests. In this regard, central governments must play a regulatory as well as a facilitative role in

---

9 Developing an internationally accepted methodology to evaluate the carbon dioxide absorbing capacity of forests, as well as social and economic indicators, is also an urgent need.

sustainable forest management, and should concentrate on building the capacity of local authorities so that they will be better able to facilitate community resource issues. In some areas the regional élite class can easily manipulate local governments in some countries. The intervention into forest management of these élite seeking special concessions is likely to lead to undesirable results in resource management, from the perspective of local people; thus central governments should regulate the involvement of local governments. The legal foundations of community forestry should include the provision that individuals or associations may access and use forestlands by entering into contractual agreements with the government. Appropriate forest zoning can be a base for sustainable forest use. Protected conservation areas should be allocated enough size to give them an appropriate management base, allowing for the proper balance between the various functions of and interests in the forest.

As observed in the impact of China's new forest policy on timber trade from Russia to China, when reforming or establishing policies related to forest use, central governments should enact appropriate measures to mitigate the impacts of their policies on forest management in neighboring countries and areas.

7.4 Reform the Legal and Administrative Bases for Sustainable Forest Management

Our study suggests that insufficient legal and administrative bases in central and local authorities are closely connected with forest loss. The shortage and/or lack of recognition of the rights, knowledge, and participation of local communities in forest management and forest development have brought about unsustainable forest use and local conflicts, leading to forest loss. Thus reform of the legal and administrative bases in relevant authorities should introduce the elements of community forestry and participatory forest management. For example, effective means can include providing early and efficient coordination to resolve inter-ministerial conflicts, developing legal mechanisms which recognize traditional land-use practices and systems of customary tenure in order to protect the rights of indigenous peoples, and using transparent procedures for granting forest concessions in consultation with all the affected parties—especially local communities and authorities. In addition, it is advisable that lands with traditional social, economic, and cultural significance are carefully delineated and excluded from forest concessions. Conducting long-term monitoring of logging operations, as well as log transport and exports, are essential tools in ensuring sustainable forest use. Through these means, national governments, provincial authorities, and the local communities can co-operate closely, and it can be effective to encourage voluntary activities and provide the necessary training to monitoring groups at the village level.

Because of the significant impacts of large-scale forest fires in forest loss, an effective legal and administrative base for proper fire control is also urgently needed. It should be ensured that administration of national and local forest fire control be strengthened and coupled with national fire control policy and programs, and that well-balanced forest fire control measures be allocated, with attention to both advanced technology and practical equipment in the system, as well as sufficient financial and human resources. As for fire control, in local-level restrictions on development activities, intentional burning—for clearing land, in particular—should be enforced more strictly. As observed in Indonesia, insufficient funds and time encouraged the use of uncontrolled intentional burning by sub-contractors responsible for land clearing in the field, so new regulations should take local reality into account so that they can work in a way that minimizes fire risk.

7.5 Regional Coordination to Realize Forest Conservation

It is obvious that the strong demand for forest products means that market forces, economic policies, and forest policies in timber-importing countries are linked with forest use and forest loss in exporting countries. Thus timber-importing countries should do their duty for world forest conservation through the international community, which in turn should place increased emphasis on community involvement and participation in its approaches to forest conservation, natural resource management, and land-use planning. In order to ensure this, principle pledges of loans or grants made by the international community should be carefully conditioned on the basis of respect for human rights and sustainable management of natural resources in agreements that are informed and transparent to the public. Compliance with such written conditions should be closely monitored in order to avoid corruption; officers and government officials must be held accountable for their policies. Assisting governments in developing community forestry or joint forest management systems is also effective for moving in this direction, thus encouraging local communities to continue to value forest resources through increased involvement in forest management. In order to apply these recommendations, the international community should make the best use of past experience of forest loss in countries such as the Philippines and Thailand. In a move towards the participation of more people in forest management, the Philippines are on the leading edge in this area compared to other Asian countries. The Philippines and international society should disseminate information about their experiences to other countries so the same mistakes are not repeated. The lessons learned in Thailand could also contribute to understanding the impacts and results of rapid deforestation on communities and a national economy.

The international community should support initiatives that provide accurate information on forestry and land-use issues at the local level, particularly concerning human rights, indigenous rights, and forest management within a country. In connection with this direction, supporting education for journalists on forest-related issues is also essential in order to avoid "stereotypical" reporting on the causes of deforestation.

The absence of appropriate coordination on transboundary issues was also observed among the factors involved in forest loss, therefore it is urgent to strengthen regional coordination in order to realize forest conservation. International cooperation in fire control and the expansion of international aid should be promoted with the full support of the developed countries in the region. With respect to burned sites, restoration should proceed under full consideration of local welfare and ecological diversity, because these areas tend to be converted later into plantations for economic benefit, and without proper restoration the biodiversity value of these sites will deteriorate in the long term, leading to serious adverse effects on ecological systems.

8. CONCLUSION

With the pressing need for solutions, based on local realities, to overcome forest loss, the recognition of the underlying causes of forest loss discussed above has gradually worked its way into the process of intergovernmental policy dialogues on forest conservation, and consequently, small but concrete changes have recently emerged. Examples include the inclusion of community-based forestry in national forest policies in Thailand and the Philippines, the expansion of voluntary certification on forests and wood products, and the rise in awareness and action against illegal forest products both in producing and consuming countries. In the last decade various institutes have studied and proposed national and local guidelines for community forestry or participatory forest management. The application to local and national forest management, however, has been very limited. Guidelines for the management of degraded tropical forests were recently compiled by the International Tropical Timber Organization, including several principles on the participation of local people (ITTO 2002). Thus integration, examination, and revision of proposed guidelines through practice should be included in the next stage of efforts.

REFERENCES

Allen, J. C. and D. F. Barnes. 1985. The causes of deforestation in developing countries. *Annals of the Association of American Geographers* 75(2):163–184.

Amelung, T. and M. Deihl. 1992. *Deforestation of tropical rain forest: Economic causes and impact on development.* Tubingen: Mohr.

Araya, A. 1998. *Plywood industry in Indonesia.* Tokyo: Nihon Ringyo Chosakai.

Bottomley, R. 2000. Structural analysis of deforestation in Cambodia, with a focus on Ratanakiri province, Northeast Cambodia. In *A step toward forest conservation strategy (2) Interim report 1999,* 441–472. Hayama: IGES Forest Conservation Project.

Brown, K. and K. Pearce, eds. 1994 *The causes of tropical deforestation: The economic and statistical analysis of factors giving rise to the loss of the tropical forest.* Vancouver: UBC Press.

Food and Agriculture Organization (FAO). 2001. *State of the world's forests 2001.* Rome: FAO.

Hirsch, P. 2000. Addressing the underlying causes of forest loss and forest policy changes in the Mekong region. Proceeding of the Third IGES International Workshop on Forest Conservation Strategies for the Asia and Pacific Region, 158–168. Hayama: Institute for Global Environmental Strategies.

Inoue, M. 2000. Indonesian forest policy and the role of NGOs. In *A step toward forest conservation strategy (2) Interim report 1999,* 283–294. Hayama: IGES Forest Conservation Project.

International Tropical Timber Organization (ITTO). 2002. *ITTO guidelines for the restoration, management and rehabilitation of degraded and secondary tropical forests.* Yokohama: ITTO.

Kummer, D. M. 1992. Tropical deforestation: A literature review. In *Deforestation in the post-war Philippines.* Chicago: University of Chicago Press.

Kuroda, Y. 2000. Development and resource politics in post-war Japan. In *A step toward forest conservation strategy (1) Interim report 1998*, 427–429. Hayama: IGES Forest Conservation Project.

Myers, N. 1980. *Conservation of tropical moist forests.* Washington, D.C.: Committee on Research Priorities in Tropical Biology of the National Research Council, National Academy of Sciences.

Nagata, S., M. Inoue and H. Oka. 1994. Forest utilization patterns in the course of economic development (in Japanese with English summary). In *An inquiry of the U-shape hypothesis of forest resources.* Tokyo: Nobunkyo.

Nectoux, F. and Y. Kuroda. 1999. *Timber from the South Seas: An analysis of Japan's tropical timber trade and its environmental impact.* Lausanne: WWF International Publication.

Okamoto, S. 2000. The growth of oil palm plantations and forest destruction in Indonesia. In *A step toward forest conservation strategy (2) Interim report 1999,* 361–380. Hayama: IGES Forest Conservation Project.

Repetto, R. and M. Gillis, eds. 1988. *Public policies and misuse of forest resources.* Cambridge: Cambridge University Press.

Rundle, T. and J. Roper. 1996. Regional patterns and historical trends in tropical deforestation, 1976–1990: A qualitative comprehensive analysis. *Ambio* 25:160–166.

Seki, Y. 2000. The structural context of post-war forest loss and changes in forest policy in the Philippines. In *A step toward forest conservation strategy (2) Interim report 1999,* 315–334. Hayama: IGES Forest Conservation Project.

Sheingauz, A. 2000a. Their causes and consequences, forest fires in Primorskiy and Khabarovskiy krais. In *A step toward forest conservation strategy (1) Interim report 1998,* 456–466. Hayama: IGES Forest Conservation Project.

———. 2000b. Outlook of underlying causes of deforestation and forest degradation in the Southern part of the Russian Far East. In *A step toward forest conservation strategy (1) Interim report 1998,* 431–455. Hayama: IGES Forest Conservation Project.

Siscawati, M. 2000. Underlying causes of deforestation and forest degradation in Indonesia case study on forest fires. In *A step toward forest conservation strategy (1) Interim report 1998,* 553–563. Hayama: IGES Forest Conservation Project.

Southgate, D. 1990. The causes of land degradation along "spontaneously" expanding agricultural frontiers in the third world. *Land Economics* 66 (1): 93–101.

UNECE and FAO. 1992. *The forest resources of the temperate zones: Main findings of the UN-ECE/FAO 1990 Forest Resource Assessment.* New York: United Nations.

UN Economic and Social Council. 1996. Implementation of forest-related decisions of the United Nations Conference on Environment and Development at the national and international levels, including an examination of sectoral and cross-sectoral linkages. Programme element I.2: Underlying causes of deforestation and forest degradation. Report of the Secretary-General E/CN.17/IPF/1996/2. New York: United Nations.

Yamane, M. 2002. China's recent forest-related policies: Overview and background from the perspective of economic growth and forest conservation. In *IGES Forest Conservation Project policy trend report 2001*, 1–12. Hayama: IGES.

Yamane, M. and K. Chanthirath. 2000. Lao cypress forests: Causes of degradation and the present state of conservation in Lao P.D.R. *International Review for Environmental Strategies* 1(1): 119–133. Hayama: IGES Forest Conservation Project.

Yamane, M. and W. Lu. 2001. Analytical overview on the recent Russia-China timber trade. *International Review for Environmental Strategies* 2(2): 335–347. Hayama: IGES Forest Conservation Project.

Chapter 2

TIMBER TRADE POLICY TO SUPPORT SUSTAINABLE FOREST MANAGEMENT

Shin Nagata
The University of Tokyo

Satoshi Tachibana
Forestry and Forest Products Research Institute

1. SCOPE OF THIS CHAPTER[1]

1.1 Introduction

An examination of the state of global forests in the 1980s reveals increases in forest area in many developed countries—and decreases in virtually all the developing countries. Although we should not overlook new kinds of forest decay, especially in Europe, the situation in developing countries is so serious that we would summarize the overall state of the forests in the developing world as "devastated."

This cross-sectional view can be demonstrated through a time-series setting. At the onset of economic development the natural forests exist in abundance. At first, people live in harmony with their rich forests, utilizing and collecting necessities from them. As they begin to cultivate land, forestland is converted into farmland. Agricultural production enables the population to grow, accelerating deforestation. Industrialization hastens the speed of deforestation, because the increased human population can achieve

[1] This chapter is a result of the sub-theme, Timber Trade, in the first phase (1998–2000) of IGES's Forest Conservation Project. The study was conducted by Prof. Shin Nagata, in cooperation with the following collaborators: Prof. Yeo-Chang Youn, Prof. Ruperto P. Alonzo, Dr. Mihoko Shimamoto, Dr. Satoshi Tachibana, Mr. Nobuyuki Yamamoto, and Dr. Yasushi Minowa.

M. Inoue and H. Isozaki (eds.), *People and Forest—Policy and Local Reality in Southeast Asia, the Russian Far East and Japan, 33–47.*

greater agricultural production through the use of chemicals and mechanization. But in the course of economic development, the pressures that cause deforestation decrease. Historically, perhaps because of different comparative advantages, once they become industrialized, economies have generally imported forest products in exchange for the export of industrial products. Thus they do not need to use as much of their own forests as sources of timber and are able to begin restoring them. They change the focus of their forest policies from development and exploitation to reforestation and protection, and increase the recreational uses and protection activities of their forests. We admit that the above description is greatly oversimplified, but there is, nevertheless, truth to this story.

In the course of their economic development, many of today's developed countries imported timber from other countries to lessen the pressure on their own forest resources. Now, many deforested developing countries could, theoretically, import timber from the restored forests of developed countries. We cannot overlook the importance of timber trade policy on forest resource management.

Within the last two decades the liberalization of trade has been promoted internationally.[2] This direction, however, is unreliable for promoting sustainable forestry and forest management, because deforestation, particularly in tropical developing countries, has progressed without slowing down—at least for the two decades under the free trade paradigm.

1.2 Scope and Methodology

The purpose of this chapter is to advocate appropriate strategies for the formation of timber trade policies that lead to sustainable forest management. The study program was composed of four sections. First, in order to connect forest resources and timber trade, the framework and data availability for forest resource accounting were studied. Second, the timber trade structures and policies of major timber-trading countries were clarified from their historical perspective, partially utilizing time-series econometric analyses. Third, sustainable forest management was discussed from the viewpoint of the measures of forest and timber certification (some results are described in Chapter 6).

---

[2] The Uruguay Round Trade Agreement has led to the following results including wood-related products: (a) The tariff reduction agreement has reduced global tariffs by an average of one-third from their base rates. (b) Commodity sectors where tariffs were eliminated by the major trading partners of the agreement include beer, brown distilled spirits, pharmaceuticals, steel, construction equipment, agricultural equipment, medical equipment, toys, furniture, and paper and paper products. (c) The agreement called for tariff reductions to be made over five years in equal annual staged reductions, although in some cases, such as pulp and paper, the reductions are being implemented over ten years. (d) The agreement of the pulp and paper sub-sector in the zero-for-zero initiative for wood products was made among the United States, the European Union, Canada, Japan, Korea, Finland, Austria, New Zealand, Hong Kong, Singapore, Australia, Brazil, and Chile. (e) In the lumber and wood products sub-sector, the Uruguay Round did not achieve the zero-for-zero initiative for wood products. The United States, Canada, the European Union, Hong Kong, New Zealand, Singapore, and Sweden supported the agreement, but Japan did not. (f) In the furniture sub-sector, the Uruguay Round achieved a zero-for-zero agreement between key countries, with tariffs to be eliminated over five years.

Finally, a spatial equilibrium model was created to simulate timber trade in the Asia-Pacific region and to help assess various policies and measures.

Recently, the structure of the global timber trade has changed drastically from trade in logs to trade in processed wood products. This change in part reflects the efforts of the environmental movement worldwide and the exhaustion of useful forest resources. As mentioned in the United Nations Food and Agriculture Organization (FAO) report, *Forest Resources Assessment 2000*, global forest resources decreased steadily each year in the 1990s, while many timber-exporting countries of the developing world registered a high rate of deforestation. Considering the shifts in the timber trade, becoming more constrained by the state of forest resources, we believe that an orderly timber trade is necessary in order to preserve and sustain our world's forests.

2. FOREST RESOURCE ACCOUNTING

Agenda 21, the action program adopted at the United Nations Conference on Environment and Development held in Rio de Janeiro in 1992, explicitly states that environmental accounts should be developed by all members (Yamamoto 2000a). The United Nations, through its statistical secretariat, the United Nations Statistical Division (UNSD), responsible for the standards of the System of National Accounts (SNA), provides a satellite system to the SNA, called the System of Integrated Environmental and Economic Accounts (SEEA). The purpose of SEEA is to show the impact of economic activities on the natural environment.

SEEA consists of several different parts, with both physical and monetary accounts. The first step is to link environmental data, expressed in physical units, with the national accounts system. The second step is to create monetary accounts, which are based on the physical accounts, in order to evaluate impacts in economic terms, allowing the estimation of, for example, the costs of the damage caused by emissions. Furthermore, an accounting system requires that data compilation should be made in terms of balanced accounts.

A close study of natural resource accounts has been made in Nordic and western European countries (Yamamoto 2000a; 2000b). The development of natural resource accounts started in Norway as early as the 1970s. Since Nordic countries have large forest areas and produce a considerable output of forest products, forest resource accounting has been one of the main fields of study since its early days. Forests are one of the crucial themes in natural resource accounts even now.[3]

What are forest resource accounts? In brief, they can be defined as consistent accounting frameworks that adjust forest-related resource and environmental

[3] Forest resource accounting will play an important role in sustainable forest management, and it should include valuation of various environmental services produced from tropical forests. In this regard, it would be necessary to develop a theory and methodological system of tropical forest environmental accounting, which could be eventually part of a national economic accounting system. Related studies were started in Japan in the latter half of the 1990s (Koike and Fujisaki 1997; Kurabayashi et al. 2002).

information. As shown in Figure 1, forest resource accounts are comprised of four accounts: (1) forest accounts, (2) forestland accounts, (3) forest product accounts, and (4) forest management accounts (Yamamoto 2000a).

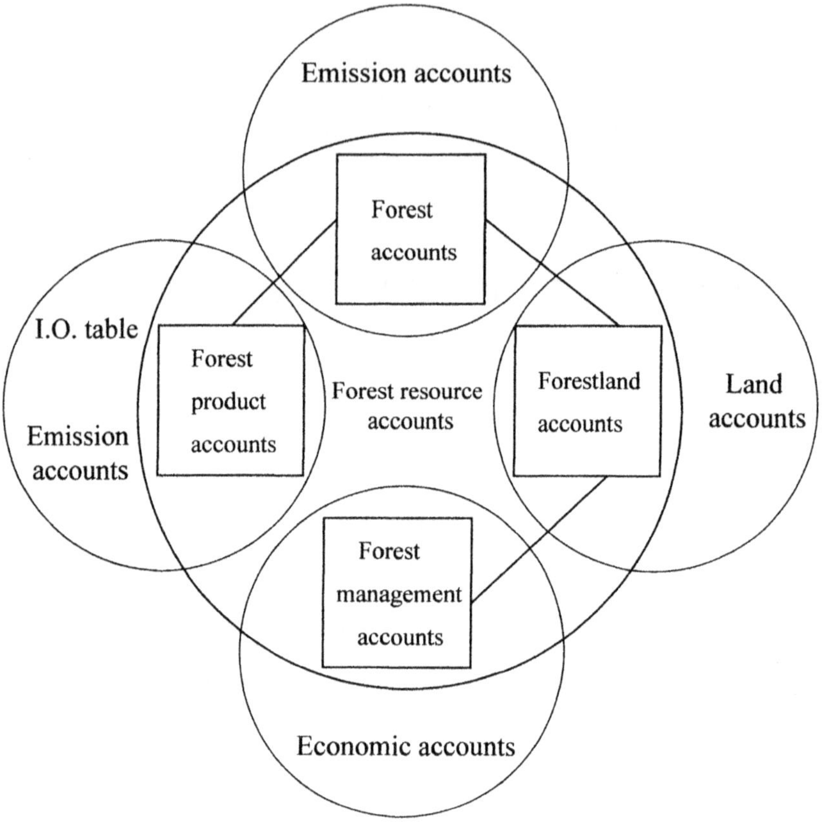

Figure 1: The structure of forest resource accounts.
Source: Yamamoto (2000a).

Forest accounts represent the initial stock, period flow, and final stock of trees growing in forests—the tree volume. What must be noticed is that increases and decreases of tree volume are caused not only by tree growth and removal, but also by changes in forestland area. In order to clarify this linkage, we can adopt forestland accounts.

Forest product accounts use weight, volume, and other physical terms to describe the various uses—from raw logs, timber, and various wood products to waste. Forest resource accounts include two parts: a forest sector and product balance table and a forest mass balance table. When we combine forest accounts and forest product accounts into one united account, we can describe a consistent system—from forest growth to the abandonment of wood residues.

Forest management accounts use monetary terms to show the flow of funds used in managing forests. Various monetary flows, such as subsidies, taxes, and investments, are described in these accounts. All monetary flows represented are real money that is transacted in actual markets.

Within forest resource accounts, forestland accounts serve to directly link forest accounts with forest management accounts and provide for their consistency. Inputs and outputs for forest management necessarily reflect the current state of forests and forestlands. In these systems, the most difficult question is how to link forest accounts with forest management accounts, because it requires the linking of land accounts and economic accounts, that is, the linkage of physical and monetary accounts. In order to construct forestland accounts, it is appropriate to utilize geographic information systems (GIS) or land information systems (LIS).

3. COUNTRY CASE STUDIES OF TIMBER TRADE POLICY: TIME-SERIES ANALYSES

The actual situation in Southeast Asia varies from place to place. In the Philippines and Thailand, which have already lost large parts of their natural forests, reforestation projects by governments are still unable to recover and maintain their forests nationwide. Unfortunately, reforestation does not immediately return money for the local poor. Therefore, other kinds of economic incentives in these countries are necessary to encourage tree planting and tending by the private sector. In the many countries in Southeast Asia, both domestic log markets, which do not compete in the international market, and domestic forest-related industries, which are mostly small-scale, will need encouragement to grow. In this section we mention the brief history and current state of some representative countries and their exporting and importing characteristics.

3.1 Indonesia

In Indonesia, forest exploitation began actively in the second half of the 1960s with the export of logs to acquire foreign exchange (Tachibana 2000a; Tachibana 2000c). According to several interviews with key personnel, up to the end of the 1970s their logging practices were regarded as sustainable, because loggers stamped a number on both the harvested log and its stump under Indonesian selective cutting rules and regulated timber distribution.

The exploitation of natural forests is ongoing, and shortages of materials to satisfy the production capacity of the processing industries, especially the huge plywood industry, are being witnessed (Massijaya and Kartodihardjo 2000). During the transition period from the Suharto era,[4] the country needed to liberalize its domestic industrial structure, but there was concern about excessive intervention from the International Monetary Fund (IMF) and multinational corporations. For example, in 1997 the IMF required the government to reduce the tariff rate on log exports, and then in 1998 it recommended the auction of forest concessions (i.e., opening up the concession market to foreign companies). The result of the IMF requirements would be the opening up of the forest resources (material) to over-exploitation.

The development of processing industries has proven beneficial for Indonesia from the viewpoint of creating employment, earning foreign exchange, efficiency improvements in terms of recovery rates, and utilizing forest resource for themselves. Therefore, the domestic processing industry, still in its infancy, should be protected from excessive outside competition and monopolies.

3.2 Malaysia

In peninsular Malaysia, the forest-related policies are based on the federal administrative framework, such as the National Forestry Policy (1978) and the First and Second Industrial Master Plans of Malaysia (1986–). Forest exploitation of peninsular Malaysia first began with the conversion of forests into rubber and oil palm plantations in the early twentieth century (Tachibana 2000b). As a result, peninsular Malaysia had plentiful rubber wood and oil palm plantations up to the 1980s, and a log export ban since the 1960s helped preserve their natural forest resources and support their domestic small wood-processing enterprises (Tachibana 2000a). Near the end of the 1980s they started producing furniture and other products using processed rubber wood materials. Since then peninsular Malaysia has promoted furniture, medium-density fiberboard (MDF), and molding productions as part of their industrializing strategy in the Master Plan. Based on Malaysian statistics, the current forest resources of peninsular Malaysia are regarded as being in the amelioration stage, making it important for them to continue promoting wood-based industries, especially high value-added products such as furniture and MDF. Under the increased planting of oil palm instead of rubber wood in the 1990s, it may be necessary to examine the next generation of materials to be used for wood products, e.g., palm.

In the state of Sabah, the forest-related industry is independent of federal policies. Forest exploitation there has expanded as a spearhead in Southeast Asia since the 1960s, because of their high quality logs (Tachibana 2000a). Sabah has limited economic resources and its state revenue was heavily dependent on forest-oriented royalties of around 70 percent in the 1980s, especially on export logs.[5] But as a result of rapid forest exploitation in the 1970s and 1980s, Sabah's forest resources have drastically decreased. According to Malaysian statistics, the forest resource is still deteriorating,

---

[4] Suharto was the president of Indonesia between 1968 and 1998.

[5] Based on econometric analysis using time-series data, in Sabah, the forest-oriented revenues have a significant influence on timber exports. It means that when state revenues were insufficient they probably increased timber exports in order to get more revenue by providing short-term timber licenses. See Tachibana (2001).

even though its speed has slowed. The scarcity of wood materials is demonstrated by decreases in old growth forests and increases in secondary forests. To promote wood-based industries, such as plywood and sawn timber, Sabah restricted log exports at the end of the 1980s by executing log export quotas from 1989 to 1992 and banning the export of logs from 1993 to 1996. In regard to diminishing timber exports, the share of forest-oriented revenue has declined from 51 percent in 1990 to 20 percent in 1999. The state needs to diversify its wood-based industries in order to improve its economy, using wood materials as an important source of income. It is also important for the state to intensively and immediately plant suitable trees. The decrease of forest resources in Sabah is a serious problem, which needs to be solved for the sake of their local economy, society, and environment.

In Malaysia's state of Sarawak, the forest-related industry is also independent of federal policies. Exploitation of its forests began actively in the 1980s, later than other states, because of its lower timber quality. The export restrictions of Indonesia and Sabah had strong influences on Sarawak's pattern of forest exploitation and timber exports. It seems now that Sarawak has switched to a relatively sustainable forest management policy, following advice from the International Tropical Timber Organization (ITTO) in the beginning of the 1990s. Thus Sarawak still has comparatively rich forests today, because of its short period of active forest exploitation and the way it was managed. According to Malaysia's statistics, the state's forest resources are larger than in the 1960s. Sarawak created a log export quota in 1992, and then tightened it in 1993, to encourage its wood-based industries. It developed some timber processing zones, and forest-oriented state revenues increased by around 40 percent in the 1990s. At the same time, the state promoted wood-based industries, such as logs, plywood, wood residue products, and pulp and paper, and then increased its exports of processed wood products. Its timber processing zones will surely play an important role in the development of wood-based industries and the state economy at the beginning of the twenty-first century. In order to encourage wood-based industrialization, it is indispensable for Sarawak to sustain its forest resources through sustainable logging practices and reforestation. At the same time, as pointed out by local and international non-governmental organizations (NGOs) over the last decade-and-a-half, it is necessary to build partnerships between aboriginal people and the use of forests in order to create the "forest-oriented and forest-based society" of the future.

3.3 The Philippines

It is important to grasp the past, present, and likely future state of the forest resources in the Philippines in order to analyze its forest products industries, but it is quite difficult to get sufficient data about forest areas and growing stock (Shimamoto 2000a). Time-series data about woodland areas defined by the national government exist, but they do not necessarily mean that trees are actually standing. The only way to comprehend the real resource situation is by integrating the fragmented data available.

The total land area of the Philippines is about 30 million hectares. The forest covered about 50 percent of the total land area in 1950, but had dropped to only 18 percent by 1996. Records show that the major depletion of forest resources in the

Philippines began shortly after World War II, and that the largest part of the domestic old growth forest resources has already disappeared. In 1992, all logging in virgin forests was banned; so, since 1991, old growth areas of dipterocarp forests stopped decreasing in the statistical data, but other natural forests continue to be degraded.

With respect to log production and trade, the peak of log exports in the Philippines was seen in the first half of the 1970s. Although log production exceeded 10 million cubic meters (m^3) early in the 1970s, after that it began to decrease drastically. In 1980, production was 6.4 million m^3; in 1985, it became 3.6 million m^3. Log exports also dropped, from 8.4 million m^3 in 1970 to 0.76 million m^3 in 1980. In 1986, log exports were finally banned, and in 1987 the government announced that it would not issue any new TLAs (the Philippine's timber concessions). At the same time, imports of logs began to increase. In 1996, the amount of imports exceeded the production quantity. The largest share of imported logs was sawlogs and veneer logs, and pulp and pulpwood imports totaled around 240,000 m^3. Major countries exporting to the Philippines are Indonesia, Papua New Guinea, and New Zealand, but other Pacific Rim countries export as well. The Philippines also imports from African and Latin American countries—more than 10,000 m^3 in 1995. The delivered (cost, insurance, and freight, or CIF) price of imported logs from these countries in 1995 was around U.S.$100 per m^3, but now there are rather large differences in the price among these countries.

Domestic producers of forest products are now at a critical point of survival. Because the supply from natural forests has been depleted, they cannot compete with inexpensive, high-quality imported goods with their old facilities and techniques. Producers face the following issues.

The first issue is tariff reduction. Producers of forest products in the Philippines have strong concerns about reductions in the tariff rate for imported forest products, because the high tariffs had protected the producers from international competition. In the case of plywood, the tariff rate was 50 percent until 1995. The Philippines, however, signed the Common Effective Preferential Tariff (CEPT) Agreement with the Association of South East Asian Nations (ASEAN) in 1992, which requires member countries with tariff rates above 20 percent to reduce the tariffs to that level within a five- to eight-year period after 1993. In accordance with this agreement, from 1996 to 1997, the tariff rate on plywood decreased to 30 percent, and in 1998 it dropped to 20 percent. The gross profit rate of plywood has been extremely low recently. If the tariff rate goes down to 10 percent, domestic plywood companies will be endangered. As for lumber, the tariff rate was 20 percent until 1997, and scheduled to be reduced to 10 percent in 1998. The Philippines Wood Producers Association requested a postponement on the reduction to 10 percent until 2000 to allow producers time to restructure the industry.

The second issue is the future supply of inexpensive domestic wood materials. One of the most important issues for the domestic lumber and plywood producers is how they can obtain inexpensive materials, either from domestic forests or foreign markets. Supplies from foreign natural forests are shrinking gradually, meaning that domestic producers can look forward to increases in the supply of inexpensive wood material from domestic plantation forests. The following estimates demonstrate that the domestic plantation forests will be able to supply the needed wood material. According to the

National Forestation Development Office (1997), the reforestation cost of a governmental project in Luzon in 1996 was 20,463.28 pesos per hectare. The estimate of the average yield of fast-growing species is 160 m^3 per hectare, which will be harvestable in eight years. Under these assumptions the reforestation costs will be 127.9 pesos per m^3. The harvesting and transportation costs are 750 pesos per m^3 for the plantation forests, meaning that the supply price will be 1,230.8 pesos per m^3, or $29.5 per m^3, at present exchange rates. These are the estimates of government projects; in the case of industrial plantations the costs may be slightly lower. But domestic producers still have to wait several years for the increased supply of domestic wood material. Another problem to be solved is how to use available plantation species for plywood face panels.

Now that the natural forests have been completely depleted, the Philippines are in a transition period, where the forest products industry is becoming more reliant on plantation forestry. One possible scenario is that the domestic forest products industry will be supported until the degraded forestlands are recovered through reforestation. This situation is similar to that of Japan. It will be necessary to reconsider the tariff reductions and the free trade system of forest products based on this point of view.

3.4 The Republic of Korea

The changes in North American forest policy for biodiversity conservation reduced the supply of coniferous logs in the international market and induced an increase in the international price of coniferous logs (Youn 2000). Consumers in the log market of the Republic of Korea subsequently changed their source of log imports from North America to New Zealand and Chile. The elasticity of Korean coniferous log demand with respect to its own price was –0.404, while that with respect to tropical hardwood logs was –0.257 (Youn 2000). The consumption of timber in Korea will continue to rise, while the domestic production of logs will increase only slightly. Therefore, the supply of timber from foreign countries will remain important to Korea.

The import of tropical hardwood timber has been also affected by international environmental policy, which stresses the conservation of tropical forests (Youn 2001). The conservation of tropical forests reduced the supply of tropical timber, raising the price. The higher relative price of tropical timber (to that of softwood) effectively reduced the demand for tropical timber in Korea, and induced an increase in demand for softwood. The higher relative price of hardwood, expected in the coming decades, implies that long-term forestry investment should be switched from the present softwood-oriented policy to one oriented to hardwood production.

4. FOREST CERTIFICATION SCHEMES[6]

Initially, the Forest Stewardship Council (FSC) focused on tropical forests, but these days they have been also including temperate forests as targets for certification. The main reasons behind the change may be (a) that tropical countries fear that constraints on tropical forest products lead to unfair trade restrictions, (b) that, as the FAO pointed out, one of the most important causes of tropical deforestation was not from timber production but from conversion to agricultural lands, and (c) the fact that timber from temperate forests constitutes about 80 percent of the world timber trade.

Although the Asian region has a large number of countries and forest areas, there are presently only a few forest and timber certifications. As for FSC certification, there were seven countries with the certification in September 2000. Their share of the world total is 2.9 percent for the number of forest certifications and 0.8 percent of the total area certified.

Past certification records reveal that the Rain Forest Alliance rises above the rest in terms of the number of certifications, and the Société Générale de Surveillance (SGS) in terms of the area of certifications. One of the reasons why certifying institutions have certain tendencies is that they have different policies. For example, the Rain Forest Alliance takes communal, non-industrial, and private clients, while SGS takes industrial clients, probably because the former is a non-profit NGO, while SGS is a profit-seeking business enterprise.

Forest certification is a "soft" policy instrument, not a command-and-control type. It operates through incentives in the markets. To make certification schemes effective, certification should be carried out as a spontaneous process by independent institutions. All the participants—from production to distribution stages—should share the benefits in order to make the scheme efficient and effective. Theoretically, the benefit comes from the higher price premium which consumers pay for the fact that they get products from appropriately managed forests. This incentive program stands at the heart of certification schemes.

There are two merits to certification schemes. The first is the price premium. Consumers agree to pay more for environmentally sound products. An important point here is that they should recognize the certification process as reliable and that they should support forest management systems which are certified through the scheme. The second is the savings in additional costs that would occur if the government was forced to introduce more stringent restrictions for the sake of environmental protection.

Currently, the most important markets for certified products are Europe and North America. In Asia, they pay attention to the sustainability of forest resources, but not enough to purchase certified products. Because most Asian developing countries do not have sufficient forest management codes and institutions, the FSC's mutual certification scheme, which has a high level of criteria and indicators of institutional forest

---

[6] In Chapter 6 the current state of forest certification in the world and some cases in the Asia-Pacific region are revealed and analyzed worldwide. It also has some brief explanation of certification organizations.

management, is worth considering for introduction as a scheme for promoting sustainable forest management (Minowa 2000; Tachibana 2000b).

Certified products do not yet make up a large share of the world market. Total global roundwood production is 53 billion m^3. Only around 0.6 percent (318 million m^3) is certified through the FSC, and only a fraction is traded as certified products. One of the main reasons stems from inefficiency in tracking the "chain of custody" (CoC). Even though forests themselves are certified, unless the entire process of production and distribution is appropriately absolved until final consumers get the products, the certification scheme functions incompletely. If the chain breaks at one place, the entire certification fails. Certified products are generally hardwoods used to make relatively small products, such as guitars, furniture, and wooden fittings. The demand for certified softwood products will increase in the future. We think that the roles of timber-importing countries, such as Japan and Korea, will be more important for encouraging related market development and sustainable forestry.

5. SPATIAL EQUILIBRIUM MODEL OF TIMBER TRADE IN THE ASIA-PACIFIC REGION

The Asian Forest Products Trade Model (AFPM) is a spatial equilibrium model, developed by reconstructing the FAO's Global Forest Products Model (GFPM), which analyzes dynamic changes of production, consumption, and trade in the Asian region (Shimamoto 2000b, 2001). Utilizing this AFPM we simulated the Asian forest products trade market for about twenty years to come.

5.1 Logs

From 1997 to 2020 the production volume of logs in the region will grow 1.77 percent per year, whereas the consumption will grow 1.83 percent per year on average. It seems that the resource base of natural forests in Southeast Asia has decreased, especially in the island areas. As for trade flows, exports will increase 0.75 percent per year on average from 1997 to 2020, and imports will increase 1.84 percent per year. It also indicates that imports from outside the region will grow more than exports.

Indonesia and Malaysia will maintain their positions as major log producers, but their production will not grow so large; their growth rates of production per year on average from 1997 to 2020 are 1.29 and 1.42, respectively. In Japan, log production growth is 0.72 percent per year, which is lower than that of the other Asian countries, while its imports will increase by 1.75 percent per year on average during the period. Major exporting countries in the Asian region are Papua New Guinea and Malaysia, and the exports of Papua New Guinea, in particular, will grow. Outside of Asia, the United States, Russia, and Chile will be major log exporters.

5.2 Sawnwood

In the simulation, an average growth rate of sawnwood production in the Asian region is 1.2 percent annually from 1997 to 2020, while the growth rate of consumption is 1.6 percent. In the calculation, the exports from the exogenous countries were set to increase 2.7 percent annually and imports into these countries were set to decrease 2.9 percent annually during the period.

In the Asian region, China will increase sawnwood production at a rate of 2.27 percent annually to respond to the growth in demand. The other major producers are Indonesia and Malaysia, but the annual growth rates will be moderate, at 1.95 percent and 1.60 percent, respectively. Vietnam will grow at a high rate of 3.21 percent, whereas Thailand is expected to have a drastic reduction in the production of sawnwood, because of resource degradation, with a big increase of 3.19 percent annually in imports. Furthermore, with the exception of Malaysia as a major exporting country, the export volumes of Southeast Asia will decrease from 1997 to 2020.

Major importers in the Asian region will be Japan, China, and Korea. Korea will show 2.81 percent of an annual growth rate of import volume, and Japan, 2.44 percent. Japan's production will decrease 0.8 percent annually.

5.3 Plywood

In the simulation period, the production volume of plywood will grow annually at 2.77 percent on average, and consumption, at 3.31 percent. China, especially, has a high growth rate of 4.7 percent per year, because of its rapid growth in gross domestic product (GDP). As for the other countries, Vietnam will increase rapidly with a high growth rate of 6.28 percent. On plywood, Indonesia and Malaysia will take the position as major producers in Southeast Asia, but they will shift their production into more highly processed goods like fiberboard, pulp, and paper.

Indonesia will keep its position of plywood exporter, though its annual growth rate in the simulation is –0.97 percent. The major importer in the Asian region consists of Japan, China, and Korea. Their average annual growth rates of imports are –2.52 percent, 1.85 percent, and 3.84 percent, respectively.

5.4 Paper

An annual growth rate of newsprint production is 3.6 percent in the period, and the consumption growth rate is 3.19 percent. The growth rate is rather high, because GDP elasticity is set to 1.07 for developing countries. In particular, Vietnam and Thailand will increase paper consumption by 7.97 percent and 7.71 percent annually, respectively. The major exporter in the Asian region is Korea, while export in exogenous regions is carried out on a much larger scale. Asian countries are exposed to their export pressures for newsprint. The major importers in the region are Japan, Hong Kong, Malaysia, China, Thailand, and Singapore. With the exception of Japan, import volume in each country will increase by 2 to 3 percent per year.

The growth rate of production of printing and writing papers is 3.14 percent, and the growth rate of consumption is 3.48 percent annually. The major producer of these products is Japan, and exporters are Korea and Indonesia. Major importers are Hong Kong, Malaysia, and Singapore. Exogenous countries are not such major players in either exports or imports of printing and writing paper.

An annual growth rate of production of wrapping paper and paperboard in the period is 1.54 percent, and a growth rate of consumption is 1.73 percent annually. This low rate is due to the low GDP elasticity, especially in developing countries. GDP elasticity in developing countries and developed countries is 0.36 and 0.55, respectively. The major producers in the Asian region are China and Japan, whereas the major exporter is Hong Kong. In Southeast Asia, Indonesia and Thailand are the main producers and exporters.

6. CONCLUSION AND POLICY RECOMMENDATIONS

Clarifying the relationship between the social economy and the natural environment in the forest sector is one of the key steps towards constructing policies necessary for sustainable forest management. To this end, forest resource accounting should be used. First, environmental information suitable for a forest resource accounting framework and information on social economic factors should be collected. Next, forest resource accounts should be constructed properly, namely, into (1) forest accounts, (2) forestland accounts, (3) forest product accounts, and (4) forest management accounts.

Controlling the forest products trade under the principle of forest sustainability and mitigating the degradation of natural forests are other key steps. In terms of concrete actions, trade restrictions may be needed to foster the domestic markets of forest products, which may encourage afforestation. Additionally, stricter controls, monitoring, and punishment for unlawful acts in the forestry and forest products trade should be encouraged. Monitoring forest resources and constructing resource accounts to clarify the relationship between human activities and forest resources should be promoted, as well.

Forest and forestry certification schemes may be one possible solution to connect consumers' preferences and sustainable forest resource management. To develop appropriate schemes, market research, which covers the flow of certified products in the national markets both in importing and exporting countries of southeastern Asian, should be conducted, as well as case studies of the certified forests of individual, enterprise, and public entities (through interviews and questionnaire surveys). In southeastern Asia, national and local governments and environmental NGOs should take the initial steps. Promoting measures for forest certification and the trade of certified wood products should also be investigated. Buyer's groups, like in Europe and America, can play an important role.

Developing an effective strategy for the certification process and timber trading in each region or country is also a key step in the scheme. Up to now we can say that many forestry industrialists are hesitant to participate in organizations with environmental protection characteristics. This is one of the reasons why participation rates in FSC certification schemes have not increased dramatically. Another is that FSC criteria and indicators are too detailed for forest owners and industrialists to understand and too rigid to apply in the field. Therefore, the following actions should be promoted in Southeast Asian countries as well in order to make forest certification schemes more accepted. The first is to construct homogeneous evaluation criteria and indicators among certifying institutions, and to follow them in carrying out certifications. The second action is to construct domestic criteria and indicators, developing those of the FSC. The third action is to establish Asian certifying institutions.[7]

As the simulation results using AFPM reveal, in the Asia-Pacific region there is strong demand for wood-based products, and it seems to be more than the forest resources currently at hand in the region can provide. So timber trade is expected to continue to play a very important role in meeting the demand. In addition, the demand will increase unevenly in the region. Some countries, like China and Vietnam, seem to vigorously demand wood-based products. The structure of the wood-based industry will change, as well. Current exporters of sawn timber and plywood, especially Indonesia and Malaysia, will shift their production to more highly processed goods. This dynamic change will affect trade structure, as well. In terms of policy recommendations, orderly trade is needed here, since timber trade influences forest resources, in most cases, negatively. Because forest resources are scarce global resources, we should not rely only on trade, but focus on expanding our resource base. We should better manage and control our scarce forest resources.

In terms of recommendations for Southeast Asian countries, we must mitigate deforestation rates and then enrich forest resources. Even for the countries where forest resources are now increasing in volume, their utilization is still developing, either for timber or other material uses, or for recreational uses. The following concrete steps in this direction should be taken.

1. Construct forest-planning schemes in accordance with sustainable forest management.
2. Promote forest resource utilization, both in terms of material and recreational use.
3. Propagate understanding among the public of the importance of forest resources.
4. Encourage forestry which contributes to sound forest management and the sustainable utilization of forest resources through such measures as tax and subsidy schemes, encouraging the recruitment of forestry workers, promoting forest road construction and maintenance, and enhancing mechanization.
5. Promote sustainable forest management in the forest products industries.
6. Shift a large part of responsibility for forest management to the public sector.
7. Revitalize rural mountainous villages through the encouragement of the forestry industry.
8. Carry out research towards the fulfillment of sustainable forest management.

---

[7] So far, established certifying institutions are based in Europe and the Americas.

REFERENCES

Koike, K. and N. Fujisaki, eds. 1997. *Forest resource accounting*. Tokyo: Institute of Developing Economies.

Kurabayashi, Y, K. Koike and N. Yamamoto, eds. 2002. *The progress in environment and resource accounting approach: A principle to the global environmental issues*. Yonago: Imai Syuppan Co. Ltd.

Massijaya, M. Y. and H. Kartodihardjo. 2000. Forest resources and the forest products industry in Indonesia. In *A step toward forest conservation strategy (2): Interim report 1999*, 291–311. Hayama: IGES Forest Conservation Project.

Minowa, Y. 2000. Sustainable forest management and forest certification in Malaysia. In *A step toward forest conservation strategy (2): Interim report 1999*, 247–258. Hayama: IGES Forest Conservation Project.

National Forestation Development Office. 1997. *Updated comprehensive site development cost and production standard*. Manila: DENR.

Shimamoto, M. 2000a. Forest resources and the forest products industry in the Philippines. In *A step toward forest conservation strategy (1): Interim report 1998*, 392–408. Hayama: IGES Forest Conservation Project.

———. 2000b. Spatial equilibrium model. In *A step toward forest conservation strategy (1): Interim report 1998*, 409–416. Hayama: IGES Forest Conservation Project.

———. 2001. Provisional version of Asian forest products trade model (AFPM) and examination of WTO demands for free trade in forest products. In *A step toward forest conservation strategy (3): Interim report 2000*, 98–124. Hayama: IGES Forest Conservation Project.

———. 2000a. Forest exploitations and timber export restrictions in Southeast Asian countries (in Japanese). *Study for Regional Policy* 3(1): 49–71.

———. 2000b. Forest-related industries and timber trades of Malaysia. In *A step toward forest conservation strategy (1): Interim report 1998*, 375–391. Hayama: IGES Forest Conservation Project.

———. 2000c. Impacts of log export restrictions in Southeast Asia on the Japanese plywood market: An econometric analysis. *J. for. Res.* 5: 51–57.

———. 2001. Impacts of log export bans on forest-related industries and the timber royalty system in Sabah State, Malaysia. In *A step toward forest conservation strategy (3): Interim report 2000*, 64–79. Hayama: IGES Forest Conservation Project.

Yamamoto, N. 2000a. Forest resource accounts and trade model. In *A step toward forest conservation strategy: Interim report 1998*, 344–358. Hayama: IGES Forest Conservation Project.

———. 2000b. Using environmental resource accounts to link economy with environment in the forest sector. In *A step toward forest conservation strategy (2): Interim report 1999*, 225–245. Hayama: IGES Forest Conservation Project.

Youn, Yeo-Chang. 2000. Korean timber markets and policy. In *A step toward forest conservation strategy (1): Interim report 1998*, 359–374. Hayama: IGES Forest Conservation Project.

———. 2001. Korean forest-based industries and forest product markets. In *A step toward forest conservation strategy (3): Interim report 2000*, 80–97. Hayama: IGES Forest Conservation Project.

Chapter 3

PARTICIPATORY FOREST MANAGEMENT POLICY IN SOUTH AND SOUTHEAST ASIA

Makoto Inoue
The University of Tokyo, Graduate School of Agricultural and Life Sciences
Institute for Global Environmental Strategies[1, 2]

1. INTRODUCTION

Usually, the field of forest policy is segmented into (1) forest inventory and planning, including forest classification; (2) management of production forest, consisting of natural forest management and man-made forest management, including reforestation; (3) management of protection forest and degraded land, including afforestation, to prevent soil erosion and natural disasters; (4) management of biodiversity in conservation areas; and (5) control of circulation and trade of timber and non-timber forest products (NTFPs).

There is general consensus that to achieve sustainable forest management, the participation of local people is essential in every domain of forest policy. Participation of the people can lower transaction costs and help avoid social conflicts over forest utilization, which themselves increase management costs (Figure 1). Thus, the participation of people should not be regarded as an opposing factor but rather as a prerequisite for sustainable forest management in segmented domains of forest policy.

---

[1] The original draft of this work was written by the author, based on the results of field studies conducted by the members listed in the following footnote, after receiving comments from Dr. Hiroaki Kakizawa, Dr. Jin Sato, Mr. Noriyoshi Kitamura, Dr. Percy E. Sajise, Dr. Do Dinh Sam, and Dr. Pearmsak Makarabhirom, as well as from various other stakeholders, such as governmental officers, NGOs, researchers, and international organizations at workshops held at Jakarta in June 2000 and at Vientiane in August 2000.

[2] Field studies were conducted by Mr. Martinus Nanang, Dr. Yosei Oikawa, Mr. Herman Hidayat, and myself in Indonesia; Dr. Atsuko Hayama and Dr. Yoshiki Seki in the Philippines; Mr. Kimihiko Hyakumura, Mr. Khampha Chantirath, and Mr. Khamvieng Xayabouth in Laos; and by Dr. Toshiyuki Tsuchiya and Mr. Le Quang Trung in Vietnam.

M. Inoue and H. Isozaki (eds.), *People and Forest—Policy and Local Reality in Southeast Asia, the Russian Far East and Japan, 49–71.*

The aim of this chapter is to analyze participatory forest management in South and Southeast Asian countries and make recommendations to improve national forest policy in selected countries.

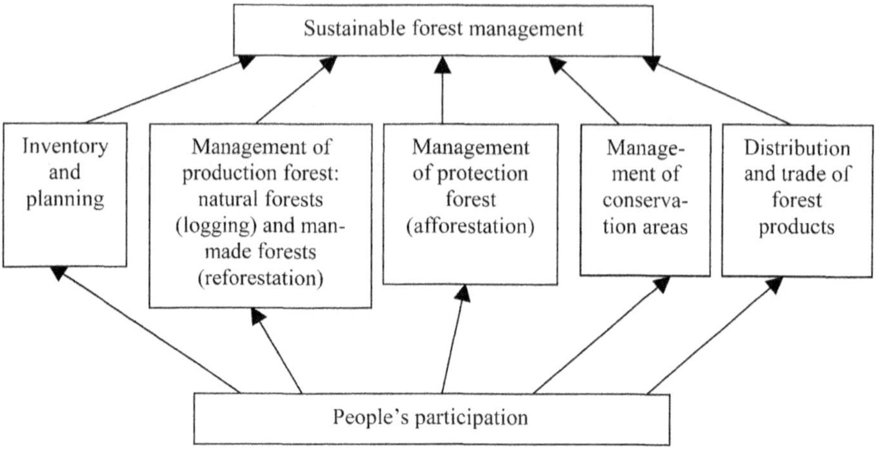

Figure 1: Importance of people's participation in every domain of forest policy.

2. PARTICIPATORY FOREST MANAGEMENT IN SOUTH AND SOUTHEAST ASIA

2.1 Participatory Forest Management in the Concept of Social Forestry

In the late 1970s professional foresters in the tropics noticed that they could not manage the forest sustainably under the principles of conventional or industrial forestry, where the local people were considered to be obstacles or constraints to forest management. Social forestry was recognized as an important norm or principle to produce successful sustainable forest management, even though industrial forestry has been dominant in practice. Many scientists and non-governmental organizations (NGOs) came to prefer the term "community forestry" to "social forestry," however, because the concept of social forestry was created as a conciliatory measure and a top-down approach was often adopted in social forestry projects.

Originally, social forestry and community forestry were defined similarly as any situation that intimately involves local people in forestry activity for the purpose of rural development (Pardo 1985). These days, however, it seems that social forestry involves a wider range of comprehensive participatory activities, and community forestry implies collective activities rather than individual activities, such as farm forestry.

In general, social forestry consists of two major components. One of these is participatory forest management in the forestry sector. The other includes (1) development of infrastructure such as roads, meeting places, schools, and clinics; (2) agricultural extension; and (3) generation of income sources for rural development, etc. If the latter activities prevail without the component of participatory forest management, however, the activities are not necessarily called social forestry, and the more general term "rural development" is used. It is evident that participatory forest management is the core of social forestry (Inoue 2000a).

Considering the unpopularity of using the term "social forestry," it might be better to use "participatory forest management" instead.

2.2 Analytical Framework

In order to compare policies, three concepts are applied as an analytical framework or valuation basis: legal status of land, activities of forest management, and main actors of forest management.

2.2.1 Legal Status of Land

Generally, possible legal possessors of land and forest are individuals, organizations, villages, outsiders, and governments, such as districts, provinces, and the state. Participatory forest management systems can be adopted regardless of the legal status of the land. Provisionally, we recognize land with differing legal status to be individual land, organizational land, village land that is owned by both a formal village or indigenous people's community, outsiders' land, and national land that is owned by the local and national government.

2.2.2 Activities of Forest Management

Even though there are many forest-related activities, we can categorize them into four activities, which are shown below.

1. **Man-made Forest Management,** including tree planting, consisting of reforestation and afforestation on degraded land.
2. **Harvesting** or natural forest management for timber production in commercially valuable forests.
3. **Conservation,** including the collection of fuelwood and non-timber forest products (NTFP) and small-scale recreation and eco-tours.
4. **Protection** of the forest from any kind of utilization. Wildlife preservation is included here. We can assume that the management of protected areas, including national parks, consists of both protection and conservation activities.

In most cases, economic benefit may be the major incentive for the main stakeholders. It often occurs that local people join in activities of man-made forest management, harvesting, and conservation, but it is more difficult for them to find incentives to participate in protection activities. When the government employs local people to protect the forests, however, the income they receive becomes the incentive.

2.2.3 Main Actors of Forest Management

There are many actors who are directly involved in local forest management. The terms "local people" and "local community" are too general to develop a strategy for participatory forest management. It is useful to classify the main stakeholders who have responsibility for and take the initiative in forest management into the following six categories:

1. **Individuals** or peasants and the members of their households in the village. Their forest management can be called "peasant forestry" or "farm forestry."
2. **Functional Groups** such as forest users' groups, cooperatives, schools, temples, women's unions, and elder's groups. Their forest management can be called "functional group forestry."
3. **Fundamental Groups** such as groups of relatives, natural villages, and indigenous cultural communities. Their forest management can be called "fundamental group forestry."
4. **Executive Bodies of the Formal Village** that are entrusted to manage the forest by the villagers. Their forest management can be called "village forestry" that includes centralized community forestry.
5. **Outsiders and Corporations** who get forest concessions. Their forest management can be called "private forestry" for the purpose of earning profit.
6. **Local and National Governments**. Their forest management can be called "public forestry."

Forest management by functional groups, fundamental groups, and executive bodies of a formal village can be regarded as "community forestry," since they are based on "collective management."

2.3 Features of Participatory Forest Management

Comparative analysis applying the framework revealed that the land still belonged to the state, and the right to use the land was granted to the local people in most South and Southeast Asian countries (IGES 2001). Other findings relevant to the activities and main actors of forest management are summarized below:

1. The functional groups, fundamental groups, and executive bodies of formal villages carry out the activities of harvesting and conservation collectively.
2. It is ascertained that most collective management by local people is not implemented by fundamental groups but by functional groups.

a. Typical examples:
   i. Forest users' groups in Nepal.
   ii. Village forest protection and management committees (VFPMC) in India.
   iii. Cooperative and other groups in Indonesia.
   iv. People's organizations (PO), associations, and cooperatives in the Philippines.
   v. Village forestry associations (VFA) in Laos.
   vi. Certain organizations in Vietnam.
b. Exceptional cases where fundamental groups are regarded as main actors:
   i. **Certificate of Ancestral Domain Title (CADT) in the Philippines**—The CADT grants indigenous cultural communities (ICCs) with the rights of land ownership. The CADT was clearly named in the Indigenous People's Rights Act (IPRA) issued in 1997. Prior to the act, the Certificate of Ancestral Domain Claim (CADC) was permitted in 1993 and integrated into community-based forest management (CBFM) in 1995.
   ii. **Land/forest allocation program in Laos**—This permits the village to manage the whole territory within the village boundary.
   iii. **Customary forest (*hutan adat*) in Indonesia**—The community that practices customary law (*masyarakat hukum adat*) can have the rights to (1) collect forest products for daily needs, (2) undertake forest management in accordance with prevailing customary laws, and (3) be empowered to improve its welfare. This was defined in the new Forestry Law in September 1999.

3. In most cases, successful participatory forest management has focused on the activities of man-made forest management, conservation, and protection, and mainly conducted on secondary forests and degraded lands, while the governments and timber concessionaires conducted the activities of harvesting to manage commercially valuable forests.
   a. Examples of successful activities on secondary forest and degraded land:
      i. **Community Forestry in Nepal** – Legislation and policies allow the forest users' group to carry out activities of man-made forest management and conservation, allowing for a fair degree of devolution of decision-making. The user group is permitted to sell and distribute forest products and fix prices without consulting the Forest Department. The district forest officer is to provide technical assistance to the user group in order to prepare two essential documents: the constitution of the user group and the work plan of the community forests (Chakraborty et al. 1997).
      ii. **Joint Forest Management (JFM) in India** – The village forest protection and management committee (VFPMC) is involved in forestry activities, including protection and man-made forest management, but objectives are set by the Forest Department and decisions are made on the basis of the objectives (Fisher 1999).

Virtually all management decisions ultimately rest with the government or Forest Department (Lynch and Talbott 1995).

b. One exceptional example of activities in commercially valuable forest was a trial conducted by the Forest Management and Conservation Project (FOMACOP) in Laos, supported by the World Bank and Finland. The trial was called **Village Forestry**, where the government gave the village forestry associations (VFA) full rights to manage and harvest commercially valuable trees in national production forests. The central government withdrew its support, however, and declared all logging operations would be put under the control of the government in its Decree No.11 in May 1999.

4. Swidden agriculture or shifting cultivation is usually not permitted as one of the activities of conservation. An exceptional example is **community forestry in Cambodia**, where a group of residents in one or more villages may continue to practice existing swidden agriculture within the areas described in a community forestry agreement during specific periods of time, as explained in the community forestry management plan.

5. It is interesting that many programs assume that collective management can also be suitable for man-made forest management, including tree planting, even though individual management seems to be suitable for planting activities rather than collective management in terms of economic incentives. Examples are:

   a. **Individual forest (*hutan rakyat*) in Indonesia**—This program is controlled by the provincial government and practiced on private land. The people hold a certification of land ownership outside the national forest. The main activity of the program is re-greening or afforestation, and *Paraserianthes falcataria* is planted by many people.

   b. **Socialized industrial forest management in the Philippines**—This program allows individuals or families and associations or cooperatives to participate in forest plantation development in forest areas, ranging in size from 1 to 10 hectares and from 10 to 500 hectares, by providing them security of tenure through the issuance of a Socialized Industrial Forest Management Agreement (SIFMA).

   c. **Tree planting by individual villagers in Laos**—People living in the villages can plant trees such as teak and fast-growing species on allocated land.

   d. **Tree planting on allocated land in production forests in Vietnam**—Individuals, households, and organizations can be allocated land and obtain red book certificates in production forests. The land area allocated to them varies from place to place. For example, each household gets three to five hectares of land on average in a mountainous region, while there are households getting more than 50 hectares in other regions.

   e. **Joint forest management in India and community forestry in Nepal,** as mentioned above.

6. The government controls protection activities. The local people are excluded or sometimes involved in the activities under government supervision.
   a. A typical example is the **Protection Agreement in Ecological Rehabilitation Zone of Special-use Forests in Vietnam.** The Special-use Forest Management Board (SFMB), having a red book certificate, concludes protection agreements with households, which acquire "green book certificates," that prohibit them from intercropping but permits them to plant trees. As for tree planting (Plantation Program), each household earns one to two million Vietnamese dong per hectare and has an obligation to protect the planted trees for three years on two to four hectares of forestland, on average. As for natural regeneration (Protection Program), each household annually earns 40,000 to 50,000 Vietnamese dong per hectares for the protection of 10 to 20 hectares of forest, on average.
   b. One exceptional example is the **management of protected areas in the Philippines.** According to the National Integrated Protected Areas System Act, or NIPAS Act (Act No.7586, 1992), the Department of Environment and Natural Resources (DENR) shall have no power to either evict indigenous cultural communities (ICC) from their present occupancy or resettle them to another area without their consent. ICCs can manage their surroundings within the restrictions as a result of an agreement between the local people and the government. In the same way, the "tenured migrant" who has actually and continuously occupied an area for five years prior to its designation as part of a protected area is eligible to become a steward of a portion of it. Their activities are governed by guidelines, however, as prescribed in the management plan as well as by the prohibitions set out in the Act.

3. POLICY RECOMMENDATIONS FOR PARTICIPATORY FOREST MANAGEMENT

3.1 Procedures and Standpoints to Elaborate Policy Recommendations

A sub-team of the Participatory Forest Management Project conducted a comprehensive process that led to the preparation of policy recommendations in the first phase of IGES's Forest Conservation Project. First of all, in the target countries (Indonesia, the Philippines, Vietnam, and Laos) we identified the **external constraints** on community participation in forest management by clarifying the gaps and contradictions between national land, forest policies, customary land rights, and forest management by the local people. Second, we identified the **internal constraints** that are immanent in the local communities in terms of economic, social, and cultural aspects. Third, **possible main actors** were identified by evaluating the local realities and national forest policies. As well, we identified the lessons learned from public participation in developed countries. We elaborated these draft policy recommendations by considering how to overcome the internal and external constraints, and recommended the main actors to carry out them (Figure 2).

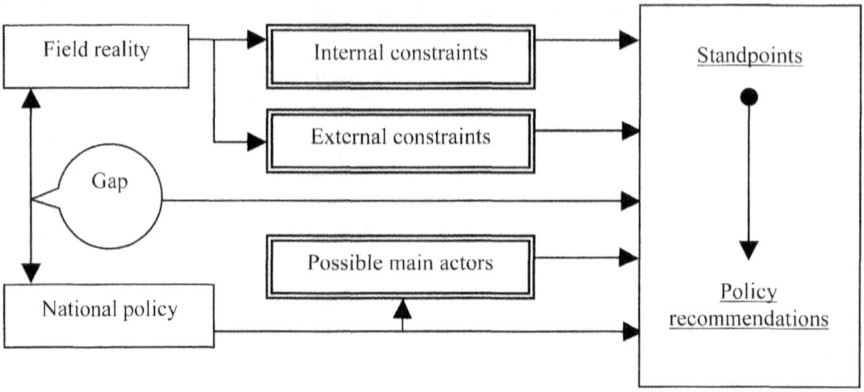

Figure 2: Procedures to elaborate policy recommendations.

Before summarizing the results, we outline here the standpoints as the premise for elaborating the recommendations.

Standpoint 1—People's participation is very important for success in sustainable forest management, with lower transaction costs, as well to avoid social conflicts over forest utilization, which themselves increase management costs.

Supplementary explanation: Most stakeholders recognize that the forest should be managed in such a way that economic benefits, social justice, and ecological sustainability can be achieved without excluding other stakeholders. The main problem to overcome is the distribution of economic benefits among stakeholders such as local people, cooperatives, timber companies, and governments.

Standpoint 2—In the tropics, the concept of community participation is more useful today than the concept of public participation.

Supplementary explanation: People's participation consists of public participation, which refers to the participation of the larger society, including city dwellers and citizens. Community participation refers to the participation of a smaller subset of society or the local community.

Standpoint 3—Our concern is to show what an ideal forest management *system* is in terms of community participation, rather than to indicate concrete *procedures*.

Supplementary explanation: Our recommendations should be examined from the perspective of feasibility under present social and political conditions.

Standpoint 4—The participatory top-down approach should not be included in the strategy for participatory forest management.

Supplementary explanation: The term "participation" in this context does not reveal the actual level of participation because the meaning of the term varies widely. The spectrum of participation could be put into three categories as follows (Inoue 2000b):

1. The **participatory top-down approach** is a blueprint approach where residents are considered to be wage laborers, volunteers, fund providers, etc.
2. The **professional-guided participatory approach** is a relatively flexible blueprint approach where drafts of the plan, made by professional planners, are examined by the residents and citizens and modified through discussion, workshops, etc. The local people and the government share authority.
3. The **endogenous bottom-up approach** is a learning process approach where professionals act as facilitators only. The local people have decision-making rights.

Formal institutional arrangements can be part of the endogenous bottom-up approach when customary law is developed in the local community, because the residents can manage their resources well by themselves. On the other hand, application of the professional-guided participatory approach seems to be reasonable when the customary law has not been developed or has already lapsed. In any case, the participatory top-down approach should be avoided, because it is usually considered by local people to be nominal and fake.

Standpoint 5—Both collective forest management and individual-based forest management are considered to be included in participatory forest management for the time being.

Supplementary explanation: It is true that collective forest management by fundamental groups, functional groups, and the executive body of the formal village is the core of participatory forest management. It can safely be said that individual- or household-based forest management on national land, called "peasant forestry" here, is also recognized as a form of participatory forest management. On the other hand, peasant forestry on private land may not be regarded as a form of participatory forest management but as "private forestry." In order to promote participatory forest management, however, it seems better that we provisionally consider it to be a form of participatory forest management.

Standpoint 6—The policy recommendations should be based on recognition—true in both developed and developing countries—of the importance and validity of participation.

Supplementary explanation: Key issues in community involvement include the following: defining the community; willingness and ability to have dialogue; presenting scientific information in a form which can be easily understood; reconciling local, national and, perhaps, global interests; and political will to respect and enforce the conclusions (Salim and Ullsten 1999). We believe that these ideas are appropriate for civil societies worldwide, including developing countries.

Standpoint 7—Diversity of legal status of land and main actors is an important aspect of participatory forest management in order for local people to make good choices in accordance with the local conditions.

Supplementary explanation: We believe it is beneficial to develop strategies for facilitating participatory forest management on national land, village land, organizational land, and individual land and for facilitating participatory forest management by individuals, functional groups, fundamental groups, and executive bodies of the villages. It is desirable that the policies of each country cover every combination of land ownership with the main actors.

Standpoint 8—The activities of man-made forest management, including planting trees, is likely to be practiced mainly by individuals. The activities of conservation should be conducted collectively by village communities and fundamental groups. The government is likely to bear the responsibility for the activities of protection, sometimes entrusting the local people with day-to-day activities.

Supplementary explanation: Forest management consists of four activities, as mentioned above. In terms of an economic incentive, individuals are most likely to prefer the activities of man-made forest management on individual land (private land ownership). In terms of opportunity cost for patrolling the forest, collective management may have the advantage in conservation.

Standpoint 9—The policy recommendations should be elaborated by making use of the results of our research. It is not necessary for them to be so comprehensive as to cover all aspects of participatory forest management.

Supplementary explanation: Even though some of our strategies have already been detailed, a local-specific strategy based on field studies would be valuable in itself.

Standpoint 10—Our policy recommendations should be elaborated from the viewpoint of the local people.

Supplementary explanation: Priorities for forest conservation differ when viewed from local, national, and global perspectives. All are legitimate and should be taken into account (WRI et al. 1992). We recognize that all stakeholders have a vested interest in conserving their forests and that the process of facilitating participatory forest management inevitably affects all vested interests to some extent.

Standpoint 11—The policy recommendations are to consist of several sets of objectives and necessary actions aimed at local people, the governments, NGOs, and international organizations.

Supplementary explanation: One function of the policy recommendations is linking the local reality to the national policy. In each action proposed here, the intended actors for each action are clearly specified.

3.2 Summarized Results of the Field Study: Current State and Constraints

3.2.1 Indonesia

3.2.1.1 Customary Land Rights and Forest and Land Management by Local People

Field studies in East Kalimantan, characterized by rich forests, in Southeast Maluku, characterized by strict customary law, and in Central Java, characterized by high population density and prevailing private land, identified the following facts:

1. Forest and land utilization is controlled by customary law, called *hukum adat,* throughout Kalimantan, even though the rules vary from community to community.
2. In East Kalimantan, the Dayaks, or indigenous people, recognize private property rights and communal property rights.
3. In East Kalimantan, under the traditional land category in Dayak villages, reserved forest (*tana mawa*), utilization forest (*tana belahan*), sacred land (*tana to'*), graveyard (*tana patai*), and other primary forest (*tana' kaso*) are considered communal property, while swiddens (*tana luma'*) and agroforest (*tana lepu'un*) planted by individuals are considered private property.
4. In East Kalimantan, the customary rights of the village to primary forest and natural resources are rather loose; customary rights of individuals and households are rather tight.
5. In Southeast Maluku, customary land (*petuanan*) is divided into three categories: private land, lineage (*marga*) land, and village land.
6. In Southeast Maluku, people are prohibited from entering certain lands during a certain period to protect resources, such as sago palm, under the local resource management system (*sasi, yutut*).
7. In Central Java, the land is categorized into homesteads planted with various perennial and annual crops (*pekarangan*), mixed gardens mainly planted with perennial crops (*kebun*), dry fields mainly plànted with annual crops (*tegalan*), paddy fields (*sawah*), hillside dry fields (*pereng*), swiddens (*ladang*), and state lands (*alas*), including grasslands (*padang rumput*) and forests (*hutan*).
8. In Central Java, the land consists of private land, national land, and village land (*tanah bengkok*) that are mostly used as paddy fields. Communal forests do not exist.
9. In Central Java, the phenomenon of "gardenization" can be seen. This is an evolution of land use in a way which expands tree gardens planted with various perennial crops.
10. People have been using the forests for swidden agriculture, sources of diet (hunting, fishing, collecting nuts, fruits, and wild vegetables), and materials for construction and handicrafts (iron wood, rattan, *Shorea* spp., etc.).
11. The poor people are more dependent on a variety of forest products, and hence are more vulnerable to sudden changes in the provision of such products and their price.

3.2.1.2 External Constraints on Community Participation

1. The rights of the local people to utilize and manage the forests have been neglected by the government.
2. Logging or timber companies, including the national forestry corporation (Perum Perhutani), usually apply top-down decision-making, and local needs are often neglected.
3. The government classified forestland in terms of its main expected functions, but the actual state of land utilization and socio-economic aspects were totally

neglected in this classification, because the main criteria for the classification used were the degree of slope, fragility to soil erosion, and strength of rainfall.
4. In Central Java the recent economic crisis triggered many young migrant workers to return to their home villages, because they had lost their jobs in Jakarta. Since then, some have started illegal logging and illegal cultivation.
5. Especially in Java, organized illegal logging by outsiders can be a great obstacle to sustainable forest management by local people as well as by the national forest corporation.

3.2.1.3 Internal Constraints on Community Participation

1. In villages in East Kalimantan where a customary forest management system does not exist, loose norms cannot function as a protective wall against external pressures.
2. In such villages weak collaboration makes it difficult to develop any form of village-wide forestry program, and high competition for forest products, particularly timber, has caused the people to be more careless about sustaining the forests.
3. In two villages of East Kalimantan, village leadership based on the law of village government does not generate community participation and tends to disintegrate the communities by serving the governmental interests.
4. In Southeast Maluku, the younger generation tends to cut trees in the customary forests to get money for commodities.
5. In Central Java, illegal logging can be seen in the national forests (managed by the national forestry corporation, Perum Perhutani) probably conducted by local people who own little or no farmland and cannot work outside the village. They may be logging to obtain firewood to be used to produce brown sugar from coconut palms.

3.2.1.4 Main Actors for Participatory Forest Management

1. In villages in East Kalimantan where a customary forest management system does not exist, **households** play an important role in swidden agriculture, rubber tree gardening, rattan forestry, and candle nut forestry. **Individuals** play a role in logging, hunting, fishing, and gathering. Collective management of the forests seems to be difficult, because customary rules concerning village property are too loose and the village leadership does not support the collective work.
2. In villages in East Kalimantan where a customary forest management system exists, the **village community** can manage the communal forests as it has in the past.
3. In Southeast Maluku, the **village community**, headed by male elders, plays an important role in customary forest management; **individuals** and **households** play important roles in agriculture and hunting.

4. In Central Java, **households** play an important role in forest management in the Tumpang Sari plantation area on national land as well as in the tree gardens on private land.

3.2.2 The Philippines

3.2.2.1 Customary Land Rights and Forest and Land Management by Local People

Field studies were conducted in the municipality of Banaue (province of Ifugao), an area recognized as ancestral domain with relatively well-preserved forests, and in the northern Sierra Madre mountain region in the province of Isabela, where commercial logging activities have operated up until recently. The studies identified the following facts:

1. In Banaue, the landscape consists of eight major land forms: terraces for rice production, drained fields for vegetable production, swidden areas for sweet potato production, low grasslands, high grasslands, community forests, private forests, and housing lots.
2. In Banaue, the lower elevations of the community forests (*inahalan*) are allocated for swidden agriculture, and the higher elevations are preserved as watershed forests, where swidden agriculture is customarily prohibited.
3. In Banaue, there has not been any regulation of resource extraction in the community forest. As a result, most of the trees suitable for woodcarving and house construction have already been taken and many woodcarvers have to find trees outside Banaue.
4. In Banaue, private forests (*pinugo, muyong*) or man-made forests scattered among the rice terraces secure water for irrigation and protect against landslides and erosion.
5. In Banaue, people make use of forest products such as firewood, construction material, and woodcarving materials for their livelihood.
6. In Sierra Madre, villagers usually categorize the land as "lowland" or "alienable or disposable land" where land title was officially recognized, and "upland" or "public timber land" where land titles were not recognized by the government.
7. In Sierra Madre, however, customary laws for utilizing forest resources were not evident to researchers, because communities (*barangays*) here were formed recently, after the Second World War.

3.2.2.2 External Constraints on Community Participation

1. A major external constraint on participatory forest management appears to be an attitude that imposes government-designed forest projects on upland communities, where local people are requested to follow them and act as tools of the government.

2. In most cases, these projects do not consider local conditions, such as land use, people's dependence on forest resources for their livelihood, and the local value systems.
3. Requiring local people to participate in government-designed reforestation projects brings about adverse effects, including resistance and conflicts between the government and local people, as well as among the local people themselves.
4. NGOs may be expected to play substantial roles, not as instruments of the government, but as catalysts to empower the local people in formulating solutions to existing problems on their own.
5. Most small-scale loggers think the forests are still controlled by the government, because logging in all CBFM sites was suspended in 1988.

3.2.2.3 Internal Constraints on Community Participation

1. In Banaue, few internal constraints of social and cultural aspects towards forest management can be identified.
2. In Banaue, even though the people have a desire to improve the stand quality in private forests by planting trees, particularly trees for woodcarving, they do not intend to do so if they have to pay for seedlings.
3. In Banaue, planting seedlings for woodcarving materials in the community forests is hardly expected, because it may be difficult to find witnesses of the plantation activities in the community forest, which is necessary in order to claim one's ownership of the trees planted.
4. In Sierra Madre, there are conflicts between the cooperative or an executive body of the CBFM program and illegal loggers and between the cooperatives and new migrants. Even the members of the cooperative expect benefits from the cooperative rather than self-sacrifice, because the cooperative is a kind of functional group.
5. In Sierra Madre, the policies of the cooperative sometimes fluctuate, because the requests of the government and the villagers at times contradict each other.

3.2.2.4 Main Actors for Participatory Forest Management

1. In Banaue, the blood kinship group is regarded as the main actor of participatory forest management in the private forests, because close blood kinship groups actively manage the co-owned private forests.
2. In Banaue, the village community can be an actor to loosely regulate the use of community forests.
3. In Sierra Madre, the cooperative seems to be the best actor for natural forest management.

4. In Sierra Madre, individuals seem to be the best actors for reforestation activities.

3.2.3 Laos

3.2.3.1 Customary Land Rights and Forest and Land Management by Local People

Field studies were carried out in Vang Vieng District and Sang Thong District, both in Vientiane Province, and in Phalanxai District, designated as a national bio-diversity conservation area (NBCA) of Phou Xang Hae in Savanakhet Province. These field studies identified the following facts:

1. In Vang Vieng, the Lao Theung (specifically the Khamu), who live at middle altitudes in the mountains, classify their land into several categories: dense forest (*patae bree kut*), old fallow (*patae reng kae*), young fallow (*patae reng kha nhom*), former swidden (*patae re tu*), present swidden (*patae re*), protection forest (*patae bree haksa*), cemetery (*patae raman*), utilization forest (*pataebree kui xay*), and house lot (*patae koun*), etc. The Lao Loum, who settled in the lowlands, categorize their land in a similar way.
2. In Vang Vieng, customary private rights are permitted on young fallow, old fallow, former swidden, present swidden, and house lots, etc., in the village territory.
3. In Vang Vieng, protection forest, cemetery, and utilization forest are managed collectively by the village community in accordance with governmental instructions.
4. In Sang Thong, customary rights to use forest resources for individuals and families have been granted within village boundaries.
5. In Sang Thong, the villagers have conserved a few patches of communal forests for protection of the water catchment, prevention of soil erosion, and maintenance of cemeteries. Villagers stated that these communal forests are very useful to them, although no clear rules and regulations relating to them exist.
6. In both Vang Vieng and Sang Thong, many forest products collected by the local people, consisting of wood and non-wood forest products, comprise their main cash income sources.
7. In Phalanxai, part of the village territory overlaps with a new NBCA, but the land allocation program was able to provide paddy fields to most of the people whose swidden agriculture was prohibited in the area of the NBCA.
8. Especially for the poor, forest products are important in daily life.

3.2.3.2 External Constraints on Community Participation

1. Decrees or implementation ordinances to enforce the Forestry Law have not yet been issued. According to the government officer who participated in the workshop, the government issued Decree No. 198 in 1998, but we have not yet examined it.
2. When converting from an existing land classification allowing use by the local people to being official land under the Land Law and Forestry Law, an important point to clarify is whether the present land and forest utilization and ownership is officially approved or not.
3. Application of the official land-use classification to the land, such as swidden land, customary conservation forestland, and dense forestland, is said to be difficult.
4. The swidden land at present includes land under cultivation, fallow, and grassland. Officially, however, land regarded as degraded at present is classified as degraded forestland, land regarded as young bush fallow is classified as regeneration forest, and land regarded as old forest fallow is classified as village-managed protection forest, conservation forest, or production forest.
5. The problem concerning the land to be classified as degraded forestland is the high possibility of afforestation on degraded forestland, although the local people are harvesting non-timber forest products (NTFP) even from grassland.
6. The problem concerning the land to be classified as regeneration forest (village-managed) is the probability of neglecting the fact that the local people hold customary tenure rights for all swidden areas, including fallow land.
7. Classification of swidden land as degraded and regenerated forestland has been planned on the presupposition that swidden agriculture should be abandoned, even though most of the local people still make their livelihood this way.
8. The problem that has arisen from the classification of felling-prohibited forestland and dense forestland as protection, conservation, and production forest is the probability that customary forest utilization is not permitted, even though the local people harvest forest products from the forestland covered with every type of vegetation.
9. Criteria for demarcating the core zone and buffer zone in NBCAs are not clearly defined. The local people do not understand the restrictions of forest utilization in these zones.
10. In reality, production forests, agricultural land, and even house lots are included in the buffer zone of NBCAs. This fact is inconsistent with the purpose of NBCAs to conserve biodiversity.

3.2.3.3 Internal Constraints on Community Participation

1. A lack of flat land suitable for sedentary agriculture and the land's low productivity force the local people to practice swidden agriculture on the degraded uplands.
2. The non-agricultural economic sectors are not developed enough to provide adequate income sources; the market system is also not well developed. As a result, local people are forced to depend on forest products.
3. The local people do not understand their rights and duties in managing forests in the village territory. As a result they sometimes do not enrich or regenerate the forest areas after land or forest allocation has occurred.
4. Actual forest utilization by the local people does not change, even after establishing NBCAs.
5. Customary forest utilization cannot automatically be regarded as sustainable.

3.2.3.4 Main Actors in Participatory Forest Management

1. Village communities or organized villagers are regarded as the main actors in village forestry programs, and can be considered as co-agents for joint forest management.
2. Village forest volunteers serve as executive bodies for patrolling the village forest for conservation and protection.
3. Individuals or households are regarded as main actors for tree planting on degraded land.

3.2.4 Vietnam

3.2.4.1 Customary Land Rights and Forest and Land Management by Local People

Field studies were carried out in Son Duong District, Tuyen Quang Province, parts of which are designated as buffer zones of the Tam Dao National Park, and Mai Son District in Son La Province. These field studies identified the following facts:

1. It seems that community forests or forests managed in a traditional manner, based on communal relationships, do not exist, because local people, especially the Kinh, may have forgotten the customary forest management system.
2. In Son Duong, the most important forest product is firewood; less important ones are birds, bats, squirrels, and medicinal plants.
3. In Son Duong, the poor collect firewood every day in the mountains; the middle class collect it three or four times a week; the rich have already stopped

collecting it and substitute branches from the plantation forest or fruit trees and residues of rice or maize for firewood.

4. In Son Duong, most of the people do not recognize that they live in the buffer zone of the national park, and they do not know the location of the park borders. Nevertheless, they understand the concept of a national park, where some practices are prohibited, such as tree felling, swidden agriculture, and hunting.
5. In Mai Son, timber is taken from natural forests for building houses and making beds, cupboards, tables, chairs, etc. Firewood is taken from natural and planted forests or gardens for cooking and heating. Bamboo is taken from natural and planted forests for building material. Other forest products include birds and animals, rattan, bamboo shoots, and medicinal herbs.
6. In Mai Son, dependence on forest products varies by ethnic group. The Kinh depend little on natural forest products; the Muong and Thai usually collect timber and firewood; and the Mong, Kho Mu, and Xinh Mun depend almost entirely on forest products.

3.2.4.2 External Constraints on Community Participation

1. Bureaucracy and centralized, top-down decision-making at the local level can pose as obstacles to the participation of the local people.
2. Although forests should be classified into three categories, there are no authentic criteria and indicators for forest classification.
3. The budget and human resources required to implement the program of land and forest allocation are limited. As a result, the local authorities cannot conduct this work effectively.
4. Local authorities have not paid attention to the fact that for local people swidden agriculture has been essential during transitional periods.
5. Arrangements and agreements on jurisdiction between the local authorities and the national government seem to be insufficient.
6. There is no effective system or program to promote participatory forest management by fundamental groups and villages, even though villages can undertake contracts to protect natural forest in national parks.
7. The national park system conflicts fundamentally with the livelihoods of the local people.

3.2.4.3 Internal Constraints on Community Participation

1. People believe that the collection of forest products is legal, even though it is illegal in ecological rehabilitation zones of special-use forests and critical protection forests.
2. The custom of exchanging ideas and experiences is not yet mature, which makes it difficult for people to acquire new ways of thinking and doing, such as legitimate participatory forest management.

3. The linkages among households have been very loose, and groups of households have not worked together to accomplish common goals.

3.2.4.4 Main Actors in Participatory Forest Management

1. The forest management board can be regarded as an actor to manage protection forests and protected areas.
2. Traditional village communities (*thon*) can be regarded as actors in natural forest and plantation management.
3. Collective associations, such as women's unions, youth unions, peasant unions, and ex-servicemen's unions, have the potential to play roles in participatory forest management because they are reliable.
4. Households, churches, temples, and other organizations, including the army, can be regarded as actors to manage small-scale, natural secondary forests and plantations.
5. Forestry companies and joint ventures can be regarded as actors in industrial plantation.
6. State forest enterprises can be regarded as main actors to manage natural forests and plantations.

3.3 Objectives and Actions Recommended for Indonesia

Recommendations for participatory forest management in Indonesia are indicated here. Please refer to our Web site <http://www.iges.or.jp/en/fc/phase1/output-e.htm> to view the recommendations for the Philippines, Laos, and Vietnam.

Indonesian forest policy is now being reformed drastically in accordance with the movement of democratization and decentralization of power. In light of this, proposals made by various stakeholders, including our recommendations, can be put onto the platform of policy discussion, even though some of the actions recommended here have already been achieved.

Objective 1: Ensure the Participation of Local People in General

Action 1-1—The government and NGOs should work together to establish the mechanism of a "green" safety net to secure the minimum level of forest conservation.

This will provide the foundation for participatory forest management in terms of forest conservation, because a green safety net is defined as the minimum regulation required by the national government to ensure sustainable forest management and forest conservation.

Action 1-2—All the parties should recognize and have a high regard for the perspective of conservation as perceived by the local people.

This will ensure full participation of local people. The compulsion of outsiders' perspectives of conservation will spoil the self-confidence of the people.

Action 1-3—The local government should play an important role and should take the actual state of land use and socio-economic conditions into consideration in forestland classification.

This will enable the government to identify the forest areas that should be managed by the local people. At present it seems to be difficult for the government to ascertain the actual state of land use under certain socio-economic contexts. In the process of land classification, the government should actively involve the local people and NGOs by employing participatory rural appraisals (PRA), etc. Although the importance of this action has been pointed out for more than a decade, ongoing efforts are still necessary.

Action 1-4—The national government should revise the Forest Village Social Development Program (PMDH) as a basis for facilitating community participation.

This will improve the livelihood of the local people, which will encourage them to participate in forest management.

Action 1-5—NGOs should cooperate with the provincial governments to encourage them to use a bottom-up decision-making process in the management of protected areas and the demarcation of forest areas. NGOs should cooperate with the district government (Kabupaten) to be involved in the activities of reforestation and re-greening, the management of protection forests and private forests, the control of hunting and collecting of non-wood forest products, and extension activities.

This will accelerate the people's participation in accordance with the devolution of authority for forest management to each level of local government.

Action 1-6—All the parties should recognize the importance of power sharing, as well as that of benefit sharing, between the local people and other stakeholders.

This will encourage the local people to have stronger will to manage their forests.

Action 1-7—The government should elaborate the guidelines for defining who are the "local people" that have the rights to manage their forests, in cooperation with NGOs and the local people themselves.

This will avoid disputes among the people.

Objective 2: Operationalize the Management of Customary Forest (*hutan adat*) as Prescribed in the New Forestry Law

Action 2-1—The government should evaluate the customary law fairly in terms of collective forest management, in cooperation with various stakeholders such as local communities, local governments, NGOs, and academics.

This will guarantee the fair judgment of the customary law. Under the new Forestry Law, the rights of the local community can be recognized under the condition that the customary laws do not contradict the national laws and local regulations.

Action 2-2—NGOs and governments should persuade the local people to modify their customary forest-use practices, where the need exists to develop appropriate technology for sustainable forest management.

This will help the local people avoid being excluded from official approval of customary forests.

Action 2-3—The government should issue a decree to show the process for designating the customary forest, even in the conservation areas.

This will facilitate customary forest management in and around the conservation areas and will secure sustainable forest management. These results can be expected, because most of the customary forest areas must be covered with relatively rich forests that have a high possibility of being designated as conservation areas.

Action 2-4—As a next step, the government should consider the release of the customary forest (*hutan adat*) from state-owned forest under certain regulations in the future.

This will solve the latent conflicts and accelerate the decentralization and devolution of forest management. It will provide the basis for the complete integration of a community-based forest management system (*sistem hutan kerakyatan: SHK*) with the management of customary forests (*hutan adat*).

Objective 3: Facilitate Collective Forest Management

Action 3-1—The government should give priority to the permission of community forestry over other permissions for natural forest management, man-made forest management, mixed plantation management (prescribed in August 1999), and tree-felling for the purpose of developing oil palm plantations and transmigration areas.

This will allow local people to obtain concessions on good quality forests; otherwise they will be allocated degraded forests only.

Action 3-2—The local people should organize themselves, discuss the rules for forest utilization, and conclude agreements for forest management in cooperation with external agencies such as NGOs and the local government.

This may reform the local leadership and provide all villagers with incentives to participate in the process, even in areas where community-based forest management does not yet exist. Although the importance of this action has already been pointed out, the former political regime made implementation impossible. Now is an opportune time to take action.

Action 3-3—The government and the state forest corporation (*Perum Perhutani*) should give permission for community forestry in Java.

This will effectively prevent illegal logging by outsiders because of the protection provided by the local people.

Objective 4: Facilitate Individual- or Household-based Forest Management

Action 4-1—The government and the state forest corporation share the profits from planted trees between the local people who participate in the *Tumpang Sari* (or the *Perhutanan Sosial* program) and the national forestry corporation, Perum Perhutani, in Java.

This will provide the participants with economic incentives to manage the forest until trees (such as teak and pine) are harvested.

Action 4-2—The state forest corporation, Perum Perhutani, should plant tree species suitable for use as fuelwood by the local people on sites of an improved *Tumpang Sari* or *Perhutanan Sosial* program in Java.

This may reduce illegal logging for fuelwood for making brown sugar from palm trees and encourage the people to take part in the management of fuelwood plantations on national land.

Action 4-3—The government and NGOs should help the local people to patrol the forest areas they manage.

This will prevent illegal logging by outsiders.

Action 4-4—The government should introduce an individual- or household-based sharecropping forestry program in the degraded production forest areas on national land in outer Indonesia.

This will motivate the local people to plant trees, even on national land, similar to the individual forest (*hutan rakyat*) program on private land in Java. Individuals or households could sub-contract with the executive bodies responsible for community forest management under community forestry.

Action 4-5—The state forest corporation (*Perum Perhutani*) should devolve the rights of forest management on parts of national land to the local people.

This will promote the process of "gardenization" practiced by the local people.

Action 4-6—NGOs should help the local people to obtain land ownership for tree plantations.

This will provide the basis for introducing the individual forest programs (*hutan rakyat*) on private land.

4. REMAINING ISSUES

The policy recommendations shown here will be useful in overcoming the gap between local reality and national policy. The next step, however, is to develop guidelines and construct mechanisms to actualize these recommendations. This is one of the objectives of the second phase of the IGES Forest Conservation Project.

REFERENCES

Chakraborty, R. N., I. Freier, F. Kegel and M. Mascher. 1997. *Community forestry in the Terai region of Nepal: Policy issues, experience, and potential.* Berlin: German Development Institute.

Fisher, R. J. 1999. Devolution and decentralization of forest management in Asia and the Pacific. *Unasylva* 50 (199):3–5.

Institute for Global Environmental Strategies (IGES). 2001. *Report of the first phase strategic research: Forest conservation* (March). Hayama, Japan: IGES.

Inoue, Makoto. 2000a. Institutions and actors for participatory forest management in Southeast Asian countries: Forest sociological approach (in Japanese with English summary). *Journal of Forest Economics* 46 (1):19–26.

———. 2000b. Participatory forest management. In *Rainforest ecosystems of East Kalimantan: El Nino, drought, fire, and human impacts,* 299–307, ed. Edi Guharidja, Mansur Fatawi, Maman Sutisna, Tokunori Mori and Seiichi Ohta. Tokyo: Springer-Verlag.

Lynch, Owen J. and Kirk Talbott. 1995. *Balancing acts: Community-based forest management and national law in Asia and the Pacific*. Washington, D.C.: World Resource Institute.

Pardo, R. D. 1985. Forestry for the people: Can it work? *Journal of Forestry* 82 (12):733–741.

Salim, Emil and Ola Ullsten. 1999. *Our forests, our future. Report of the World Commission on Forests and Sustainable Development*, 124. Cambridge University Press.

World Resource Institute (WRI), World Conservation Union (IUCN) and United Nations Environment Programme (UNEP). 1992. *Global biodiversity strategy,* 23. Washington: WRI, IUCN, and UNEP.

Chapter 4

SUSTAINABLE AND PARTICIPATORY FOREST MANAGEMENT: LEGAL AND ADMINISTRATIVE SUPPORTING MEASURES, AND FINAL RECOMMENDATIONS

Hiroji Isozaki
Iwate University[1]

1. BASIC ELEMENTS REGARDING PARTICIPATION OF LOCAL PEOPLE

With the extensive diversity in the countries of the Asia-Pacific region, the legal and administrative systems related to forest conservation also vary widely. This diversity makes it a challenge to concisely describe the overall situation in the region. While some countries are attempting to change their economic systems, others face economic crisis. Though many problems exist in each country, there has also been progress in solving them. In particular, many countries are currently changing their forestry laws or have already changed them in recent years. Legal reforms focus on several issues, a major one involving the promotion of participatory forest management.

In terms of forests, at the global level, it can be said that promoting community forestry is one of the most important issues to be addressed, and a number of measures

[1] This chapter was written in cooperation with Kiyoshi Komatsu and Makiko Yamauchi, based on the results of collaborative work under the Legal and Administrative Supporting Measures sub-theme of the Forest Conservation Project of the Institute for Global Environmental Strategies with Prof. Toru Iwama, Prof. Wang Xi, Ms. Loudes E. Tolentino, Dr. Ari Nakano, Dr. Pearmsak Makarabhirom, Mr. Mas Ash Santosa, and Dr. S. Sothi Rachagan. A draft of the text was examined at two international workshops, one held in Jakarta in June 2000, and the other in Vientiane in August 2000, as well as at an international seminar held 16–18 January 2001 in Tokyo. This chapter therefore represents the input of various stakeholders, including governmental officers, non-governmental organizations, researchers, and international organizations.

M. Inoue and H. Isozaki (eds.), *People and Forest—Policy and Local Reality in Southeast Asia, the Russian Far East and Japan, 73–89.*

related to forest conservation have been implemented to date, including international treaties adopted at the regional and global levels. These treaties, which recognize that the participation of local people is important, share two elements.

The first element is the composition of participants, with many stakeholders in forest conservation. Stakeholders can include anyone who participates in the forest management process—ranging from local community members to people who live in urban areas, non-governmental organizations (NGOs), and companies. It is useful to think of the people involved in four categories: the general public, the affected or concerned public, local people and communities, and indigenous people.

Another key element is the level of participation, which can be divided into three categories: access to information, participation in the decision-making process, and access to means of redress. The level of participation is based on how people participate in the forest management process and how their participation may be ensured.

Almost all treaties that require the contracting parties to ensure stakeholder participation, however, do not clearly identify the necessary legal and administrative measures to ensure their participation. Considerable leeway is permitted regarding the measures for ensuring their participation. Decisions on which measures should be adopted depend on each contracting party.

This report examines the legal and administrative systems for forest conservation and participation at the national and international levels, based on research conducted by the IGES Forest Conservation Project from 1998 to 2001. Results of the research help provide the perspective required to understand the current situation in the Asia-Pacific region. In addition, this report indicates current problems that must be overcome. Finally, based on the research, some recommendations are offered.

2. LEGAL AND ADMINISTRATIVE SYSTEMS AT THE INTERNATIONAL LEVEL

2.1 Current State of International Instruments Related to Forest Management

There is no consensus in international society on building a new international treaty, such as a convention on forests. Discussions still continue, and the results will depend on each country's willingness to participate.

Sustainable forest management is recognized as a key concept in documents that resulted from discussions at the Intergovernmental Forum on Forests (IFF), the Intergovernmental Panel on Forests (IPF), and other frameworks for forest conservation created since the Rio Summit in 1992. Accordingly, sustainable forest management might be the focus of the proposed convention.

Many other legally binding international instruments already exist. Those concerned in some way with sustainable forest management include the Convention on Biological Diversity (CBD), the Kyoto Protocol of the United Nations Framework Convention on

Climate Change (particularly its Kyoto Protocol), the Convention on Wetlands of International Importance Especially as Waterfowl Habitat (Ramsar Convention), the Convention Concerning the Protection of the World Cultural and Natural Heritage (WHC), the Convention on International Trade in Endangered Species of Wild Fauna and Flora (CITES), and the International Tropical Timber Agreement (ITTA). Moreover, there are three non-legally binding instruments: Principles on Forests, Proposals of the IPF, and Proposals of the IFF.

Some countries have expressed concern about duplication between existing instruments and a new convention on forests, and also point out the need to implement existing instruments effectively before establishing new ones.

On the other hand, it could be said that already too many instruments deal with sustainable forest management, and inconsistencies between them have been noted. For example, the possibility exists of a conflict emerging between the CBD and the Kyoto Protocol due to differences in their respective purposes. Plantations are important for both these instruments as well as for sustainable forest management. The CBD, however, aims to conserve biological diversity, whereas the Kyoto Protocol values increased carbon sinks. If plantations are expanded in order to increase sinks, problems will arise from the perspective of conserving biological diversity. Since contradictions may become obstacles, when plantations are being established, for example, some countries feel that new international instruments, including legally binding ones, are necessary.

In order to analyze the relationships among the international treaties related to forest management, the basic principles of international law should be recognized. An important principle is *lex posterior derogat pripori*, which states that if a country ratifies many international treaties and there are contradictions between previous treaties and newer ones, the newer ones are effective in that country, with some exceptions. Another principle is *lex specialis derogat lege generali*, which states that if a country ratifies many international treaties and there are contradictions between general international treaties and specific ones, the specific laws are effective in that country, with some exceptions. According to these principles, existing international treaties do not provide sufficient reasons to reject the establishment of new laws, even if existing laws address the issue.

2.2 Principles in Each Existing International Treaty Related to Forest Conservation

There are several important principles included in the existing treaties, which can be said to be legal principles on forest conservation and could be included in a "Convention on Forests." The principles can be extracted from six international treaties: CITES, WHC, the Ramsar Convention, ITTA, CBD, and the Convention to Combat Desertification (CCD). The Kyoto Protocol can affect forestry policy, but it is not yet certain what action will be required with respect to carbon sinks. Therefore, the Kyoto Protocol cannot indicate the principles for forest conservation at the present time.

The principles can be also found in "soft" law instruments related to forest conservation. According to experiences with international environmental law, several principles are defined in non-legally binding instruments at first, and then these principles are included in legal instruments. Hence, non-legally binding instruments should be analyzed in order to clarify the principles on forest conservation. Regarding forest conservation, there are three important soft law instruments, such as the Principles on Forests, the IPF's Proposals for Actions, and the IFF's Proposals for Actions. In particular, the IPF's Proposals for Actions includes some important principles. In addition, resolutions or decisions made at the conferences play an important role in the implementation in some treaties. In particular, the parties to the Ramsar Convention adopted an important resolution on "wise-use" guidelines.

The results are shown in Box 1. These principles can be divided into three parts. The first part shows general principles of forest conservation. These principles can be applied not only to forest issues but also to other issues. Therefore, they can be said to be common principles of international environmental law. Due to their nature, the principles do not describe concrete measures for sustainable forest management. Rather, they describe concepts of international environmental law. The second part shows concrete measures for forest conservation at the international and national levels. These principles indicate concrete and necessary measures in order to achieve sustainable forest management. In other words, these principles detail the obligations required of governments.

There are differences in the nature of the obligations required by each treaty. CITES and the WHC, which were adopted earlier than other international treaties, limited their objectives to controlling the trade of endangered species and protecting valuable ecosystems. On the other hand, the international treaties adopted after the Rio Summit deal with more comprehensive issues than the earlier two treaties, which included regulation of some activities related to forest management. The two more recent treaties set up policy processes to deal with problems related to forest conservation by entire societies.

According to these principles, governments should regulate land use, depending on the character of ecosystems, and create comprehensive policies, taking into consideration the social, cultural, and economic aspects of forests. In order to ensure appropriate decisions are made, information relevant to forest management is necessary; hence, inventory, monitoring, research, and access to information are required. Needless to say, environmental impact assessment (EIA) is required in the decision-making process. In addition, cooperation in forest conservation by all stakeholders is indispensable to the success of sustainable forest management. Hence, coordination among ministries regarding forest conservation is required, as is public participation. Moreover, for all stakeholders, capacity building is necessary for the management of forests in a sustainable manner; this will allow all stakeholders to equitably share the fruits of forest management.

Box 1: Principles included in existing international treaties on forest conservation.

General Principles	• Common concerns of humanity • Rights of future generations • Sovereign rights over natural resources under jurisdiction • Precautionary • Prevention • Principles of sustainability • International cooperation • Principles on subsidiarity • Common but differentiated responsibilities
Principles on International Level	• Information exchange • Consultation
Principles on National Level	• Consideration of environmental impact (monitoring EIA) within jurisdiction on decision-making process open to foreign people • Legislation (new legislation, revision of existing laws) • Institutional arrangement among concerned ministries and authorities (establishing strategy, policy/nomination of focal point) • Regulation of land use in order to avoid negative impact on ecosystems (protected area/zoning/permission for certain activities) • Regulation of international trade on goods or wildlife relevant to forests in order to avoid negative impact to ecosystems (permission for the export/import only of goods or wildlife which are produced by processes managed under principles of sustainable management) • Rehabilitation/restoration of destroyed/degraded forest • Respect for traditional management system • Sharing information/ensuring access to information • Sharing benefit equitably/ensuring access to resources • Research • Training • Public awareness • Participation of stakeholders

2.3 Provisions on the Rights of Communities in Forest Management

The International Labor Organization's Convention No. 169, Concerning Indigenous and Tribal Peoples, states that "the rights of ownership and possession of the peoples concerned with the lands which they traditionally occupy shall be recognized. Government shall take steps as necessary to identify the lands which the people concerned occupy, and guarantee effective protection of the rights of ownership and possession. Adequate procedures shall be established within the national legal system to resolve land claims by the peoples concerned."

Principle 22 of the Rio Declaration affirms the vital role of communities in environmental management and development, but provides no guidance on how to ensure effective participation.

Article 8(j) of the Convention on Biological Diversity requires parties to "respect, preserve, and maintain the knowledge, innovations, and practices of indigenous and local communities embodying traditional lifestyles relevant to the conservation and sustainable use of biological diversity." This provides a framework for international legal protection for certain types of local community forestry management systems, but the section stipulates that it's "subject to legislation," which weakens its effectiveness.

Chapter 32 of Agenda 21 is directed at the interests of farmers, which are identified as "all rural people who derive their livelihood from activities such as farming, fishing, and forest harvesting." It calls upon national governments to give effective land tenure to these groups. It also notes that the absence of legislation on land rights has been an obstacle in taking action against land degradation in many farming communities in developing countries.

The Convention to Combat Desertification recognizes the rights and interests of community-based resource users, as well as the participation of these groups, as essential for sustainable resource management. Article 10 stipulates that "effective participation at the local, national, and regional levels" in policy planning and implementation should be provided for.

These instruments indicate not only the need for participation by local people, and assure their rights, but also the recognition by international society of the importance of participation by local people. In order to ensure the rights of local people, the European Council adopted a new convention in 1998, the Convention on Access to Information, Public Participation in Decision-making and Access to Justice in Environmental Matters (Aarhus Convention). It requires public participation in decisions on specific activities (article 6); public participation concerning plans, programs, and policies relating to the environment (article 7); public participation during the preparation of executive regulations and/or generally applicable legally binding normative instruments (article 8); and access to justice (article 9).

3. LEGAL AND ADMINISTRATIVE SYSTEMS AT THE NATIONAL LEVEL

3.1 Legitimacy Based on Property Rights or the Right to Use Lands and Forests

3.1.1 Rights to Forest and Land Use

In order to ensure the legitimacy of participation, several measures can be used. One is to ensure participation by authorizing property rights or forest-use rights. Some countries enacted legislation which authorized tenure rights for local communities and people. Other countries authorize individual rights to use forests and land.

The constitution of Papua New Guinea stipulates that land and forests are owned by their customary owners. Custom is defined as "the custom and usages of indigenous inhabitants of the country existing in relation to which the matter arises, regardless of whether or not the custom or usage has existed from time immemorial." Hence, almost all land and forests are owned by local people or kinship groups, based on custom. Exploitation of land and use of forests requires their consent. The Forestry Act, which came into force in 1992, is legislation directly related to forest management. Under the Act, the state reserves the exclusive right to reach forest management agreements with landowners, once they approve the conditions. The National Forestry Board selects a forest company and recommends that the minister grant a timber permit. When consensus is not achieved, landowners cannot exploitation the land. Accordingly, the government cannot grant a timber permit without the approval of the local people and community.

In China, on the other hand, according to the constitution, land (including forests and other natural resources) is owned by either the state or the collectives. A similar system exists in Vietnam and Laos. In the case of China, the forestry law encourages people to participate in afforestation activities by rewarding contributions, which can take several forms such as voluntary tree planting, eco-agriculture, cooperation between nature reserves and local communities in managing nature reserves, wasteland contracts, and individual afforestation contributions.

Wasteland contracts rent bare land in mountainous areas to farmers and to unemployed people in urban areas. Based on the contract, tenants are given the right to manage and use the lands protected under the forestry law. Those who are given the right to use bare land can plant trees in the area and generate benefits from the plantations. Contracts have fifty-year terms, and the benefits of activities approved in the contract are tax-free during the term. Throughout these measures, the forestry law encourages participation in afforestation activities and ensures the benefits for the people.

3.1.2 Problems

Even though people have tenure rights, or the right to use land and forests, the lack of a system providing opportunities for participation is an obstacle. In Papua New Guinea, their national forest plan, which provides a detailed policy on forest management at the national and provincial levels, does not require public participation in the planning process. In addition, a lack of awareness and understanding of the necessity for sustainable forest management prohibits its realization. And although the consent of landholders is indispensable for forest management agreements, meaningful consultation between the government and landholders is rare. The landholders are consulted without advice from experts—lawyers in particular. Inconsistencies among national forest policies and other relevant policies, such as environmental policies, exist within the government, and the political will to commit to sustainable development is often lacking. Inappropriate advice that ignores the real situation also prevents sustainable forest management. In China, although awareness of the need for forest conservation is increasing, recognition of it at the local level is still insufficient. Moreover, insufficient consultation between government and landholders is an obvious problem.

3.2 Legitimacy of Rights Authorized in Environmental Law or Other Relevant Laws

3.2.1 Legislation Related to the Participation of Local People in Conserving the Environment and Protecting Indigenous Peoples

In the Philippines, customary rights can be recognized in other ways. The old civil code stipulated: "Where no statute is exactly applicable to the point in controversy, the custom of the place shall be applied, and in the absence thereof, the general principles of the law." This stipulation was changed when the civil code was revised in 1950. Despite this change, judges still may recognize custom as it was in the old code. Moreover, the Philippines have enacted legislation related to the participation of local people since 1992. In the National Integrated Area System Act, local people can participate in the management processes of protected areas. Ancestral lands within protected areas and customary rights to these lands can be recognized. In 1997, the Indigenous Peoples Rights Act was approved by Congress to ensure the rights of indigenous peoples. Although the legislation does not generally ensure participation, it opens the door for the participation of local people in each issue, such as management of protected areas and protecting the rights of indigenous people.

Environmental legislation provides greater opportunities to participate than those existing in forestry laws. For example, laws related to forestry in Indonesia did not take into account the voices of local people or communities before the economic crisis. (After the economic crisis, forestry laws were revised, and the new legislation included articles on the participation of local people.) Legislation on environmental conservation, however, allows participation in the decision-making process. In particular, the EIA system provides local people and communities with opportunities to participate. Even though there are many problems with the current EIA system, it is a good case where opportunities to participate were provided and the interests of the local community were protected. Hence, EIA has become an indispensable tool for ensuring participation.

3.2.2 Environmental Impact Assessments and Participation

EIAs are recognized as important tools for ensuring the participation of local people in environmental management systems. In the Asia-Pacific region, many governments have recognized the importance of EIA and local participation, and as a result, many EIA systems in this region include provisions concerning them. Fortunately, in some cases, EIAs have provided local people with opportunities to express their opinions on forest management to their governments. At the same time, however, many problems with existing EIA systems have been pointed out.

3.2.3 Influence of International Society

Since the 1970s, several countries (e.g., the Philippines and Thailand) established framework laws for the conservation of the environment which included some provisions regarding EIA systems, but in some cases no prescriptions were given on how to implement them. One of the reasons these countries established the laws was to take action on the declaration of the 1972 United Nations Conference on the Human Environment. Most of these countries made these provisions to demonstrate to other countries, especially developed countries, and multilateral aid organizations their commitment to address environmental issues.

It has been pointed out that this legislation achieved its purpose simply by its existence. With few exceptions, these provisions were not implemented for some time; however, the situation has been changing since the UN Conference on Environment and Development (UNCED) held in Rio de Janeiro in 1992. Most of the countries (e.g., Malaysia, the Philippines, Thailand) concerned began to implement these provisions after UNCED by prescribing EIA procedures. Similar progress was also made in some economies-in-transition. In Vietnam, training and case studies began during the 1980s, and in 1993 the government established a law on environmental protection and made a decree prescribing procedures for implementation. Considering these cases, one could conclude that international society has affected attitudes toward the environment in these countries. In addition, it has been pointed out that multilateral and bilateral aid organizations have been playing important roles there as well. As these organizations fund or implement a large number of large-scale projects, their funding activities sometimes have a considerable impact on the environment. Each organization has prepared its own EIA procedures or guidelines in order to avoid negative impacts on the environment and criticism from environmental NGOs. In the case of Laos, these procedures and guidelines have affected the country's policies. Although the country does not yet have an official EIA system in place, in fact, EIAs are implemented on an informal basis. Because the procedures and guidelines of the organizations noted above provide good examples of EIA, the government of Laos refers to them in EIA activities. As this case shows, these organizations can provide useful models and references for developing countries.

3.2.4 Importance of Regional Cooperation

Although the Association of South East Asian Nations (ASEAN) Agreement on the Conservation of Nature and Natural Resources was adopted in 1985, it is not yet being enforced due to the reluctance of member countries to establish legal obligations to conserve the environment, and the low priority given to environmental conservation in each nation's policies. Like the example above, however, international society can affect these countries' attitudes, and it appears that UNCED has also influenced ASEAN countries.

In 1994 ASEAN adopted its Strategic Plan of Action on the Environment, which requires ASEAN countries to take actions such as "1.1—Continue to support the documentation of regional EIA experiences, leading towards the harmonization of procedures," "1.2—Initiate activities that will make use of natural resource and environmental accounting studies and approaches," and "1.3—Establish procedures that would initiate the integration of environmental concerns in the various ASEAN programs and activities," in order to "support the development of a regional framework for integrating environment and development concerns in the decision-making process."

There are many similar national strategic plans in ASEAN countries. Although a strategic plan is not a legal instrument, countries are free to make improvements in their national policies. It is possible that even non-legal instruments can be effective in ASEAN countries and be good tools for improving their EIA systems.

3.2.5 Current Situation in Each Country

Various types of EIA systems exist in the Asia-Pacific region, reflecting the diversity of each country's situation. Almost all the countries in this study (Indonesia, Malaysia, the Philippines, Thailand, and Vietnam) had provisions regarding EIA in their legislation. In the case of Laos, although no legislation yet exists concerning EIA per se, EIA is included as a procedure for approval of foreign investments. In any case, these countries are presently carrying out EIA in some form. As mentioned above, various types of EIA exist, but there are some common characteristics. It has been observed that many countries tend to extend the scope of their EIAs. In Vietnam, almost all projects must conduct an EIA in order to obtain approval from government, other than a few exceptions where serious environmental impacts aren't expected. In Thailand and Indonesia, the scope of EIAs has been growing. Almost all countries require project proponents to execute EIAs before starting projects, and the results of EIAs are referred to when the government makes decisions on whether or not to approve a project.

3.2.6 Ensuring Local People's Participation

Most developed countries recognize the importance of local people's participation in decision-making and have tried to ensure it. This recognition also affects international society, and evidence exists that this is true in Southeast Asian countries, where many provisions in legislation refer to local people's participation (Indonesia, Malaysia, and the Philippines). In Vietnam, where national law does not yet require EIAs, attempts are made to include public participation at the city level (i.e., Ho Chi Minh City, Hanoi).

A case in Malaysia demonstrates the effective participation by the local people. In the Penang Hill development project, many local people made comments that an environmental assessment was needed, and an NGO involved in a review process claimed that there were many flaws in the assessment process. In the end, the review committee did not accept the environmental assessment and the development project was canceled. While this case is a good example of participation by local people, many problems with the process were also pointed out.

3.2.7 Case of Environmental Impact Assessment

In a case on environmental impact assessment in Malaysia, local people initiated a lawsuit seeking the extension of an exemption from EIAs issued by the government. The Environmental Quality Act enacted in 1974 as federal legislation included a provision on EIA, though without many specifics. The subsequent Environmental Quality Order was enacted in 1987 for the purpose of defining EIA processes. In the case of projects listed under the Order, project proponents were required to implement EIAs before starting a project and, in the process, were required to provide opportunities for local people to participate.

The Malaysian federal government declared that projects proposed in the state of Sarawak were exempt from the duty to implement EIAs. Since 1993, the Bakun Dam project had been proposed for construction in Sarawak, and a serious impact on the environment was anticipated. At the time the proponent announced the project, EIA procedures were just being prepared. According to a declaration, the project proponent did not need to implement an EIA, but three residents claimed that they had the right to obtain a copy of the EIA concerning the project and that they also had the right to representation. The first trial ruled that "the EQA was enacted to be applicable to the entire nation. Subsidiary legislation was permitted to give full effect to the EQA. Under the guidelines prescribed by the project proponent…it cannot be made without some form of public participation…for this is a right vested with the plaintiffs." This judgment, however, was reversed by an appeals court.

3.2.8 Problems

Cases exist in the Asia-Pacific region where the public was allowed to participate in EIAs. Whether positive or negative, this indicates that these countries are gaining experience in the implementation of EIAs. Looking beyond these cases, one can identify current problems and challenges which Southeast Asian countries face with EIAs. Coordination between central and local governments and among the relevant ministries in central governments has been a problem; overlapping jurisdiction may allow some parties to escape the obligation to implement EIAs. To properly implement them, experienced people are necessary; however, a shortage exists—not only of scientific experts but also facilitators of local participation. In addition, there are shortages in local government staff, and implementation of EIAs is sometimes too costly. In some cases, this has become an obstacle to effective implementation. Besides this, many people are not aware of the existence of EIA regulations, and even if they are aware, they may not know how EIAs can be utilized. These shortcomings can be a hindrance to public participation. Some cases indicate problems with appropriateness and fairness in EIA processes controlled by the government. One problem here is that the decision to start an EIA process depends on the government administration. Another problem is the absence of judicial procedures to monitor or check on decisions made by the administration.

4. CONFLICT RESOLUTION IN SUSTAINABLE FOREST MANAGEMENT—WITH SPECIAL REFERENCE TO INDONESIA

As mentioned above, there are several ways to ensure the participation of local people. It is necessary for the protection of their rights to set up mechanisms for remedying injustices when their rights are violated. The courts are one important tool. Recently, other tools have appeared, such as alternative dispute resolution (ADR) systems. In any case, appropriate conflict resolution mechanisms are necessary in order to protect the rights of local people.

4.1 Analysis From Country Studies in Indonesia

Four cases, which occurred in remote areas of Kalimantan and Sumatra, are analyzed. Generally, their settlement patterns and processes were not satisfied, even though the government conducted administrative action or promised to take follow-up measures. In most cases, the parties concerned were the central government, as licenser, local government, as the field supervisor apparatus, plantation entities, as forest concession holders, and traditional or local communities.

From cases in Indonesia, the causes of conflicts that occur in forest management can be summarized as follows:

1. Various problems in the field of forest management have occurred as a result of the implementation of forest concession rights granted by the government.
2. For example, the location of forest concession rights often overlap with traditional and residential areas of the local community. Exploitation based on forest concessions has threatened the life of the local community by cutting

trees and creating plantations without any compensation payments to the locals.
3. Inappropriate allotment of forest concession rights is the major cause of conflict.
4. Delays in the government's responses to community objections have caused conflicts to accumulate, become more complicated, and more difficult to settle. Many of the worst cases have been completely neglected.
5. Decision-making processes are not transparent. The decision-making process for allocating forest concessions and distributing natural resources barely involves local communities and people who live in the area in question. Therefore, the interests of the local community are not recognized and are almost always neglected. This weakens the position and influence of local people and communities, both politically and economically, compared to the company or entity holding the concession right. Neglect of the interests of local communities is based on the lack of legal recognition of their rights.
6. The involvement of parties that have no relevancy to the case has been another problem. Military intervention is one example.
7. Another important factor in the settlement process is the lack of neutral and independent mediators that are not indebted to parties in the community, the concession holders, or the government. Neutral third parties with the skill and capacity to settle conflicts can greatly influence the settlement process.

4.2 Conflict Resolution According to the Current Legal System in Indonesia

In 1993, a decree by the government's forest ministry authorized the harvesting of forest produce, including timber, by traditional communities within concession areas if they obtained permission from the concession holder and authorization from the minister of forests (Ministry of Forest Decree No. 251). This development is a potentially significant, but limited, step toward negotiating partnerships with forest communities and ensuring that incentives exist for local people to promote sustainable forest management. On the other hand, generally speaking, Indonesia's current legal forest tenure system still works against the nation's forests and the livelihoods of many local forest communities. In most cases, the remoteness of areas under timber concessions can overwhelm the government's ability to collect reliable data, set boundaries, and monitor concession-holders.

In 1999 a new law on the management of productive forests was issued to reform the enforcement of the law and to enhance the participation of the local people.[2] Chapter

---

[2] Direct laws and regulations on public participation are as follows: Law No. 5/1967 (Basic Regulations on Forestry), Law No. 5/1990 (Conservation of Bio-Nature Resources and its Ecosystem), Law No. 24/1992 (Spatial Planning), Law No. 5/1994 (Ratification of the UN Convention on Bio-diversity), Law No. 23/1997 (Environmental Management), Govt. Reg. No. 28/1985 (Forest Protection), Govt. Reg. No. 69/1996 (Spatial Layout), Presidential Decision No. 32/1990 (Management of Protected Areas).

6 of the new law stipulates that the indigenous people have the right to harvest forest products to meet their daily needs. In addition, the involvement of indigenous people and local communities in the process of forest industrialization is given a high priority. The people are also allowed to participate by sharing their input on the use of the forests. This minimizes and reduces conflicts over forest management.

Through a growing debate in Indonesia over forest tenure issues, the importance of public participation and benefit-sharing has been recognized.

4.3 Necessary Mechanisms For Conflict Resolution

In designing conflict resolution mechanisms for sustainable forest management, the four factors below should be considered.

1. The rights of local and traditional communities and people, including tenure over forestland and rights of access, ownership, control, and harvesting, must be recognized, guaranteed, and stipulated clearly enough to be invoked at court or in public forums, when problems based on the rights come into question.
2. The local community and people should be included in the decision-making process of regulations, legally binding normative instruments, as well as plans, projects, and policies on forests. The procedure should include reasonable time-frames for the different phases, allow sufficient time for informing the public, and provide for early participation.
3. Access to information by local people should be guaranteed, especially at the early stage. If a request has been ignored or inadequately answered, that person should have access to review procedures. Once a conflict occurs, the settlement process should be established once invoked by a concerned party, and the process should be transparent.

In the settlement mechanism, three components should be included: an objection mechanism, allowing the public to respond or object to the government; a reliable dispute settlement mechanism or ombudsman, which is truly independent; and the option to choose to conduct the settlement through a legal forum or out of court.

5. RECOMMENDATIONS: MEASURES FOR SUSTAINABLE FOREST MANAGEMENT AND EFFECTIVE PARTICIPATION OF LOCAL PEOPLE

From analysis of existing legal systems related to forest management at the national and international levels, some obstacles to implementing the necessary legal systems have been identified. In order to overcome these obstacles, the following recommendations are proposed.

1. Enhance compliance with existing treaties (see Box 2)

There are many existing international legal instruments related to forest conservation, yet there is still not a consensus for establishing a new international convention on forest conservation. Therefore, it is necessary to consider measures on forest conservation at the international level, while referring to three points.

a. It is important to enhance compliance mechanisms for existing instruments.
b. Coordination among these instruments and dispute settlement will be important for existing compliance mechanisms.
c. When developing new compliance mechanisms, it is important to keep in mind the differences in the characteristics of instruments that exist for nature conservation compared with instruments concerned with other issues.

2. Ensure local people's participation

Pubic participation is indispensable for the sustainable use and management of natural resources. It is clear that it has become one of the legal principles in some international instruments, but the specific contents of the principle depend on the characteristics of the natural resources in question. Research has revealed several important principles of public participation, as follows:

a. The participation of local people and indigenous people should be respected and ensured.
b. Forest management systems should adopt the "Subsidiary Principle," a general principle of governance that means making and implementing decisions at the lowest effective level of government or organization. Not only central governments, but also international treaties and organizations should support local people and indigenous people's participation. The former should act in a subsidiary role and as a safety net, when local and indigenous people cannot solve problems by themselves.
c. Based on learned experience, several actions are necessary to ensuring local people's participation:
   i. Ensure their economic benefit.
   ii. The right to consult with the concession-holding company by providing advice from experts should be secured by law.

iii. Disclosure of information relevant to forest conservation and management systems in order to create public awareness among local people.
iv. Support of the local government by the central government, in order to ensure local and indigenous people's participation.
v. Coordination and cooperation among these organizations.
vi. Capacity building is indispensable for realizing sustainable forest management.
vii. Awareness of local people and government officials of the necessity of realizing sustainable forest management is important.
viii. Ensure access to remedies when the rights of local people are violated.

3. Necessity of dispute settlement mechanisms

Concerning dispute settlement, the following components should be considered in designing mechanisms for solving conflicts over forest management:

a. The rights of local and traditional communities and people, including tenure over forestland and the rights of access, ownership, control, and harvesting should be recognized, guaranteed, and stipulated clearly enough to be invoked in court or in a public forum, when problems based on these rights come into question.
b. Local communities and people should be included in the decision-making process for regulations and legally-binding normative instruments, as well as forestry plans, projects, and policies. The procedure should include reasonable time-frames for the different phases, allow sufficient time for informing the public, and provide for early participation.
c. Access to information by local people should be guaranteed, especially at the early stage. If her or his request has been ignored or inadequately answered, the person should have access to a review procedure.
d. Once a conflict occurs, the settlement process should be established upon the request of a concerned party. The process should be transparent.

In the settlement mechanism, three components should be included: first, an objection mechanism, allowing the public to respond or object to the government; second, a reliable, independent dispute settlement mechanism or ombudsman; and third, the option to choose or agree to conduct the settlement through a legal forum or out of court.

Most of these elements have been mentioned and discussed widely at international forums. Although the importance of these elements has been recognized at the national or sub-regional levels, few concrete measures have been taken to realize these concepts.

Box 2: Measures for ensuring compliance.

Public awareness:

Effective implementation can be ensured by cooperation between all stakeholders. Public awareness can be a strong motivation for cooperation between stakeholders.

Administrative measures:

Administrative measures give detailed and desirable procedures of the obligations or interpretations of articles in order to support smooth implementation. For example, the guidelines for implementation of a treaty can provide concrete procedures to ease implementation at the national level.

Coordination:

Coordination requires measures for ensuring the flexibility of treaties to deal with the various conditions and situations of contracting parties. This measure includes coordinating relevant treaties to avoid duplication and conflict. For example, the United Nations Framework Convention on Climate Change and the CBD stipulate adopting protocols to deal with changes in scientific knowledge and to supplement the general obligations of these conventions. The CCD has an article requiring contracting parties to adopt regional action programs to deal with the varied conditions of contracting parties. These measures ensure flexibility in overcoming problems. The CBD and the Ramsar Convention exchange a memorandum of understanding (MOU) in order to coordinate the two conventions.

Monitoring and inspection:

These measures mean that implementation is monitored or inspected by other parties or persons nominated by treaties. Reporting from contracting parties on the status of implementation can be included in this category.

Giving of benefits:

This measure involves the establishment of a system to support implementation by contracting parties by providing financial and technical support. Some treaties establish funds to support developing countries. Almost all treaties require developed countries to support developing countries.

Sanction:

The treaties punish countries for violating treaty obligations. There are several stipulations requiring sanctions for violations, but they are often unable to be implemented, because sanctions would create serious political conflicts, and such conflicts are not desirable.

Conflict resolution:

Almost all treaties have provisions or mechanisms for resolving conflicts, in line with the purpose of the treaty. There are several types of conflict resolution measures, such as mutual negotiation, conciliation, arbitration, and the International Court of Justice.

Reservations:

Some treaties permit contracting parties to create reservations against some articles or annexes. For example, CITES has an article allowing reservations against annexes. The effect of this article on the treaty can be criticized, but it cannot be denied that reservations help to increase the number of contracting parties, in some cases.

PART 2

POLICY AND LEGISLATION IN TERMS OF LOCAL PARTICIPATION

Chapter 5

PUBLIC PARTICIPATION IN INTERNATIONAL ENVIRONMENTAL AGREEMENTS

Kiyoshi Komatsu
Institute for Global Environmental Strategies

1. INTRODUCTION

As described in previous chapters, the participation of local people is recognized as an important element in sustainable forest policy. Many activities are related to local participation, but many obstacles are often encountered when trying to implement them. One obstacle is the failure to clearly define participation leaving it open to be interpreted in many ways.

Such confusion can be seen in international environmental law. There are many international instruments that recognize and describe the importance of participation, but they have certain limitations, even though some are legal instruments. One problem is that the definitions differ greatly. Depending on the purpose and context of each instrument, different aspects of participation are emphasized, so the word "participation" has taken on several general interpretations. In order to ensure effective participation, proper descriptions of these interpretations are necessary.

In this chapter, the various descriptions of participation related to forest conservation in international instruments are analyzed in order to identify their implications. Although these instruments have already been mentioned in the previous chapter, before analyzing them, an outline of their current status is reviewed here in order to elaborate the reasons for the necessity of the participation of local people in these affairs.

Next, descriptions in both non-legal and legal instruments are analyzed to clarify the elements of participation described, as well as to point out their limitations. Based on these analyses, measures for facilitating and promoting the participation of local people are further elaborated. These can be considered supplementary to the measures mentioned in Chapter 4 and seen as suggestions from the point of view of international law.

M. Inoue and H. Isozaki (eds.), *People and Forest—Policy and Local Reality in Southeast Asia, the Russian Far East and Japan, 93–103.*

2. OUTLINE OF INTERNATIONAL INSTRUMENTS RELATED TO FOREST MANAGEMENT

2.1 Processes for Adopting a "Convention on Forests"

Many activities related to forest management have been conducted by international organizations, governments, and non-governmental organizations (NGOs) at the global level since the United Nations Conference on Environment and Development (UNCED) in 1992. One of the most important of these is found in two policy dialogue processes: the Inter-governmental Panel on Forests (IPF) and the Inter-governmental Forum on Forests (IFF). The character of each is unique, including their original objectives: one was to make proposals for actions to tackle the many problems of forest management; the other was to build consensus for adopting a new international legal instrument on sustainable forest management at the global level, a so-called Convention on Forests.

The adoption of a convention on forests was one of the main agenda items at UNCED, but it was not successful because of confrontation between developing and developed countries, and their inability to build consensus. In the end, the Non-legally Binding Authoritative Statement of Principles for a Global Consensus on the Management, Conservation and Sustainable Development of All Types of Forests, the so-called Forest Principles, was adopted as a non-legal instrument. Even though forest issues were dealt with in Chapter 11 of Agenda 21, the action program endorsed at UNCED, no legal instruments on forest conservation were adopted.

Since reaching consensus on a convention on forests had been an important part of the political agenda, the IPF and IFF were expected to act as temporary fora to produce a convention. Therefore, they were organized on an ad hoc basis by setting the time period of the process; each held four meetings over two periods since 1995 (the IPF from 1995 to 1997, and the IFF from 1997 to 2000). In the end, they still failed to reach consensus.

Even though they took almost five years in total, negotiations among parties came to a deadlock. Negotiators failed to achieve consensus on adopting a new legal instrument, but they agreed in the final meeting of the IFF process in February 2000 to establish the United Nations Forum on Forests (UNFF) as a new, permanent body within the United Nations to deal with issues related to sustainable forest management. This was approved by the United Nations Economic and Social Council (ECOSOC) in October 2000, and in its resolution the UNFF was required to conduct activities divided into six functions.[1]

---

[1] The UNFF's six functions include the following: (1) facilitate implementation of the IPF/IFF proposals for action, (2) ensure continuity of policy dialogue and monitor implementation of IPF/IFF proposal actions, (3) coordinate among relevant international instruments and organizations, (4) enhance cooperation among all stakeholders, (5) monitor and assess progress at national, regional, and global levels, and (6) strengthen political commitment. Coordination among existing instruments is recognized as important, and for this purpose the Collaborative Partnership on Forests (CPF) was established under the UNFF. It consists of representatives of the following international instruments and organizations: the secretariat of the Convention on

Meetings are organized annually.[2] The UNFF is to conduct activities based on the ECOSOC resolution mentioned above, and other resolutions adopted at the first substantive meeting of the UNFF, until its fifth meeting in 2005.

Although the UNFF was established as a permanent body, the adoption of a convention on forests remains an important agenda item. Again, the necessity of a convention on forests will be elaborated until the fifth meeting. Accordingly, there is still a chance that the countries will agree to start negotiations on a convention, but this is just a possibility and still uncertain.

Even though, to date, international society has failed to build consensus to start negotiations on adopting a convention on forests, there are already many international instruments at the global level related to forest conservation, such as the Convention on Biological Diversity (CBD), the Convention Concerning the Protection of World Cultural and Natural Heritage (WHC), the Ramsar Convention on Wetlands (Ramsar Convention), the Convention on International Trade on Endangered Species (CITES), the International Tropical Timber Agreement (ITTA), and the International Convention to Combat Desertification (CCD).

2.2 The Necessity of Participation

These instruments contain several principles that could be included in a convention on forests, as mentioned in the previous chapter.[3] Cooperation among all parties is one principle. But ecosystems are too diverse to stipulate universal conservation measures, meaning that these instruments cannot make the principles mandatory. Accordingly, there is discretion on the part of each contracting party in implementing these treaties.

The reason for requiring the participation of local people is connected to these characteristics. The requirement leads to all parties participating in the process of planning and implementation of relevant policies on the sustainable use of natural resources, including forests. This can include local people as one of the important participants. In addition, each contracting party is required to take appropriate measures, but due to the varied ecosystems in each country it is almost impossible for national governments to do so in a way that covers their entire country. This is why the role of local communities and local people is emphasized and their participation is so important.

Biological Diversity, the Center for International Forestry Research, the Department of Economic and Social Affairs of the United Nations Secretariat, the Food and Agricultural Organization of the United Nations, the International Tropical Timber Organization, United Nations Development Programme, United Nations Environment Programme, and the World Bank. See E/2000/35.

[2] In the first meeting, two documents were adopted: the Multi-Year Program of Work (MYPOW) and the Plan of Actions. They were expected to facilitate and promote implementation of the Proposals for Actions of both the IPF and IFF. See Komatsu Kiyoshi, "Toward New International Arrangement for Sustainable Forest Management—United Nations Forums on Forests." *Policy Trend Report 2001*. IGES Forest Conservation Project, page 76–78. 2002.

[3] See Chapter 4.

3. ACTIVITIES IN INTERNATIONAL INSTRUMENTS RELATED TO THE PARTICIPATION OF LOCAL PEOPLE

3.1 Descriptions in Non-Legal Instruments (IPF/IFF Proposals for Actions)

As mentioned above, many non-legal instruments already exist, and although implementation of them is not binding, they show some recognition on the part of international society of their importance. This can affect the building of new international treaties or national policies, and so analysis of these non-legal instruments is necessary.

As a measure to deal with environmental issues, the importance of public participation has been included in many international instruments. For example, the Rio Declaration states the necessity of public participation and describes access to information and means of redress as measures to facilitate and ensure it.[4] It also points out that women, youth, and indigenous people are indispensable actors in the process.[5]

Agenda 21 also emphasizes the necessity of public participation, highlighting the importance of access to information and environmental impact assessment (EIA) processes.[6] Moreover, it also recognized the role of indigenous communities and the importance of traditional knowledge in realizing sustainable development.[7] In this regard, Agenda 21 requires several actions: the assurance of indigenous people's rights to their lands, self-management of natural resources, participation in decision-making, financial assistance, and the provision of legal frameworks.[8]

These two non-legal instruments deal with environmental issues comprehensively, and they point the way for the trend of environmental laws and policies at the international level. Since these instruments describe many actions dealing with public participation, including that of local people, they show that international society prefers to take on policies that involve all of society in activities towards sustainable development. Several descriptions of indigenous and local people are found in them, illustrating that international society also recognizes the importance of the role of local people and the necessity of their participation, including measures to ensure their participation, such as rights of access to resources, tenure rights, participation in decision-making processes, financial assistance, and legal frameworks.

Several documents relating directly to forest conservation have been produced through international dialogue. In the Statement on Forest Principles, the importance of the participation of "interested parties" in the process of implementation and planning

---

4 See Principle 10.

5 See Principle 22.

6 See 23.2 "Section 3: Strengthening the Role of Major Groups" *Agenda 21*, page 219. 1994.

7 See 26, ibid. (N6), page 227.

8 See 26.4, ibid. (N6), page 228–229.

of national forest policies is pointed out in particular. Local and indigenous people are also specified as important actors.[9] It also describes the importance of preserving and utilizing traditional knowledge, and emphasizes the necessity of equitable sharing of benefits from the use of traditional knowledge among parties, including local and indigenous people.[10]

In the IPF's Proposals for Actions, more than 130 actions on forest conservation are laid out for governments, international organizations, NGOs, and other parties. The participation of all parties related to forest conservation is described as one of the pillars and is required in formulating national forest programs,[11] research on forests, and the development of criteria and indicators (C&I).[12] In relation to traditional forest-related knowledge (TFRK), the role of local people is also recognized as important. Included in the Proposals is a requirement that countries take several actions, such as conducting research on TFRK, ensuring indigenous people and local people's participation in national forest programs from planning to implementation, and promoting public awareness in order to preserve and utilize TFRK.[13]

In the IFF's Proposal for Actions the importance of the participation of all parties related to forest conservation is recognized as well. The participation of local people in management processes for protected areas is also emphasized, including the need for economic incentives or support for local people to ensure their participation.[14] Moreover, the establishment of a legal framework on participation, where appropriate, is described as an important measure in order to tackle underlying causes of deforestation. The IFF's Proposals for Actions, like the IPF's, also recognize the importance of TFRK, describing several actions such as preserving and utilizing traditional knowledge and ensuring the equitable sharing of benefits. Moreover, the IFF describes the necessity of establishing and implementing a framework for preserving TFRK and coordination among relevant activities.[15]

In these three non-legal instruments, the necessity of participation of all parties in forest conservation is described, but detailed procedures and processes of participation are not included. The participation of local people is emphasized in the context of preserving and utilizing TFRK, and they list several actions such as the establishment of a framework for preserving TFRK and coordination among relevant activities, including promoting public awareness. Again, they emphasize the necessity and importance of the

---

[9] See Principle 5(a) A/CONF.151/26 (Vol. III).

[10] See Principle 12 (d) ibid. (N9).

[11] See Report of the Intergovernmental Forum on Forests on its fourth session, para. 1--9 (E/CN.17/2000/14).

[12] Actions described can be categorized as follows: institutional arrangements/formulation of policy, international cooperation, research, ensuring participation, open information and information exchange, and training. See Komatsu Kiyoshi, "Outline of Brain Storming Forum on IFF and a 'Convention on Forests.'" In *A Step toward Forest Conservation Strategy (1).* IGES Forest Conservation Project, page 14–15. 1998. Also see para. 8–17, Report of the Ad Hoc Intergovernmental Panel on Forests on its fourth session (E/CN.17/1997/12).

[13] See para. 32–40, ibid. N(11).

[14] See para. 79–90.

[15] See para. 68–75.

participation of local people, but do not provide detailed processes. The development of detailed procedures is left to each country's discretion. In addition, since these are not legal instruments, they are still just indicators for policy direction.

3.2 Activities Described in Legal Instruments (Convention on Biodiversity, the Ramsar Convention on Wetlands)

Both the conventions on biodiversity and the combat of desertification have articles related to the participation of all parties—the participation of local people in particular.

Article 14(1) of the Convention on Biodiversity requires public participation in environmental impact assessment processes where appropriate. It also requires contracting parties to ensure the participation of local and indigenous people and to "promote their wider application with the approval and involvement of the holders of such knowledge, innovations and practices" in Article 8(j). In the same article, it established the Ad hoc Open–end Working Group, based on the decision of the Fourth Conference of Parties in 1998, in order to address its implementation and relevant provisions. As a result of discussions at the meeting, the Working Program on the Implementation of Article 8(j) and the related Provisions of the Convention on Biological Diversity were proposed.

At the Fifth Conference of Parties of the CBD in 2000, Decision V/16 was adopted. It included the working program in its annex, along with seven elements. Each one lists tasks as necessary measures in order to achieve the objectives. In the working program, the participation of the local and indigenous communities is required by contracting parties as central. In order to ensure effective participation, it requested that several points be taken into account.[16] Benefit sharing, monitoring, exchange, and dissemination of information are recognized as main measures. In order to implement them, several things are required. With benefit sharing, prior informed consent is important. In general, there are not many detailed measures, but the working program consists mainly of activities for developing guidelines or detailed measures. Hence, concrete measures should be clarified as a result.

Article 8(j) also stipulates that nations should "encourage the equitable sharing of the benefits arising from the utilization of such knowledge, innovations and practices." In this regard, the CBD has other working elements, including its Panel of Experts on Access to Genetic Resources and Benefit-sharing.[17] A report of the panel's first meeting listed several measures to ensure benefit sharing, including prior informed consent, contractual arrangements, information dissemination, and capacity building. As a result of work on this issue in the CBD, the new Bonn Guidelines were adopted at the Sixth Meeting of the Convention of Parties (COP) to the CBD as Decision VI/24 Access and

[16] These include the need to "(a) build on the basis of their knowledge, (b) strengthen their access to biological diversity, (c) strengthen their capacity on matters pertaining to the conservation, maintenance, and protection of biological diversity, (d) promote the exchange of experiences and knowledge, and (e) promote culturally appropriate and gender specific ways in which to document and preserve women's knowledge of biological diversity."

[17] The panel was established by Decision 8, 1998.

Benefit-sharing as related to genetic resources. In the guidelines, measures are described as possible detailed elements of procedures and mechanisms for ensuring access to genetic resources and equitable benefits sharing.

The CCD also includes articles requiring the participation of local people: Article 3(a) and (c), Article 9(1), Article 10.2(e), Article 13.1(c), and Article 19.1(a). In the CCD, the role of local people is emphasized. For example, other than Article 9.1, these articles mention "local people" or "local community" in stipulations, even if local people are described as one of the parties. Moreover, several articles describe measures necessary to ensure the participation of local people. For example, Article 10.2(e) emphasizes the necessity of ensuring access to information and technology to local people, and Article 19.1(a) points out the importance of capacity building of local people. Article 3(c) points out the importance of all parties' awareness, including local people. Article 19.1(a) describes the necessity of participation of local people as well, and it emphasizes the importance of the participation of women and youth.

In relation to traditional knowledge, the CCD also has several articles that require the participation and approval of local people.[18] In particular, Article 18.2 describes some measures necessary for preserving and utilizing traditional management systems of natural resources and land. It recognizes the importance of participation in developing an inventory of traditional technology, knowledge, know-how, and practices. It also emphasizes the importance of having the approval of local people to utilize these for commercial purposes and assurance of their benefits. Moreover, the dissemination of information, the development of new technology based on information from traditional management systems, and the adaptation and integration into modern science are encouraged and facilitated.

The Ramsar Convention does not mention the necessity of participation, but the importance of participation of concerned parties, in particular local people, had been recognized while contracting parties accumulated experiences in its application. Based on this recognition, the Ramsar Convention started developing guidelines on ensuring the participation of local and indigenous people. As a result, Resolution VII.8 was adopted at COP-7 in 1999. The guidelines for ensuring local people's participation were attached to the resolution in the annex. They show several elements for the participation of local people as follows: the establishment of the framework of participation, incentives for local and indigenous people, the exchange of information and knowledge, capacity building, and the utilization of traditional knowledge. The elements are indicated as possible elements for a system of promoting and facilitating the participation of local and indigenous people; however, there are no concrete procedures or processes. This means that they could not develop universal procedures and processes because of the diversity of ecological, social, cultural, and economical situations in the countries of each contracting party. In other words, the contracting parties were requested to develop systems or measures for ensuring the participation of local people by implicitly accommodating the situations in each country.

Even though they are legal instruments, they still have limitations. As mentioned in the explanation of the guidelines in the Ramsar Convention, it could not indicate

---

[18] Article 16(g), Article 17.1(c), Article 18.2, and Article 19.1(e).

universal measures or procedures on participation of local people at a global level, nor could the other two instruments, for the same reasons as with the Ramsar Convention.

3.3 Activities in International Organizations Related to Sustainable Forest Management (ITTO, WIPO, FAO)

The International Tropical Timber Agreement includes many documents developed by the International Tropical Timber Organization (ITTO) as a result of their policy development work. As one of the main results, their criteria and indicators (C&I) on sustainable management of tropical forests can be pointed out in Criteria and Indicators for Sustainable Management of Natural Tropical Forests (1998). It also presents indicators to measure each criterion and highlights the participation of local communities. As indicators of that criterion, several actions are listed in Indicator 7.16 regarding traditional forest management systems, consideration for the customary rights of indigenous and local communities, as well as participation in economic activities and involvement in forest management agreements.

In relation to the preservation and utilization of traditional forest-related knowledge, the Food and Agriculture Organization of the United Nations (FAO) has several documents related to the use of genetic resources.[19] One of the reasons why the FAO is connected with this issue is that the progress of biotechnology has been based on traditional knowledge in the use of national resources, usually by ignoring the interests and the efforts of local and indigenous people to preserve that knowledge.[20] In order to deal with this problem, the FAO developed the Code of Conduct for Germplasm Collecting and Transfer. In the code, collectors of germplasm are requested to take several actions, including compensating farmers.[21] Descriptions in these documents are based on the concept of "farmers' rights" that was developed by the FAO and described in Resolution 5/89, adopted in 1989. It aims to protect the benefits of farmers who have cultivated seeds over generations, but the role of the Farmers' Rights is not clear and

[19] The Code of Conduct for Germplasm Collecting and Transfer, adopted by the FAO, provides a model of procedures for issuing licenses for collecting germplasm and guidelines for collectors themselves, and clarifies the responsibilities of the sponsor of the mission of collecting activities, the curators of gene banks, and the users of germplasm.

[20] Some medicines were developed based on such traditional knowledge, and they have made profits for companies, but the efforts of local and indigenous people to maintain such traditional knowledge have been ignored in the existing system of intellectual property rights.

[21] Other actions include showing their gratitude to the traditions of local society and their property rights, responding to requests of the local community for information or assistance on germplasm, informing local people about the purpose of collecting activities, and recognition of the "rights and needs of farmers and those who manage wild and cultivated plant genetic resources." Several concrete measures of compensation are also indicated: access to new improved varieties and other products, support research activities on the conservation of germplasm by the local community, training at both the institutional and farmers' level, transfer of appropriate technology on conservation and the use of genetic resources, support for evaluation and enhancement of local varieties and indigenous germplasm, and providing scientific and technological information and other appropriate support for the local community.

still remains just a concept to encourage the protection of the benefits of local communities.[22]

In relation to intellectual property rights, the World Intellectual Property Organization (WIPO) has already had some experience.[23] In order to tackle the problems dealt with by the Code of Conduct adopted by the FAO, the necessity of modifying intellectual property rights was recognized by international society, but there are gaps between the nature of intellectual property rights and traditional knowledge on genetic resources.[24] Accordingly, WIPO, which deals with matters related to intellectual property rights at the global level, is required to bridge the gap between intellectual property rights and traditional management systems on genetic resources.[25] WIPO has dealt with this issue since 1998. And in 2001, it established the "Intergovernmental Committee on Intellectual Property and Genetic Resources, Traditional Knowledge and Folklore" in response to requests from member countries.[26] The work of the committee continues, with no consensus yet on taking concrete actions and measures on some crucial issues.[27]

The activities of these international organizations related to traditional knowledge or natural resources management systems focus on ensuring the interests of local people and are the result of deliberations by international society. There is confrontation between countries, however, in particular on the matter of intellectual property rights—

---

[22] See Dinah Shelton, "A Right–based Approach to Public Participation and Local Management of Natural Resources." *The Third IGES International Workshop on Forest Conservation Strategies for the Asia and Pacific Region*, ed. Inoue Makoto and Isozaki Hiroji, page 234. 2000.

[23] As an aspect of the preservation of cultural heritage, WIPO has already developed "The Model Provisions for National Laws on the Protection of Expression of Folklore against Exploitation and other Prejudicial Actions" in cooperation with UNESCO.

[24] For example, traditional knowledge is held by groups, such as kinship or ethnic groups, but intellectual property rights are owned by individuals. Therefore, there is difficulty in determining the owner of the rights. See UNEP/CBD/COP/5/8, page 92.

[25] Center for International Environmental Law, "Comments on Improving Identification of Prior Art," page 3. 2 August 1999; David Downes, "Using Intellectual Property As a Tool to Protect Traditional Knowledge" Center for International Environmental Law Discussion Paper prepared for the Convention on Biological Diversity Workshop on Traditional Knowledge, Madrid, November, 1997, page 7.

[26] Three main themes were set out for committee discussion: (1) access to genetic resources, (2) protection of traditional knowledge, whether or not associated with those resources, and (3) the protection of expressions of folklore. See WO/GA/26/6, page 5–7.

[27] One of the issues is *prior art*, which means that knowledge or information is held by the public before the application is filed. Moreover, applications which belong to prior art can be refused by authorities. This procedure is required by national laws and the Patent Cooperation Treaty (PCT). Hence, prior art is one of the conditions used to examine patent applications, and the information on prior art is necessary for determining whether the application is a novelty and invented or not, but there is no experience in using traditional knowledge as prior art in concerned governmental agencies. Besides, forms of traditional knowledge are much different from ordinary prior art and there is no database of traditional knowledge for the purpose of patent application procedures. Therefore, it is necessary to modify and revise existing application procedures. See WIPO/GRTKF/IC/2/6, page 3, para. 1.

one of the reasons why WIPO has continued its activities. In other words, measures related to intellectual property rights that can impact many countries is a sensitive issue. Accordingly, there are obstacles to the work conducted by WIPO.

4. CONCLUSION

As mentioned above, there are several types of elements involved in the issue of local participation in forest management. In both the Rio Declaration and Agenda 21, the participation of local people as well as other parties—the "public"—is emphasized. The necessity of public participation is also pointed out in the IPF's Proposal for Actions. The Statement on Forest Principles does not use the word "public," but it describes the "participation of interested parties, including local communities and indigenous people, industries, labour, non-governmental organizations and individuals, forest dwellers and women." This is similar to the concept of public participation. It also highlights the necessity of involving all parties in society in forest conservation. These instruments only emphasize the importance of ensuring participation; they do not provide descriptions of detailed and concrete procedures or the forms of participation.

On the other hand, the role of local and indigenous people is included in other instruments, such as the Convention on Biodiversity and the Convention on Desertification, which emphasize the role of local and indigenous people in preserving and utilizing traditional knowledge and natural resources management systems. In particular, equitable benefit sharing is highlighted as a very important issue, with relevant measures and procedures indicated. And more detailed measures or procedures are contained in the ongoing works of WIPO. Therefore, measures for the equitable sharing of benefits in these instruments are still general and contracting parties still have discretion on their implementation.

Accordingly, one could say that these instruments recognize the necessity of the participation of all parties. Measures indicated in them for ensuring participation are too general, however, to be applied directly at the national level. Therefore, each country must decide on prioritized measures according to its own situation. In particular, the development of detailed and concrete procedures of participation is needed to reflect each country's unique situation at the field level.

In other words, international instruments can only indicate frameworks for ensuring participation, but not concrete, detailed procedures, and these procedures can only be developed by each country with consideration for its own unique situation. Therefore, the intended purpose of these instruments cannot be achieved by just establishing frameworks for ensuring participation, but by developing detailed and concrete procedures.

Accordingly, ensuring participation depends on each country's ability to develop and implement mechanisms to ensure the participation of local people. In this context, the means of ability implies the need to consider several dimensions, such as the ability to develop mechanisms that include detailed and concrete procedures on participation, knowledge of legal and administrative systems of forest management, the ability to coordinate among various parties, and so on. Governments need to develop training programs to enhance the abilities of government officials, local people, and other parties. International cooperation should also be required to be reviewed from the point of view of enhancing its ability to ensure participation.

Chapter 6

FOREST CERTIFICATION SCHEMES IN JAPAN, INDONESIA, AND MALAYSIA: BACKGROUND AND REALITY

Norihiko Shiraishi
The University of Tokyo

Satoshi Tachibana
Forestry and Forest Products Research Institute

Forest certification is a market-driven, institutional system that promotes environmentally conscious, socially acceptable, and economically sustainable forestry management practices through third-party evaluation and certification. Despite the use of the word "institutional," which gives the connotation of it being a government service, it is generally a private initiative. As well, the certifiers and the companies they evaluate both actively participate in the scheme. Forest management entities, once granted certification, make use of it through advertising or by stamping or labeling their logs and other wood products with a logo that communicates the organization's achievement. When consumers choose to purchase their products, they are, in effect, supporting good forest management. Forest certification is considered a viable means for differentiating from wood illegally logged or produced through unsustainable forestry management practices.

1. FOREST CERTIFICATION SCHEMES UNDER THE FSC AND ISO

In reaction to the accelerated deforestation in developing countries in the tropics towards the end of the 1980s, concerned European environmental organizations formed a movement to boycott tropical timber. The timber industry's concern about this movement led to the idea of forest practices certification by certifying sustainable forest

M. Inoue and H. Isozaki (eds.), *People and Forest—Policy and Local Reality in Southeast Asia, the Russian Far East and Japan, 105–112.*

management and wood sourced from such forests. But the intelligibility of this notion of forest certification hindered it from realizing its full potential, as the certification methods and criteria of different certifying bodies varied, causing a proliferation of different logos and confusion among consumers. At the same time, because of a lack of consumer consciousness of the need for a strict certification process, even wood coming from unreliable sources, in terms of sustainable management, was allowed into the market. In order to deal with the situation and to come up with a neutral, just, and reliable forest certification system, influential environmental organizations, such as the World Wide Fund for Nature (WWF), came together in 1993 to form the Forest Stewardship Council (FSC). Because this came a year following the United Nations Conference on Environment and Development (UNCED), the rise of forest certification coincided with other regional initiatives, such as the Montreal Process and the Helsinki Process, in promotion of sustainable forest management.[1]

In light of its founding objectives, the certification scheme structure under the FSC was designed to encompass forest management around the world under various social, economic, and ecological conditions, emphasizing on-site field assessments and performance criteria. The term *on-site* refers to an assessment process that emphasizes the field, and not exclusively documentation. *Performance criteria* refers to the delivered practice and management standards evaluated. In effect, the intention of the scheme is to compromise on the establishment of institutions and laws on forest governance, even under conditions within developing countries that may still be at a premature stage.

In 1996, the FSC developed its "Principles and Criteria for Forest Management," which are the principles for all certification criteria, and it has accredited certification bodies and been involved in forest certification worldwide since. The area under FSC certification has seen a stable increase, and as of April 2002, covered 27.92 million hectares (ha) in over 54 countries.

The International Organization for Standardization (ISO) 14000 Series, also referred to as the environmental management system certification, is often brought into discussions in relation to the FSC's forest certification scheme. The ISO scheme lays out a continuous decrease of environmental loads by incorporating the "Plan→Do→Check→Act," or PDCA cycle, into the organizational structure. Under the ISO scheme, establishment of the management system refers to the act of compiling documentation in the form of manuals or such, incorporating goals and procedures of activities into plans and policies, and carrying these out. The ISO approach of management systems may well function in societies that have attained high "quality of life" or in organizations with high levels of structural order capable of controlling the PDCA cycle, but the nature of the ISO system presupposes the organization's capacity to define its own goals and employ corrective actions and procedures for their attainment. Thus, assessing attainment in terms of absolute measures tends to be unsuitable. Another issue is the deviance between the corporation's environmental

[1] The Montreal Process Working Group (MPWG) for the Conservation and Sustainable Management of Temperate and Boreal Forests was initiated at UNCED in 1992. The Helsinki Process developed guidelines for the sustainable management of forests in Europe, seeking to identify measurable criteria and indicators to evaluate progress.

measures and the products being marketed. Because of this, labeling practices on products with ISO certified status of the organization have not materialized.

The ISO's focus on management systems and the FSC's emphasis on performance in the production procedures do not necessarily conflict, but rather complement each other. In countries abroad there are cases where the forestry management entity has opted for certification under both the ISO and FSC schemes. It should be noted that establishment of a management system is also required as a part of the FSC certification standards.

2. DEVELOPMENT OF REGIONAL FOREST CERTIFICATION SCHEMES AND MUTUAL RECOGNITION

Since the launch of the FSC in 1993, there have been numerous attempts around the world for countries to come up with a forest certification scheme of their own. In Europe, the Helsinki Process was created in reaction to UNCED, and around the mid 1990s national initiatives arose in laying ground for environmental conservation, biodiversity laws and regulations, and forestry practice policies which incorporated environmental aspects of forest management. In timber-exporting countries, such as Finland and Norway, as well as in countries whose consumers have higher environmental consciousness, such as the United Kingdom and Germany, the proceedings of the governments in laying ground for environment and forest-related laws and regulations were met by forest industry initiatives in organizing forest certification schemes, reflecting the voices of environmental organizations and citizens. In the United States and Canada, similar movements arose and forest certification schemes initiated by the American Forest & Paper Association (AF&PA) and the Canadian Standards Association (CSA) were instituted in the respective countries, concurrent with similar developments in Europe. These regional initiatives in forest certification in Europe and North America are based on the ISO management system's approach; in other words, they are composed on the premise of conformity with laws and regulations and complementation in areas needing reinforcement by the certification requirements. A program structure is embodied which prompts management entities with higher capacity to seek even higher standards. This is to say that forest certification systems are found to be functioning holistically with the institutions and laws of the respective countries.

When absolute equivalency between two different forest certification schemes is acknowledged by both parties, and a product certified under the conditions of one of the schemes is entitled to labeling using the other schemes' logo mark, this state is called mutual recognition. Even when acknowledgement is not mutual—when one scheme recognizes the other scheme to be within its own requirements and therefore worthy of certification—this state is called endorsement. Endorsement or mutual recognition is said to be inevitable in the process of introducing locally-certified material into the global market.

The Pan-European Forest Certification (PEFC) system has taken on the role of an umbrella system of forest certification schemes in Europe. It has prompted advancement in the endorsement of forest certification systems developed nationally and of carrying the PEFC logo. It has further extended its realm beyond the regional frame to incorporate the two North American schemes as endorsees of the program, and has condensed the proliferation of certification schemes to the two general influential streams of the FSC and the PEFC. Notably, the PEFC has been expanding its areas of operation by using endorsement to the extent that it has brought itself to a competitive status with the FSC, which has taken a worldwide perspective. At this point, the PEFC's related bodies are criticizing the FSC structure for its costliness for certification and the non-attainment of the separation of the three powers (i.e., administration, legislation, and judicature). In reaction, the FSC's supporters claim that the PEFC is only applicable to a limited region and criticize the ambiguity of its performance criteria.

It should be taken into consideration, however, that the societal background of the establishment of the FSC and PEFC are each unique and that the structure of their certification schemes are more or less different, as well (Table 1). The FSC scheme, founded on the premise of combating the deforestation of tropical forests, does not take for granted the compliance of institutions and laws of the applied locality, and is characterized by performance-based assessment conducted on-site. In order to effectively compete in the market, chain of custody (CoC) tracking and labeling procedures are essential.[2] On the other hand, the forest certification schemes under the PEFC were developed in countries where compliance with institutional laws and regulations is secured, with the main purpose being to improve forest management standards. Both schemes were established with their respective imperative rationales, and the superiority of one scheme over the other should not be discussed based on the differences in system structure.

3. CASE STUDIES

3.1 Current State of Forest Certification in Japan

Among the forests and forestry arenas of Japan, the certification of the Sumitomo forestry division and forestlands under ISO 14001 in August 1999 was the earliest of all other initiatives, followed by certification under the FSC scheme by Hayami Forestry in February 2000. As of April 2002 a total of two entities had received certification under the ISO scheme, and four were certified under the FSC scheme. The ISO certification has been limited to large enterprises or local government units, and due to the characteristics of application practice within the organizational structure, observable changes have been limited. As for the FSC scheme, numerous forest owner's

[2] CoC is a system that tracks products from their origin, through transportation, storage, transformation, and commercialization until they reach the final consumer. Certification provides the consumer with a guarantee that the product comes from a certified process.

associations and publicly owned forests have acknowledged FSC as a tool for local community development and effective cooperative management, and are in the process of preparing for undergoing assessment.

Forest certification has several points of merit. The roles of local and central governments in relation to forest management can be devolved to certification, and if improvement of forest management standards can be granted, this should be an advantage for the government as well as for the public. For the forest management entity, forest certification can lead to incentives, such as enlarged market shares and a premium in sales prices. At the same time, there are other advantages that can be presupposed, such as internal risk management of the entity, effective corporate management, and empowerment of the workers.

There are movements among forestry-related bodies in Japan that are seeking to establish Japan's own forest certification scheme. This move was initiated by the prospect that Japan's reliance on import timber will increase in the future and to certify as many forestry entities as possible at a lower cost than certification by the FSC. Looking into precedent cases of countries adopting forest certification schemes around the world, however, most cases exemplify a strategy built for timber export, and strategies for promoting domestic forestry against foreign import timber through certification schemes are nil. In considering a forest certification scheme for Japan, there are numerous issues to be dealt with, such as gaining trust from the market, identifying incentives for management, and ways of handling the added costs.

3.2 Current State of Forest Certification in Indonesia

Forest certification in Indonesia is in the early stages, but some programs already exist, notably from the Indonesia Ecolabel Institute (LEI), the central government's mandatory scheme (Ministry of Forestry [MoF]), and the Association of Indonesian Forest Concession Holders (APHI) self-declaring scheme.

The LEI, which was established officially in 1998, certified a natural forest of 90,957 hectares in Sumatra in mid-1999 and a furniture company in Uniseraya in March 2002. This same forest was also certified by SmartWood under the FSC scheme. In order to share their experiences and lessons, to increase confidence nationally and internationally, and to develop capacity building for Indonesian assessors, the LEI and the FSC reached a joint certification protocol in 1999 and have been conducting joint field surveys since. Referring to criteria and indicators, the LEI scheme, which is based on production forest management units (PFMU), is quite similar to the FSC scheme. The difference between the two is in the examination process; the FSC has two steps of pre-assessment and assessment, and the LEI has three steps of pre-assessment, field assessment, and decision-making process. Furthermore, the FSC has been in the process of certifying two plantation forests in South Sumatra and East Java since March 2000, while it suspended four certified forests in October 2001 for not clarifying the volume of illegal logs in their operations.

Table 1: Structural comparison of the FSC and forest certification schemes of various countries.

	Forest Stewardship Council	Forest certification schemes of various countries
Nature of schemes	A self-completing certification scheme, aiming for applicability, even within the context of a locality with premature institutions and laws or in compliance with such.	A certification scheme functioning in hand and in mutual complementation with the institutions, laws, and regularities of the locality that presupposes compliance with such.
Nature of certification criteria	Conducts on-site audits focusing on the attainability (performance) of the management standard.	Emphasis on a management system where continuous improvement is secured; similar to ISO 14000. Compliance with domestic laws and regulations is the starting point of evaluations, with the aim of raising management standards by certification.
Motives for certification. Target forest management entities	Certification of certain selected management entities whose performances are high and who participate in certification through self-motivation.	Appears to be self-motivated, but achievement of certification is practically prescribed. Aims for certification of all forest management entities in the area.
Market. Incentives for management	Niche market. Management entities as units aim for market competitiveness by means of differentiation.	General market. A qualification to enter into the global market when forest certification is recognized. States as units aim for competitiveness in trade of domestic products.
Need for chain of custody or labeling schemes	Essential in identifying and differentiating certified products.	Necessary only in the event that forest certification becomes a norm. Necessity is relatively low for the time being, as long as forests are certified.
Establishment of forest reserves	Necessity of establishing forest reserves in areas identified as of high conservation value stated in principles and criteria. This requirement may go beyond the extent of domestic laws and regulations.	Domestic laws and regulations are the premise, and certification requirements do not exceed these limits.
Possibilities of mutual recognition	The FSC's nature of social relevance and public quality invites such possibilities, but currently no certification schemes match that of the FSC's.	In comparing the respective criteria, similarity levels are not high, but there is a commonality in that the emphasis is on process.

Indonesia's ministry of forests just launched a verification process in 2002 in collaboration with international organizations, including the International Tropical Timber Organization (ITTO) and the Center for International Forestry Research

(CIFOR). This scheme, which targets all 375 forest concessionaires and has an independent body to be supported, seems easier than the LEI scheme at this stage. The APHI scheme, which has been trying to meet the ITTO's criteria and indicators since 1993, also re-started in 2002, supported financially and technically by the ITTO.

Based on interviews with key personnel in forest certifier organizations and forest industry trades in Indonesia, there is a possibility of convergence of the three schemes into the LEI scheme in the near future. It means that the LEI scheme surely plays an important role in promoting sustainable forest management in Indonesia. There are some difficulties, however, in promoting its original certification. (1) To take both the FSC and LEI certifications together doubles the expected costs for forest holders, because it is necessary for them to realize mutual certification. (2) An expansion of illegal logging and trade indicates that there are some problems in timber distribution in Indonesia, including the scale and behavior of the wood processing industry. It is necessary to develop and fulfill a tracing procedure system with a label or stamp in order to promote legal timber distribution directly and sustainable forestry indirectly. (3) The incentive to obtain and implement forest certification is smaller, because certified wood does not produce enough benefits in the international timber market of the western Pacific Rim region. Producers are closely observing the trends of the timber market among major importing countries such as Japan right now. If Japan develops a strong market for certified timber, they may pay attention to forest and timber certifications without delay.

3.3 Current State of Forest Certification in Malaysia

In Malaysia, the National Committee on Sustainable Forest Management, established in February 1994, formulated a total of 92 activities to operationalize the ITTO's 5 criteria and 27 indicators at the national level (Forestry Department Headquarters Peninsular Malaysia 2001). At the same time, the ten state forestry departments in peninsular Malaysia formed a working group in order to support the committee. The committee also identified a total of 84 activities to be implemented at the forest management unit (FMU) level under the ITTO's 6 criteria and 23 indicators for sustainable forest management (SFM). These cover resource security, the continuity of timber production, conservation of flora and fauna and other forest resources, an acceptable level of environmental impact, socio-economic benefits, planning, and adjustment to experience. The respective state forestry departments have also formulated management specifications for effective monitoring and evaluation, which formulated 206 management specifications at the national level and 191 at the FMU level. The states of Sabah and Sarawak are in the process of forming a plan of their own management specifications.

The Malaysian criteria and indicators (MC&I) for SFM, coordinated by the Malaysian Timber Certification Council (MTCC), were completed in 1999 for forest management certification in assessing forest management practices in all FMUs, using the standard of the Malaysia-Netherlands joint program mentioned below. Currently, Malaysia has 64 indicators, 200 activities, and 170 standards of performance for SFM at the national level, and 56, 171, and 150 at the FMU level, respectively, under the

ITTO's seven criteria, which consist of enabling conditions for SFM, including forest resource security, forest ecosystem health and condition, flow of forest produce, biological diversity, soil and water, and economic, social, and cultural aspects. The MTCC had certified around 2.3 million hectares in forests and 26 enterprises in CoC by the end of 2001.

A pilot study on timber certification was initiated in mid-1996 under the ambit of the Malaysia-Netherlands joint working group. The ad hoc working group produced 29 indicators, 87 activities, and 49 standards of performance for forest management certification at the FMU level under six criteria, with the exception of forest ecosystem health and conditions in the above seven criteria of the ITTO. Three participating states (Terengganu, Pahang, and Selangor) were chosen as the FMUs. At the initial stage, four Malaysian exporters and five Dutch importers participated in the project. Assessment was carried out in accordance with the MC&I. Furthermore, a chain-of-custody assessment was put into practice in compliance with the Société Générale de Surveillance (SGS) qualifier program. As a result of surveillance and exercises in 1998, there have been improvements in forest management since the first audit. This pilot study later expanded to five states in peninsular Malaysia.

In the process of Malaysian forest certification, they have been trying to develop FSC-compatible MC&I so as to build mutual forest certification, but it has not been developed for the term, following just three FSC certifications of 55,083 hectares in natural forest in the state of Sabah, 12,434 hectares in plantation forest in the state of Selangor, 9,725 hectares in natural forest in the state of Ipoh, and 47 CoC enterprises by the middle of 2002. The FSC certified 52 enterprises under the CoC scheme between August 1997 and August 2000, whereas five enterprises have been suspended up to now. Based upon the information from the MTCC and as part of the collaboration efforts between the MTCC and FSC, the multi-stakeholder National Steering Committee was formed in 2001 in order to determine the terms and conditions for further cooperation and to revise the MC&I to make them compatible with the FSC principles and criteria. It is an important issue for the MTCC and Malaysian SFM to develop a mutual system of forest certification with the FSC in the next phase.

REFERENCE

Forestry Department Headquarters Peninsular Malaysia. 2001. Briefing notes on the Malaysian criteria and indicators for forest management certification.

Chapter 7

FOREST POLICY DEVELOPMENT IN LAOS

Noriyoshi Kitamura
United Nations Food and Agriculture Organization

1. INTRODUCTION

1.1 Scope of this Chapter

This chapter has three objectives: first, to introduce the development of general forest policy in Laos after the socialist revolution in 1975, based on official documents and interviews; second, to analyze implementation of the policy and legislation, based on official documents, budgets, programs, and interviews; and finally, to analyze the background and process of policy formulation, change, and the promulgation of various forest related legislation, based on data and interviews. Some difficulties exist that limit this chapter's scope to a chronological overview and limited analysis of major forest related events and legislation. For example, official documents, especially of decision-making processes, implementing programs, and budget allocation, are unavailable or inaccessible, and statistics concerning forestry activities are very limited and unreliable, if even available. More detailed analysis by sub-sector, including logging, environmental conservation, and forest management by local people, also remains to be done.

1.2 Structure of this Chapter

The next section provides a brief account of the natural and socio-economic situation of Laos, followed by the role of its forestry sector.

Section 3 is the main part of this chapter. The Lao People's Revolutionary Party (the Party) has the leading role in the management of the political system (Lao People's Supreme Assembly 1991). The Party Congress, held every five years, is the highest decision-making forum and prepares a five-year socio-economic development plan.

M. Inoue and H. Isozaki (eds.), *People and Forest—Policy and Local Reality in Southeast Asia, the Russian Far East and Japan, 113–126.*

Forest related sections of the Congress's report and the five-year plans act as the forest policy or policy directions, and major events and legislation are introduced and analyzed against this backdrop in the five-year period following each Congress. This chapter also provides some observations concerning forest policy development, approaches, and governance in the last section. All documents of Congress and legislation referred to in this chapter are in the Lao language, and the author is solely responsible for translation.

2. FORESTRY IN THE NATURAL AND SOCIO-ECONOMIC CONTEXT

2.1 Natural and Socio-Economic Conditions

Land-locked Laos covers 236,800 square kilometers and has a population of 5.2 million, growing at 2.3 percent per annum. Its land is mostly mountainous, except for flood plains along the Mekong River, and agriculture remains the major sector of the economy, contributing 52 percent of gross domestic product (GDP) and employing over 80 percent of the labor force. Its per capita income and social indicators are among the lowest in the region, with nearly 40 percent of the population living below the national poverty line. Disparities between the urban and rural populations has been widening, and subsistence farmers in the mountainous northern and eastern parts of the country are largely engaged in slash-and-burn cultivation, and remain poor and isolated from major markets and social services.

2.2 The Role of Forests and Forestry in the Economy, Society, and Environment of Laos

Laos is endowed with natural riches, especially forest and water resources, particularly in comparison with neighboring countries. Forests and the forest sector play very important roles in the macro-economy and subsistence livelihoods of rural society in Laos. In 1990 forest cover was 47 percent of total land area, and another 44 percent was in some form of degraded forestland. The total land area and forest area per capita are by far the highest in the region.

Forestry contributes 7 to 10 percent of Lao GDP and 15 to 20 percent of non-agricultural GDP. It contributes more than 30 percent of total export value and even more of net foreign exchange. Log royalties, as a share of government tax revenues, were 20 percent in the mid-1990s, but have decreased to 8 to 9 percent in recent years, due mainly to lower yield volumes.

About 80 percent of domestic energy consumption is wood-based, mostly in the form of fuelwood and charcoal. Most of the rural population, about 80 percent of the total, is heavily dependent on forests for food, fuel, medicines, tools, houses, and so on.

They also generate cash income from the sale of non-timber forest products, including cardamom, rattan, and others. Shifting cultivation is the main form of agricultural production in the mountainous areas, especially in the northern and eastern parts of the country. Due to the increasing population, worsening soil conditions, and government land-use and allocation policies that restrict swidden lands, shifting cultivation is no longer sustainable and poses a major threat to soil conservation and the remaining forest areas.

Two of the strategies used by the government to attain economic development are to increase food production through irrigation and generate revenues through hydropower facilities. The conservation of forests in watershed areas is essential for the development and maintenance of these systems. Moreover, Laos occupies 33 percent of the lower Mekong basin area and still maintains relatively rich biodiversity in the region. Forest ecosystems are diverse, varying from dry dipterocarp in the lowland to dry evergreen, mixed deciduous, and coniferous forests in higher elevations. These ecosystems and accompanying high diversity of wildlife are of regional and international importance.

3. POLICY, LEGISLATION, AND IMPLEMENTATION

Below is a description of forest-related policy formation, legislation, and implementation. Please see Appendix 1 for a chronological breakdown.

3.1 Formative Years (1975–1986)

At the end of the 1970s the forestry sector had the following main objectives: (1) to utilize forest resources for the welfare and development of the people and to create capital, (2) to follow exploitation with plantation, and (3) to invest the capital created in industries based on forest and agricultural products (Persson 1983). Under these objectives and the socialist idea of planned and state-led economic management, nine state forest enterprises (SFEs) were established (late 1970s). They were allocated an average of 200,000 to 300,000 hectares of rich forest for management, including planning, harvesting, planting, protection, and processing. With assistance from donors and government investment, many of the SFEs were equipped with modern heavy logging machinery and large-scale wood processing factories. Although the management and operations were very inefficient, log production picked up rapidly after 1980 (Table 1). Tree planting was stressed, but actual plantation establishment was very limited during this period (Table 2).

The first comprehensive forestry legislation passed after 1975 was the Council of Ministers (CM) Instruction No. 74 on Forest Protection on 19 January 1979. It included provisions on national resource ownership, permission of forest conversion and logging, prohibition of shifting cultivation in watershed areas, traditional uses by local people, and promotion of tree planting for restoration. The preface is interesting since it recognizes the importance of forests, not only for the national economy and people's daily lives, but also the impacts of drought and flooding caused by serious forest destruction.

Table 1: Log production volumes, 1965–1999.

Five-Year Period	Industrial Log Production (cubic meters)
1965–69	391,662
1970–74	518,416
1975–79	505,624
1980–84	786,839
1985–89	911,000
1990–94	1,987,000
1995–99	3,291,869

Source: World Bank et al. (2000).

Table 2: Forest plantation development.

Year	Planted Area (ha)	Year	Planted Area (ha)
Until 1980	1,059	1995	8,828
1981–85	1,384	1996	11,850
1986–89	1,257	1997	12,290
1990	716	1998	9,030
1991	1,359	1999	6,353
1992	901	2000	15,160
1993	2,219	2001	20,501
1994	3,798	2002	15,800

Source: Figures collected from various internal documents from the Ministry of Agriculture and Forestry.

Although the document from the 1981 Third Party Congress was not available to the author, Persson provided the objectives of the forestry sector in the First Socio-Economic Development Plan, which can be considered the forestry related section of the Congress document, and they are almost in line with the 1979 CM Instruction (Persson 1983). Given that Lao society and the economy remain highly subsistence-based and decentralized, due to the lack of transportation infrastructure and a lack of means to implement the stated policy (including financial and human resources), the importance of the Instruction was only symbolic. Exploitation of forest resources for collection of royalty revenues and foreign currency through SFEs was the only major forest policy that was fully implemented.

3.2 Fourth Party Congress and Second Socio-Economic Development Plan (1986–1990)

At the Fourth Party Congress in March 1986, something called the "New Economic Mechanism" was adopted, signaling a shift to a market-oriented economic management system under state leadership. The second priority of the development plan was to curb and eventually stabilize shifting cultivation—a major government policy since 1975—after increasing food production. The program stressed that 300,000 hectares of forests were being destroyed by shifting cultivation annually, causing serious environmental problems. But its approach to stabilizing shifting cultivation was very cautious, stressing that it could not be stabilized by order and force, but rather by providing alternative crops or livelihoods to replace it. The document acknowledged this as a most difficult and complex challenge.

Chapter 1 of the document explained that two-thirds of the population in Laos depend on shifting cultivation for their livelihood, mainly in the northern provinces and eastern mountainous areas, and that their daily lives lacked *orderliness*. This suggests that only ten years after the socialist revolution there may have been a political purpose of national unification behind the policy.

In the document, the program was four pages long, compared to only two pages on the program's first priority. Interestingly enough, after addressing the main topic of shifting cultivation, more than a half of the four pages dealt with the topic of comprehensive forest management. It said that forests need to be managed under a comprehensive idea and plan which takes into account not only log supply but also other environmental functions, and that legislation on the protection of forests and valuable non-timber forest products (NTFPs) is necessary. Regeneration and tree planting need to follow logging, and logging should not occur without a regeneration plan. Finally, it concluded that wood harvesting based on scientific plans would bring both economic and environmental benefits in the long run.

This was a strong expression of political will to manage forests in a sustainable way, and this reflects a concern about ongoing forest exploitation without control and measures for protection. Unfortunately, no legislation for forest protection or comprehensive forest management was enacted after the Congress.

3.3 First National Forest Conference (1989)

By the time of the National Forest Conference in May 1989 it was thought by most people that the country was still endowed with rich and endless forest resources, although some degree of deforestation had been going on because of shifting cultivation and industrial logging. But when the scientific land-use survey revealed that the actual area of forests was much smaller than believed, it sparked concern about the forests and growing consensus to take action. It was against this backdrop that the landmark conference was convened.

The week-long meeting was presided over by then prime minister and party secretary-general, Mr. Kaisorn Phomviharn, and attended by over 500 people from local governments and the private sector. In addition to the report by the minister of agriculture and forestry, the ministers of economic planning and national defense and the president of the National Assembly made speeches.

Mr. Kaisorn himself made as many as three speeches, one of them very long and detailed. He repeatedly stressed the importance of forests for both the economy and environment and the necessity of legislation for the comprehensive management of the forests, including strict implementation. He also addressed illegal logging and the unusual wealth of a number of logging related people, and attributed this to plots and tricks by the enemy to exploit the country's rich natural resources.

The conference produced a resolution, which is said to be the basis for forestry policy up to now. Important policy related parts of it are reproduced in Box 1.

Legislative response to the recommendations in the resolution was very quick. In October 1989 the Council of Ministers issued two decrees: No. 117 on Management and Use of Forest and Forest Land, and No. 118 on Control and Management of Aquatic Animals, Wildlife, Hunting, and Fishing. Another, Decree No 117 had only16 articles focused on clear definition of the roles and duties of the Ministry of Agriculture and Forestry (MAF) concerning forestry, the allocation of forests and forestland, and various restrictions on logging by enterprises and local people.

Decree No. 118 was even shorter, with only nine articles. It stated that wildlife belongs to the national community and under government control, then covered the roles and duties of MAF in wildlife management. Concerning wildlife protection, Article 3 defined prohibitions, including the use of destructive hunting methods such as explosives, hunting of restricted species, hunting of nearly extinct species, and hunting during the breeding season. Later in 1993, MAF issued Instruction No. 298, Control of Wildlife and Hunting (24 March) for field implementation of PM Decree No. 118, and included a list of protected wildlife species classified into restricted, control, and general categories.

Stabilization of shifting cultivation remained high on the conference agenda, and one policy tool mentioned in the resolution was to rationalize forest use and introduce alternatives to shifting cultivation for local people through forest and forestland allocation to villagers and villages. This was elaborated in PM Decree 117 as follows:

- Allocate 2 to 5 hectares of forest and forestland to each household and 100 to 500 hectares to each village.
- Allow them to manage and use allocated forest as long as wood volume increases.
- Allow them to inherit or transfer allocated forest to others.
- Recognize ownership of individuals and groups to degraded land where they plant trees, regenerate, grow crops, or raise livestock.

This new policy was first tested in the following years in some northern provinces, then introduced nationwide in 1994 with some modifications.

After establishing SFEs in almost all provinces (18 in total in 2002), provincial forest enterprises (PFEs) were created for logging and processing, or to issue logging permits to private companies in order to collect revenues to finance development projects, ignoring control by the central government. On the other hand, SFEs were operating at a huge loss to the government every year, because of lack of management skills, inability to collect money, and subsidies. Soon after the conference, most of the SFEs and PFEs were dissolved or privatized, together with most of the other state-owned enterprises, complying with the government policy of market-oriented economic management and the separation of business from government.

Box 1: Strategic directions and ideas to the year 2000.

Strategic directions and ideas to the year 2000

1. Preserve and expand present forests through development of a management and protection system.
2. Rationally use benefits from the forests, especially both the use and replenishment of economic benefits of forest resources.
3. Link restoration, protection, and expansion of forests with the need for cereal production, crop production, a new cropping system, and sedentary livelihood of people in mountainous areas. Provide a sedentary livelihood to 60 percent of the 1.5 million people currently practicing shifting cultivation by the year 2000.

Tasks for immediate implementation after the conference

Local level

1. Ban indiscriminate logging and burning of forests. In special cases, the central forestry section can consider and propose to the Council of Ministers that permission be granted for logging. Search for revenue sources other than from forests.
2. Pilot projects of sedentary livelihood, forest use, and forestland allocation to households and villages in selected places, then expand to other places based on lessons learned.
3. Take into consideration the balance between log supply capacity and the demand for processing. Inspect all wood processing factories to decide which ones may continue and which must cease operation.
4. Delineate conservation forest at the national, provincial, and district levels.

Ministry of Agriculture and Forestry

1. Prepare laws on forests and the environment.
2. Assist local offices in piloting forest and forestland allocation.

The conference resolution called for a ban on indiscriminate logging, and PM Decree No. 117 stipulated that logging take place only in production forests that are inventoried, along with economic and technical plans approved by the MAF, full payment of log royalties to the government, and compensatory tree planting. With foreign assistance attempts were made to strengthen planning and management capacity, but it is not clear if production forests were defined and delineated or if economic-technical plans were prepared and implemented nationwide. Judging from the issue of another logging ban in 1991, technically sound management was not introduced.

3.4 Fifth Party Congress and the Third Socio-Economic Development Plan (1991–1996)

At the Fifth Party Congress in March 1991, forestry was again highlighted in the political report made by Mr. Kaisorn, the party secretary-general and president of the state. The tone of the report was similar to the previous one and went into detail, calling for the classification and delineation of all forests, the definition of clear responsibilities and benefits for management of each forest type, acceleration of forest and forestland allocation, and assistance to the mountain people to establish sedentary livelihoods to replace shifting cultivation.

After the Party Congress, Prime Minister (PM) Decree No. 67, which banned logging, was issued in August. In the preface, the illegal issuance of permits for logging and wood processing factories, document forgery, and deception of concerned officers were mentioned as some of the main causes of rampant illegal logging. After abolishing the SFEs, a logging quota system was introduced. After setting a national quota, the central government gave the provinces a logging quota for performing log sales contracts with buyers, but the provinces abused this system to create additional funds for development projects, and there were many irregularities concerning log sales contracts and actual logging, grading, and sales. This clearly reflects what was going on in the field, despite the resolutions of the Forestry Conference, the Party Congresses, and the decrees.

PM Decree No. 67 declared a country-wide logging ban and ordered the Ministry of Agriculture and Forestry to prepare measures to make forest management, logging, processing, and trade follow and respect the laws and regulations, as well as to survey and classify forests into categories of protection, conservation, regeneration, and economy (production). The logging ban was well implemented, with log production in 1992 about half of the previous year, and it is said that most were old logs or felled trees carried over from the previous year. But a year later (1993) the level of logging had returned to above the 1991 volume.

In 1993 two important decrees were issued: PM Decree No. 164, Establishment of National Conservation Forests, called national biodiversity conservation areas (NBCAs) in English; and No. 169, Management and Use of Forest and Forest Land. The first was truly landmark legislation concerning biodiversity conservation in Laos. Although only on paper, it established 18 national conservation forests covering 2.8 million hectares, 12 percent of the country's total land area. These conservation forests, with the addition of two more in a later year, were the fundamental basis for biodiversity conservation in the country that still exists today.

PM No. 169, issued only five days after No. 164, was a response to the logging ban decree and, as its title indicates, was new, comprehensive forestry legislation to replace CM No. 117 (1989). It was also the basis for the 1996 Forestry Law. This decree consists of 54 articles, including definitions of forest, ownership, forest categories, management of forests based on contract, and prohibition of development and forest activities in protection and conservation forests. The five forest types—protection, conservation, production, regeneration, and degraded forests—defined in the Forestry Law were already shown in this decree, but there were some inconsistencies between the two. One is that the relationship is not clear between the national conservation forests established by PM Decree No. 164 and the conservation forest defined in No. 169. The other is the difference in the level of authority for designating conservation forest. No. 169 says that the MAF designates all five forest types, but the national conservation forests are established by the prime minister.

Although this decree was repealed by the 1996 Forestry Law, it played an important role in the development of the framework for land and forest allocation, formally started in 1996. Chapter 4, on forest management contracts, defined three types: the collective forest management contract for villages, the family forest management contract, and the afforestation business contract. The collective forest management contract is between a district and a village on the use and management of forests, and the family forest management contract is between a district and a household on the use of degraded forest land for agriculture, tree planting, and raising livestock, with an incentive for households that a long-term use right would be given after three years of successful implementation of the contract. These two management contracts are combined into a set of backbones for land and forest allocation, which has been implemented as the main institutional vehicle to replace shifting cultivation.

In regards to logging, PM Order No. 16, on the reform of the control system of tree harvest (7 October 1994), introduced a two-way quota system. It set logging quotas for the export of both logs and processed products for the three state-owned enterprises under the control of the defense ministry, and another quota for the provinces for domestic use. The three military affiliated state-owned enterprises were made responsible for the north, central, and southern regions, respectively, and they took over logging machinery and factories from the SFEs. The background of this reform is not clear, but in the preamble of the prime minister's order the private companies were accused of various irregularities, and the provinces were accused of not following the rules and regulations set out in the government orders and instructions. This quota system continued to the year 2000, when PM Order No. 11 was issued to return to the previous system of all quotas given to the provinces. The largest of the three, Borisath Phathana Khed Phudoi (BPKP), expanded its operations, using revenues from logging, to own 58 affiliated companies. At the same time, its loans from state-owned banks occupy more than 40 percent of the total non-performing loans lent by the banks (IMF 2002), and it is now under the control of the finance ministry for reform.

In October 1994 the government issued PM Decree No. 186, on the delineation and allocation of land and forest for tree planting and protection, provides a basic legal framework to promote tree planting, including exemption from a land tax on tree plantations with more than 1,100 trees per hectare, ownership of planted trees (use, harvest, sale, transfer, and inheritance), and exemption from paying royalties. This decree was needed to secure investment in tree planting by foreign and domestic companies, and was subsequently replaced by the Forestry Law.

3.5 Sixth Party Congress and the Fourth Socio-Economic Development Plan

At the Sixth Party Congress, in March 1996, forestry was given less emphasis, compared with previous Congresses. The following points were set as priority directions in the fourth plan:

- Wide implementation of land and forest allocation for local people to protect and use forests.

- Continuation of shifting cultivation stabilization with a focus on watershed areas, rich forest areas, and conservation forest areas, while paying attention to the provision of sedentary settlement and permanent occupation.
- Promotion of tree planting of various species for domestic use and commodity production.
- Strict control of the logging and wood business.

The National Assembly approved forest-related laws one after the other: the Forestry Law (1996), the Land Law (1997), and the Environment Protection Law (1999). The Forestry Law, which remains the governing legal document today in Laos, is comprehensive and gives fairly clear directions on many aspects of forestry, except for two. First, it says nothing about log harvest contracts and sales. Second, it says management authority shall be delegated to the provinces, districts, and villages, but it is very unclear about the extent of delegation. Since 2000, several sets of MAF regulations for implementation of the Forestry Law have been issued by consolidating related older decrees and instructions.

Two other important legislative documents enacted in 1996 are PM Order No. 3, on the continuation and expansion of land management and land and forest allocation, and MAF Instruction No. 822, on land and forest allocation for management. These documents provide a legal framework and implementation guidelines on land and forest allocation. As seen in Table 3, the area and number of households involved in shifting cultivation has gradually decreased but still remains high.

Table 3: Shifting cultivation from 1996 to 2000.

	1996	1997	1998	1999	2000
Area (hectares)	176,605	158,843	140,800	115,800	118,900
No. households	186,265	126,110	142,785	131,700	174,036

Source: Figures collected from various internal documents from the Ministry of Agriculture and Forestry.

The allocation of land and forests has two main components. One is the contract-based allocation of degraded land to households for crop cultivation, tree planting, or grazing. If they perform well, based on a contract (a three-year land-use certificate), they qualify for the title to the land. Second, after land allocation, the village forest is classified as use, protection, or other areas, and an agreement on the use of each forest type is signed. After several years of piloting, the present style of land and forest allocation was formally introduced in PM Decree No. 186 in 1994, and implementation was accelerated with the MAF instruction in 1996, as shown in Table 4.

Table 4: Implementation of land and forest allocation from 1991 to 2000.

	1991	1992	1993	1994	1995	1996	1997	1998	1999	2000
Villages	7	20	21	519	953	1,242	1,327	1,232	1,009	555
Households	54	296	580	10,329	60,747	82,206	40,158	95,659	90,886	21,789

Source: Figures collected from various internal documents from the Ministry of Agriculture and Forestry.

From 1996 to 2000, land and forest allocation was carried out in more than 5,300 villages (more than 40 percent of the total number of villages), and more than 330,000 households were allocated land (about 50 percent of the total agriculture households). This is one of the very few forest-related programs that had both clearly-defined policy objectives and detailed instruction for field implementation, and was actually implemented nationwide, with the objectives of promoting crop production to replace shifting cultivation through allocation and titling of production land, and protecting forestland through classification and the stabilization of shifting cultivation.

This process works better in areas where crop production has already been started and demand for land is so high that secure land tenure is needed. Taking into account, however, the low level of infrastructure development and the limited crop markets in the most remote areas of Laos, there are not many of the 5,300 villages where land and forest allocation has accelerated crop production, as officially intended. In many cases it is used as a direct measure to stabilize shifting cultivation, i.e., by allocating three or four plots per household for shifting cultivation (rice production for own consumption), while closing other areas to shifting cultivation. This measure was cited as one of the causes of poverty aggravation experienced by villagers (ADB et al. 2001).

As for forest allocation, the village forests are to be classified into the five categories defined in the Forestry Law, based on traditional use and discussions between villages and local authorities. In practice, however, most village forest areas are designated for protection purposes, with very limited areas allocated for use by villagers (Ministry of Finance 2002).

During this period logging volumes dramatically increased, as shown in Table 1, and so did the number of wood processing factories. Most of the increase came from advance logging, in anticipation of development project sites, especially hydropower dams. PM Order No. 11 (2000) allocated much smaller logging quotas to the provinces and no specific quota for the military-run state-owned enterprises. It is widely said, however, that actual logging far exceeds the official quota, because of the strong demand for logs in neighboring countries and growing domestic demand, combined with cash-starved provinces.

3.6 Seventh Party Congress in March 2001 and Fifth Socio-Economic Development Plan

The ruling party and the government set development targets for the years 2020, 2010, and 2005 at this Congress, and these were endorsed by the National Assembly. Under the targets, shifting cultivation is to be basically stabilized by 2005 and completely stabilized by 2010. Tree plantation for commodity production is to be strongly promoted, with the target area for the five-year period to 2005 set at 134,000 hectares. The third priority action is to accelerate classification and delineation of forests for protection, conservation, and production purposes.

The promotion of tree planting has long been a priority policy in Laos. As seen in Table 2, the annual planted area increased to the order of 10,000 hectares in 1996, and after a drop in investment in 1998 and 1999 because of the Asian financial crisis, it reached its highest at 20,000 hectares in 2001. Much of this increase was commercial plantations of *Eucalyptus* spp., established by joint venture companies and individuals, and partly supported by the Asian Development Bank's industrial tree plantation loan program. In order to promote tree planting by small-scale farmers, in 2001 the government started to charge a reforestation fee on top of the log royalty, with the fees allocated for promotional activities, establishing nurseries, and for seedling production for distribution to farmers.

The MAF issued the Regulation on Village Forest Management in June 2001, mostly repeating existing provisions concerning village forests, namely, the classification of village forests into the five categories, limiting the harvest of logs to household use only, and so on. For the first time, however, the regulation legalized the collection and sale of NTFPs, one of the few sources of income for rural villagers, as long as management plans are formed and approved.

On 13 October 2001, a bidding system for log harvesting and sales was partly introduced by PM Decision No. 60, Approval of Annual Plan of Tree Planting, Log Harvest, and Collection of Other Forest Products for 2001–02. It is expected that the bidding system will be developed and expanded to replace the current quota system.

In May 2002, PM Decree No. 59, on the sustainable management of production forest, was promulgated. In addition to provisions on delineation and planning of production forest, it acknowledges participation of local villages in all aspects of production forest management, including labor, planning, sales, and benefit sharing, based on contracts between villages and district authorities. Although implementation is yet to be seen in 2003, active participation of villages is expected to contribute to technically-sound logging and economically-efficient log sales, as well as to the development of remote villages by using the village share of the benefits.

4. OBSERVATIONS

4.1 Policy and Legal Framework Development

After the Forest Conference in 1989, where a set of forest policies was agreed upon, the development of a forest-related legal framework was vigorously pursued, as seen in the Appendix, including several very important decrees issued by the prime minister that were then brought together under the 1996 Forestry Law. Although there are still some inconsistencies between the law and the decrees and a lack of transparency in decision-making processes, taking into account the short time-frame and very weak institutional and human capacities, the development of forest-related policy and a legal framework marks a major achievement.

4.2 Top-down Approach and Local Use

As typical in socialist countries, where all natural resources are national property and the governments allocate their use and management, a guiding principle in forest policy and legislation is that the government decides everything. It sets logging quotas for companies, sets log prices, allocates land to farmers, draws boundaries of various types of forests, and tells local people inside these areas what they can and cannot do. At the same time, huge amounts of forest resources have been used very inefficiently, beyond official control, and local people are in short supply of forests to maintain their livelihood.

Current policy needs to be redirected, from top-down direct control to provision of a framework that enables various stakeholders to participate in planning, decision-making, and competing in resource use and processing. Especially, villagers need to be recognized as managers of the state forests and forest resources within their village boundaries, not as poachers. Some signs of redirection are being seen, such as the legal recognition of NTFP sales by villagers and the test introduction of bidding. The recent PM Decree No. 59 in May 2002 on the sustainable management of production forest is also epoch-making in this sense. Reform of the general forestry sector policy needs to follow up and strengthen these positive developments.

4.3 Governance

Governance issues exist in all sectors, not being specific to just forestry. But forestry, especially harvesting, processing, and trading, is still a lucrative business in Laos and one of the few sources of revenue for development projects and other activities. Unfortunately, the preliminary results of a 2000 land-use survey reveals an alarming rate of deforestation, especially in the central and southern provinces, which were once forest rich but harvested heavily in recent years. As was clearly recognized as early as 1986 at the Fourth Party Congress, wood harvesting based on scientific plans will bring both economic and environmental benefits in the long run. Strong determination and action at the leadership level is indispensable on this issue.

REFERENCES

Asian Development Bank, Laos State Planning Committee and Laos National Statistics Center. 2001. *Participatory poverty assessment*. Vientiane: Laos State Planning Committee.

International Monetary Fund (IMF). 2002. *Selected issues: Lao PDR*. Washington, D.C.: IMF.

Lao People's Supreme Assembly. 1991. *Constitution*, Article 3.

Ministry of Finance, Land Titling Project. 2002. *Existing land tenure and forest lands study*. Laos Ministry of Finance.

Persson, Reidar. 1983. *Forestry in Lao PDR*. Stockholm: Swedish International Cooperation Agency.

World Bank, Swedish International Cooperation Agency, Ministry of Foreign Affairs, and Government of Finland. 2000. *Lao PDR production forest policy, status, and issues for dialogue* (draft).

APPENDIX 1: CHRONOLOGY OF FOREST-RELATED LEGISLATION AND EVENTS

Late 1970s	Establishment of nine state forest enterprises
1979	CM Instruction No. 74 on forest protection (17 July)
1980s	SFEs in operation
1986	Fourth Party Congress and Second Socio-Economic Development Plan
1987–89	"New economic mechanism" reforms in monetary, pricing, exchange rates, trade, etc.
1989	First National Conference on Forests (May)
	CM Decree No. 117, Management and Use of Forests and Forestland (5 October)
	CM Decree No. 118, Management and Protection of Aquatic Animals, Wildlife, Hunting, and Fishing (15 October)
1990	Privatization of state-owned enterprises, including SFEs, except those in strategic sectors
1991	Fifth Party Congress (March) and Third Five-Year Socio-Economic Development Plan
	Constitution (15 August)
	PM Decree No. 67, Logging Ban, country-wide (28 August)
	PM Decision No. 66, Approval of Tropical Forestry Action Plan for Lao PDR (7 September)
1992	UNCED in Rio de Janeiro
	PM Decree No. 99, Land (19 December)
	Mr. Kaisorn died (December)
1993	PM Decree No. 164, Establishment of National Conservation Forests (29 October)
	PM Decree No. 169, Management and Use of Forests and Forestland (3 November)
1994	PM Order No. 16, Reform of Control System of Tree Harvest (7 October)
	PM Decree No. 186, Delineation and Allocation of Land and Forests for Tree Planting and Protection (12 October)
1996	Sixth Party Congress (March) and Fourth Five-Year Socio-Economic Development Plan
	Forestry Law (October)
1997	Land Law (April)
1999	Environment Protection Law (April)
2000	PM Order No. 11, Control of Forest Work and Wood Business (23 August)
2000–01	Several sets of MAF regulations on implementation of the Forestry Law
2001	Seventh Party Congress (March) and Fifth Five-Year Socio-Economic Development Plan
	MAF Regulation No. 535, Village Forest Management (18 June)
2002	PM Decree No. 59, Sustainable Management of Production Forest (22 May)

Chapter 8

POTENTIAL ROLES FOR NGOs IN COMMUNITY FORESTRY IN LAOS: LIAISONS BETWEEN THE SOCIALIST GOVERNMENT AND GRASSROOTS PEOPLE

Satoru Matsumoto
Mekong Watch Japan

1. CONNOTATIONS OF "COMMUNITY FORESTRY" IN LAOS

Community forestry was defined very broadly by the United Nations Food and Agriculture Organization (FAO) as "any situation which intimately involves local people in a forestry activity" (FAO 1978). At a January 1992 workshop on community forestry held in Vientiane, jointly organized by the Lao government and a network of non-governmental organizations (NGO), community forestry was interpreted as equivalent to locally-based and customary regimes of forest management (MAF-SAF 1992). The term "community forestry" is usually used to translate the Lao term *paa mai xao ban* (villager's forest).

Premrudee Daoroung (2000), advisor to the Community Forest Support Unit of the Department of Forestry for three years from the unit's commencement to 1996, observed that NGOs prefer to use community forestry to emphasize the rights and roles of local people, and recognize the close links with the people's movement in regard to their own forest. Her analysis was reinforced by the Lao government's reluctance to use the term "community forestry," partly due to its prevalent use in Thailand (Department of Forestry 1997), where community forestry emphasizes the history of the people's movement.

It is commonly understood that community forestry has two polar objectives: (1) to protect or elaborate on the rights of local people in their indigenous forest management as a social movement, and (2) to institutionalize or professionalize local forest management (Hirsch 1997; Daoroung 2000; Carter 1999). In the context of Laos, it is

M. Inoue and H. Isozaki (eds.), *People and Forest—Policy and Local Reality in Southeast Asia, the Russian Far East and Japan*, 127–138.

often said that there is no grassroots movement in the country, where the communist party has been the sole power since 1975 and freedom of expression to criticize the government policies is restricted in practice. It is too simplistic, however, to deny the existence of any social movement at the grassroots level, because there are some reports of local criticism against governmental exploitation of local natural resources (Tubtim et al. 1996).

Presumably, there are two different roles for NGOs to play in promoting community forestry in Laos. In a situation where there is official pressure to involve local people in forest management, referred to by whatever name the government uses, NGOs can utilize such governmental initiatives to actualize what the local people wish to achieve. The other presumable but more difficult role is to support "grassroots movements" by local people in taking initiatives to retain or secure community rights to control forest resources. This chapter aims to respond to this presumption, based on a review of literature and semi-structured interviews with staff of five relevant NGOs out of nine engaged in activities related to community forestry in Laos (NGO Directory Committee 2000).

2. LAND AND FOREST ALLOCATION POLICY AND COMMUNITY FORESTRY

Land and forest allocation (LFA) has emerged as one of the practical products of the development of the legal framework in terms of forestland tenure, and it has had significant influence on community forestry in Laos, as described later in this chapter. The most crucial part of LFA is its demarcation of village boundaries and the allocation of forests and non-forest lands to village authorities and villagers. Forests are normally categorized as protection, conservation, production, regeneration, and degraded forests. The degraded forests or the lands which belong to no one are allocated to villagers. This policy has been implemented as a part of the National Program for Shifting Cultivation Stabilization. It explicates that the government urge local governments to promote LFA in order to protect forests from shifting cultivation and to encourage shifting cultivators to engage in lowland agriculture, even though it is widely recognized as one of the indigenous formulas of local forest management.

Although community forestry does not necessarily include collaboration with the government in forest management, in practice it is unrealistic to ignore government-driven initiatives in a socialist country like Laos. Brown (1999, in Carter 1999) summarizes the local perspectives which are motivating factors for collaboration with governments as securing tenure and rights of resource use by local people, sustainable and long-term production, distribution of assets, local decision-making, and empowerment or control over forest management. In this sense, securing tenure and rights to resource use by local people are key motivational elements for them to collaborate with government-initiated programs, and LFA has had the greatest impact on community forestry activities supported by NGOs in Laos.

3. NGO ACTIVITIES IN LAOS

Although freedom of association is guaranteed by the constitution, local NGOs are not allowed to be established in Laos due to a lack of procedural regulations. There is one *de facto* local NGO called the Participatory Development Training Center, but it is registered as a private, non-formal education institution at the Ministry of Education.

The Lao government promulgated the Decree of the Prime Minister on the Administration of Non-Governmental Organizations (NGOs) in Laos, dated 28 April 1998, and followed it by issuing Guidelines for the Implementation of the Decree by the Ministry of Foreign Affairs on 8 July 1999. According to the decree and guidelines, the authority to permit NGOs to operate in Laos belongs to the Ministry of Foreign Affairs, and they do not contain any provisions for registration of local or Lao NGOs.

The other specific and more relevant feature of NGO activities in Laos is the relationship with the government. In general, whenever NGOs visit project areas, especially villages, they are accompanied by government officials, who usually have counterparts in local authorities. This situation sometimes creates difficulties in terms of time arrangements or authoritarian attitudes. Merits, however, can also be found, because close contact with relevant government agencies creates certain political spaces to influence policies based on the outcomes of grassroots activities.

4. NGO COMMUNITY FORESTRY ACTIVITIES

In this section, five international NGOs working on community forestry in Laos are examined. German Agro Action (GAA) and ZOA have operated projects in the northern provinces; the Japan International Volunteer Center (JVC), in the central province; and Community Aid Abroad (CAA), CUSO, and GAA, in the southern provinces.

4.1 Community Aid Abroad

The first activity related to community forestry was a training workshop held at the end of 1995 for government officials and village headmen in the Vangvieng district of the province of Vientiane, followed by discussions on community forestry among major stakeholders, both in the village and the district, and a forestry survey using "participatory rural appraisals."[1] At that time, Community Aid Abroad (CAA) was involved in only one village in the district. Through this process, CAA supported three families in the village to obtain legalized rights for access to and use of one hectare of arable community land for establishing an integrated orchard and garden. As a result, these families were able to generate additional incomes and ensure a certain level of food security. It led to the implementation of LFA in the village and encouraged the district authority to implement the community forestry plan in another location adjacent

[1] Interview with Mr. Sengthong Vongsakid, Project Manager of CAA Laos, in Vientiane on 10 August 2001.

to this village in 1997, because the local government appreciated the positive impacts in the village supported by the CAA.

Based on the request of the Thateng district authority in the province of Sekong, a similar process was preceded, including discussion among concerned stakeholders and a village survey on community forestry. Subsequently three villages in the Thateng district implemented LFA in 1997 to set up a community forest area, a village forest committee, and the necessary regulations.

Following initial positive outcomes through ad hoc community forestry activities in both provinces of Vientiane and Sekong, the CAA began a new project called the Community Forestry and Plant Genetic Conservation Project in two districts of these provinces, in order to alleviate material manifestations of poverty and to empower local communities and local authorities. The major activities were as follows (CAA 2001):

- Support for LFA was provided in two villages in the Vangvieng district to manage the watershed, important for local irrigation and water supply, and to reduce slash-and-burn cultivation by the Kmu ethnic group by allocating land to poor families and supporting their production.
- A study trip on community forestry inside and outside the country for villagers and local government officials was conducted.
- Seedlings of native hardwood tree species were planted in the degraded forests of the villages on National Arbor Day, based on the discussions of the village committees of the three target villages in the Vangvieng district.

The reason why the CAA expanded their LFA support to the other two villages in the Vangvieng district is that there was increased logging pressure from the villages adjacent to the first village, and villagers recognized the urgent necessity to involve neighboring villages in their community forestry activities.

The risks and constraints that the CAA and villagers face in the province of Sekong include large-scale logging activities and drastic population growth associated with a government resettlement program. For instance, the CAA's official district counterpart emphasizes the importance of forest conservation when he visits project villages with CAA staff, but also comes to the same villages with logging companies to exploit the community forests. It creates distrust among villagers against local officials and suspicion against the benefits of the project. Furthermore, logging pressures tempt village volunteers in charge of forestry to seek short-term profits from logging.

4.2 CUSO

From mid-1993 to April 2000, CUSO coordinated the Community Forest Support Unit (CFSU), a government body working on community forestry.[2] The CFSU was the product of a government/NGO collaboration to implement an action research and

[2] The project coordinator was away during the stay of the author in Vientiane. Therefore, this section is based on an interview with Mr. Wayne Brook, Country Representative of CUSO Laos, in Vientiane on 6 August 2001.

training program based on the conclusion of the National Workshop on Community Forest, held in January 1992 (Pahlman 1993, in Daoroung 2000). CUSO decided to terminate its support for the CFSU, however, because the Department of Forestry submitted a budget proposal of U.S.$250,000 to the Ministry of Foreign Affairs without CUSO's approval. CUSO assumes that the Department of Forestry feels that NGO projects are not necessary when it has a donor project with a huge budget like the Forest Management and Conservation Programme (FOMACOP).

In mid-2000, CUSO commenced a new project called the Sustainable Integrated Agriculture and Forestry Project (SIAF) in two southern provinces, Sekong and Saravan. The project covers 35 villages in four districts, with forest conservation as one of its components. CUSO expects that the villagers will develop a sense of ownership and become self-reliant in management activities and the conservation and protection of natural resources. In order to reach the objectives, major activities being prepared include the following:

- providing villagers with information, communication equipment, and knowledge on forest conservation,
- training village forestry volunteers and villagers on forest conservation,
- organizing workshops to exchange experiences and establish regulation on participatory sustainable use of forest resources and forest products,
- supporting LFA for the sake of management and utilization by villagers, and
- awareness-building of natural resource conservation.

In the first year of SIAF, CUSO focused on providing training to provincial and district officials and village committee members using various techniques, including participatory extension, project management, and micro-credit. Small-scale village-based projects were just started in mid-2001. Non-timber forest products (NTFPs) are one of the key areas in its community forestry component. CUSO intends to support the securing of NTFPs for food security, traditional herbal medicines, and small income generation. Even though LFA is described in the project outline as a target activity, CUSO still doubts if it is the right way to accomplish its objectives, while it admits the importance of villagers being able to manage their forests by themselves.

4.3 German Agro Action

German Agro Action (GAA) began the Community Development for Conservation Project in the Phu Xiang Thong National Biodiversity Conservation Area (NBCA) in 1998 and in the Xe Ban Nouan NBCA in 2001.[3] Both are located in southern Laos. In the Phu Xiang Thong NBCA, the number of target villages was 25 during its first phase (1998–2001) and is now 34 in its second phase (2001–).[4]

---

[3] Interview with Mr. Manfred Back, GAA Project Advisor, in Vientiane on 7 August 2001.

[4] GAA has another project related to community forestry in the province of Oudomxay. However, project staff were not available for an interview during the author's visit to Vientiane in August 2001.

As described in the project's name, the GAA has placed much emphasis on community development as a means to achieve forest conservation. A short evaluation after the first phase corroborated its positive impacts on food security through the introduction of natural composts and bio-pesticides. While the target increase of rice production during the first phase was 25 percent, the actual increase reached 40 percent. Agricultural production was diversified by promoting integrated farming with fishponds and vegetable gardens. Agricultural development contributed to encouraging villagers to engage in farming and reduced the opportunities and need for villagers to enter into the National Biodiversity Conservation Area (NBCA) forests.

The second phase is placing more attention on NTFPs. The project focuses on identifying which NTFPs villagers have taken from forests and used for their own livelihood. Resource-use sketches include which NTFPs are available in each season. Three Laotian staff of the GAA supervise the activities implemented by local officials. Based on the results from the resource-use sketches, user groups to conserve NTFPs will be organized. The project established a steering committee consisting of key officials from relevant local authorities. The committee has permitted villagers to extract NTFPs from NBCAs.

The GAA does not intend to support LFA, because it is government-initiated and too mechanical, with little discussion with villagers. In some cases, legally identified forests are not actually forests. In addition, as a part of LFA, the law requires establishing a village development plan, but the government cannot allocate enough funds for it. The GAA understands that LFA is correct in principle but contains many problems related to its methods and processes.

Training in agricultural techniques and participatory rural appraisals are also included in the GAA's project activities, because it understands that agriculture and human resource development are imperative in promoting forest conservation.

4.4 Japan International Volunteer Center

The Japan International Volunteer Center (JVC) was the first NGO to begin community forestry projects in Laos in 1992.[5] Since 1996, a community forestry component has been a part of the Integrated Agriculture and Forestry Project in the five districts of central Laos in the province of Khammouane. The JVC has worked in 25 villages and has four major activities related to forestry, as listed below:

- supporting the LFA process to transfer rights to manage forests within village boundaries,
- supporting villagers to be able to deal with company-initiated development activities inside village boundaries,
- building capacity of district and provincial officials and village forestry volunteers who have worked as key persons to promote community forestry activities at the village level, and

[5] Interview with Mr. Akira Miyoshi, Country Director of JVC Laos in Vientiane on 2 August 2001.

- exchanging information and experiences with other forestry projects both inside and outside the country.

In its assistance to the LFA process, the JVC is concerned about the appropriate allocation of agricultural land to landless people, although they believe that LFA itself is not the work of NGOs but of the government. One of the JVC's activities, highly appreciated by other NGOs and the government, is the participatory and time-consuming method it uses in the LFA implementation process, utilizing visual materials so that villagers understand LFA's processes and significance to their lives. The major objective of the community forestry component of the project is to establish villagers' legalized rights to manage and use their forests, and the JVC strives to make the legalized rights substantial in reality, not just on paper.

The JVC recognizes two crucial challenges. One is the use of production forests after LFA. In one village, they did not understand the logging situation in the village production forest, and in another village, villagers cut five trees in the village protection forest when they could not satisfy their needs with resources available in the production forest. The other challenge is how to deal with interventions from outsiders, such as a cement factory built adjacent to village-protected forests or military agriculture land established in an allocated village forest.

4.5 ZOA

ZOA completed its three-year contract with the government for the Forest Conservation and Rural Development Project in mid-2001.[6] The project was divided into three components: (1) developing sustainable agriculture and livelihood systems; (2) enabling the communities to manage, conserve, and rehabilitate local forest resources in sustainable manners; and (3) enhancing the capacity of project staff, government counterparts, and villagers. A total of 17 villages in two districts in two northern provinces, namely, Xieng Khouang and Luang Namtha, were involved in the project from April 1998 until June 2001.

Since ZOA has worked for many years to support reintegration processes for repatriates, who were in exile from Laos during or after the Indochina War due to various reasons, and community development in surrounding villages, there are many Hmong ethnic people in their project area.

The major activities related to community forestry were as follows (ZOA 2001):

- training village forestry volunteers to enable them to take leading roles beyond their past mandates, which included doing paperwork for cutting firewood and timber,
- land allocation and village forestry, applying the participatory approach to planning and implementation developed by the German aid agency, GTZ,

---

[6] Interview with Mr. Kennedy O. Cruz, Project Manager of ZOA Laos in Vientiane on 6 August 2001.

through the Nam Ngum Watershed Management and Conservation Project (NAWACOP),

- development of tree nurseries at the village level and supporting credit for income generation,
- planting trees for firewood and lumber in the villagers' gardens or village woodlots, and
- protection from forest fire by establishing firebreaks.

Due to geographical conditions in which forests in Xieng Khouang were destroyed by bombing and war during the 1960s and 1970s, ZOA emphasized the significance of tree planting in order to secure necessary fuelwood and domestic demand for lumber.

ZOA perceived villagers in the project area to be major contributors to deforestation by their unsustainable forest use. They began to assist with LFA because of their belief that land titles and a feeling of ownership would promote more sustainable forest use and management.

5. TRENDS IN ANALYSES OF NGO COMMUNITY FORESTRY ACTIVITIES IN LAOS

On the basis of the information about community forestry activities operated with support from international NGOs in Laos, it is evident that community forestry is a component of integrated projects. In particular, the integration comprises the three basic fields of activity of agriculture, forestry, and human resource development. With respect to community forestry, four major activities are found: LFA, capacity building, tree planting, and NTFPs.

5.1 Land and Forest Allocation (LFA)

Except for the GAA, four organizations support the process of LFA to some degree. They are concerned with reduction of slash-and-burn cultivation (CAA); protection of watersheds (CAA); ensuring the rights of villagers to manage and use village forests (CAA, CUSO, JVC, ZOA); allocation of arable lands to poor families (CAA, JVC); resolution of conflict over forest resource use among neighboring villages (CAA, JVC); or application of participatory methods for the LFA process (JVC, ZOA).

In regard to establishing legitimate rights of local people to manage forests, support for LFA is apparently relevant to community forestry. But as the country representative of CUSO Laos explained, there is a degree of suspicion about LFA's actual effectiveness in ensuring these rights. Similarly, the advisor of GAA Laos raised the question about how LFA is promoted, although he believes in its significance. The research outcome encourages us to revisit the question raised by Hirsch: Does institutionalized or legalized community forestry, as an official acceptance of local rights to manage forests, mean the professional and institutional co-optation of grassroots movements for community empowerment? An analysis will be conducted in the final section.

5.2 Capacity Building

All five NGOs regard capacity building or human resource development as a crucial element of their activities. Their targets are villagers, including forest volunteers, and local government officials at district and provincial levels. Major activities are training of village forest volunteers, study tours inside and outside the country, sharing experiences, and technical training for plantation and participatory development skills.

Training village forest volunteers, who are officially appointed for being responsible for village forest management, is one of the key activities. It is also a very common approach among NGOs in Laos to develop their grassroots activities centering on unpaid, village-based individuals who are usually authorized by village committees. While they are recognized as indispensable to NGO activities at the village level, a few NGOs face similar difficulties in "training" them to honor all the responsibilities of their activities. For example, a village forest volunteer has the responsibility to monitor authorized logging and investigate if logging activity complies with village regulations and takes place in village production forests. The CAA is concerned that village forest volunteers, in some cases, overlook illegal logging inside village protection forests because of "remuneration" from logging companies or local officials. The JVC is also considering replacing or re-educating its village forest volunteers.

Another common feature of capacity building is training for local government officials. Since all NGOs in Laos are to visit and work in villages with local counterparts, they go to great lengths to make them understand the concepts behind each project in order to obtain their support. In the case of community forestry activities described in Section 4.1, however, local counterparts are often responsible for managing logging concessions and even logging practices at the village level. This is closely linked with the difficulties related to village forest volunteers.

5.3 Tree Plantations

The CAA and ZOA support tree planting as a part of forest conservation activities and as a means to address deforestation. They have never defined common causes, neither proximate nor underlying, but have indicated that some of the factors involved in deforestation in their project areas include slash-and-burn cultivation (CAA), population growth induced by governmental resettlement plans (CAA), and war (ZOA). Causes may highly depend on the social and historical conditions of each local situation. In terms of community forestry, two aspects should be raised here: incentives for villagers and the meaning of the local rights over forests.

Regarding the former aspect, ZOA's evaluation report states: "In 2001 most farmers, however, did not want to continue the tree nursery activity. They consider the activity too labor intensive, while trees can only be sold for low prices." On the contrary, it reports the success of tree planting in woodlots for firewood and lumber for domestic use (ZOA 2001). This case indicates the difficulties that villagers have in planting trees for commercial purposes as opposed to domestic demands, though it is too simplistic to generalize on the basis of this one case.

On the latter aspect, the CAA's project manager explained one of the reasons for promoting tree planting as, "It is impossible for villagers and NGOs to criticize private companies or local government who execute logging in the project area. But unless we do something, villagers will only lose their forest resources in vain. The sole way in which we can help villagers is by planting trees to restore lost forest resources, not to act against companies or government."[7] Similar constraints have been reported by the JVC in relation to other types of development initiatives by companies and government. This issue will be discussed in the last section.

5.4 Non-Timber Forest Products (NTFPs)

All NGOs interviewed recognize the importance of NTFPs, although specific projects are supported by only two organizations. CUSO's concerns about NTFPs are mainly for food security and herbal medicines for domestic use. Little attention is paid to income generation at present. The GAA's activities more or less emphasize research on seasonal resource use of NTFPs to identify actual utilization. The JVC is also considering involvement in NTFPs as a major element for community forestry activities in terms of effective use of production forests.

The interviews with NGOs indicate that community forestry activities of NGOs with respect to NTFPs are still limited to assessing the local situation and supporting small activities for food security or domestic demands. This is partly because securing markets was difficult and commercial use of NTFPs extracted from natural forests was not officially permitted. However, the Regulation on the Management of Village Forests (No.0535/AF.2001) was promulgated on 18 June 2001. It prescribes that "Forestry produces may be gathered from natural forests for commercial purposes, such as mushrooms, bamboo shoots, sticklack, benzoin, cardamom, palm fruit and others to generate additional household revenues" (Article 8). The impact of this new regulation has not been predicted yet, but it is certain that a door was opened to encourage villagers to sell NTFPs for commercial use.

6. CONCLUSIONS: CHALLENGES IN COMMUNITY FORESTRY IN LAOS

The two hypothetical roles of NGOs in community forestry, discussed earlier in this chapter, are to utilize governmental initiatives to realize what the local people wish to achieve, and to work for a grassroots movement of local people who advocate retaining or securing community rights to control forest resources by their own initiatives. In conclusion, this final section explores whether or not these roles are being played by

[7] Interview with Mr. Sengthong Vongsakid on 10 August 2001.

NGOs. If they are, how? If they are not, why not? What are the challenges in community forestry in Laos?

In Section 2, four perspectives (introduced by Brown) for collaborative forest management between grassroots people and government agencies were cited: (1) securing tenure and rights to resource use by local people; (2) sustainable and long-term production; (3) distribution of assets; and (4) local decision-making and empowerment or control over forest management. Applied to NGO activities in Laos, the first perspective is equivalent to LFA, the second is comparable to various development initiatives (including agriculture support or tree planting in which the government is willing to involve NGOs), and the third is equivalent to allocation of arable lands to poor families. In this sense, it can be said that NGOs in Laos use governmental initiatives to realize what the local people wish to achieve.

The fourth perspective, local decision-making and empowerment, is related to the latter hypothetical role of NGOs and is closely linked with the remaining question raised in Section 5.1: Does institutionalized or legalized community forestry as an official acceptance of local rights to manage the forest mean the professional and institutional co-optation of grassroots movements for community empowerment? In respect to constraints pointed out by the CAA and JVC, more or less, seeds of grassroots movements can be found. The JVC has had reports from people in villages where forest allocation was completed about a plan to construct an agriculture station in the village's protection and production forests and, in another village, about a plan to construct a cement factory adjacent to the village's protection forests. The CAA also recognizes a constraint connected with logging in village forests. If the rights of local people over allocated forests were really ensured by law, such constraints would not be raised. However, while the JVC has not conceived any solutions or actions to deal with these issues at the moment, the CAA attempts solutions through more training of village forest volunteers to improve their skills and enhance their responsibilities (CAA 2001). Referring to the second hypothetical role of NGOs, they have a clear intention to work for grassroots movements of villagers in this socialist country to take initiatives to control their own forest resources, but concrete outcomes have yet to be seen.

At the same time, apprehension remains about "the professional and institutional co-optation of grassroots movements." If tree plantation activities are developed to compensate for forest resources lost to logging by private companies, as cited in Section 5.3, it may lead to "co-optation" of a local movement.

What is the most difficult hurdle in accomplishing empowerment of local people regarding community forestry? As explained in Section 4.3, the GAA supported villagers to be allowed to extract NTFPs that they need from within NBCAs. This activity can also be another type of empowerment. What is the major difference between this NTFP case and logging issues?

Gilmour and Fisher (1997) found that the people's movement in community forestry in Nepal emerged after a relatively mature and people-oriented program was established as a government program, and the reason why there was no obvious political pressure was the limited access to relevant forest resources by urban elites. There is no empirical analysis of political pressure on empowerment approaches of NGOs in Laos, but the case of Nepal can imply one possible factor which makes it difficult to empower local people in community forestry in Laos.

It is possible to say that the NGOs in Laos have enough experience and knowledge to utilize government initiatives for the benefit of the local people. The challenge that this research identifies, however, is how to empower local people or include them at the crucial stage in decision-making when conflicts occur over forest resource use among villages and more powerful stakeholders, while understanding the political constraints in the country.

REFERENCES

Brown, D. 1999. Principles and practice of forest co-management: Evidence from West-Central Africa. Draft document. London, UK: ODI.

Carter, Jane. 1999. "Recent experience in collaborative forest management approaches: A review of key issues." Paper presented for the World Bank's Forest Policy Implementation Review and Strategy Development (FPIRSD).

Community Aid Abroad (CAA). 2001. *Progress report on Community Forestry and Plant Genetic Conservation Project, Reporting period January 1999 to December 2000.* Vientiane: CAA.

Daoroung, Premrudee. 2000. Global actors, local forests. Unpublished thesis for Master of Science, University of Sydney.

Department of Forestry. 1997. National Village Forestry Strategy. Discussion paper, second draft. Vientiane, 7 March 1997.

Food and Agriculture Organization (FAO). 1978. *FAO forestry paper no.7: Forestry for local community development.* Rome: UN Food and Agriculture Organization.

Gilmour, D. A. and R. J. Fisher. 1997. Evolution in community forestry: Contesting forest resources. Proceedings of an international seminar, Community Forestry at a Crossroads, 17–19 July, at Bangkok, Thailand. RECOFTC.

Hirsch, Philip. 1997. Community forestry revised: Messages from the periphery. Proceedings of an international seminar, Community Forestry at a Crossroads, 17–19 July, at Bangkok, Thailand. RECOFTC.

MAF-SAF. 1992. Proceedings of the Community Forest Workshop, 22–26 January, at Vientiane, Lao PDR. MAF-SAF.

Manivong, Khamphay and Berenice Muraille. 1997. Partnership for forest management: Joint forest management in Savannakhet, Lao PDR. Presented at an international seminar, Community Forestry at a Crossroads, 17–19 July, at Bangkok, Thailand. RECOFTC.

Pahlman, Charlie. 1993. NGO support to community forestry activities in Lao PDR. Summary of presentation by Charlie Pahlman, CUSO Country Representative, Lao PDR, at the National Workshop on Forestry Conservation, Management and Community Participation, 28–30 April, at Vientiane.

NGO Directory Committee. 2000. *Directory of NGOs in the Lao PDR 2000.* Vientiane.

Tubtim, K., K. Phanvilay and Philip Hirsch. 1996. Decentralisation, watershed and ethnicity in Laos. In *Resources, Nations and Indigenous people: Case studies from Australasia, Melanesia and Southeast Asia*, ed. R. Howitt, J. Connell and Philip Hirsch, 265–277. Melbourne: Oxford University Press.

ZOA Lao P.D.R. 2001. *Forest Conservation and Rural Development Project final report.* Vientiane: ZOA.

Chapter 9

COMMUNITY-BASED FOREST MANAGEMENT IN THE PHILIPPINES: RETROSPECT AND PROSPECTS

Juan M. Pulhin
College of Forestry and Natural Resources, University of the Philippines Los Baños

Perlyn M. Pulhin
Ginintuang Masaganang Ani Corn Program, Department of Agriculture, Philippines

1. INTRODUCTION

In the Philippines the forests are considered the country's natural resource base and an important pillar in promoting sustainable development (DENR 1990). Despite this, past efforts have been ineffective in halting a persistent onslaught on these resources. Over the years they have degenerated, because of massive logging activities, fuelwood gathering, charcoal making, shifting cultivation, and permanent agriculture (Kummer 1992).

Historical records indicate that in 1521, when the early Spanish colonizers first landed, about 90 percent of the country, around 27 million hectares, was covered with lush, tropical rainforest (Sajise 1998). By 1900, 21 million hectares of forest cover remained—about 70 percent of the country's total land area of 30 million hectares (Garrity et al. 1993; Liu et al. 1993). At present, only about 17 percent (5.4 million hectares) of the land is forested, with only about 800,000 hectares of primary forests remaining (DENR 2002).

Adding pressure to the current situation are the more than 20 million Filipinos that occupy the uplands. It is believed that about half of them are solely dependent on forest resources for their livelihood. These settlers are characterized by a high annual population growth rate of 2.8 percent, around 0.5 percent higher than the country's average population growth rate of 2.3 percent (Sajise 1988).

M. Inoue and H. Isozaki (eds.), *People and Forest—Policy and Local Reality in Southeast Asia, the Russian Far East and Japan, 139–156.*

The inability of the state-cum-corporate forestry approach to address the problems of upland poverty, forest destruction, and undemocratic access to the country's forest resources has contributed to the emergence of a new forestry paradigm. Most popularly known as community-based forest management (CBFM), this new approach to forest management regards the forest occupants as partners in forest development and conservation (Pulhin 2001). It departs from the traditional notion that considers these communities as the culprits in forest destruction, and advances a more decentralized, participatory, and people-oriented strategy to forest management.

This chapter traces the historical crossroads of CBFM in the Philippines, reviews its categories, and presents its current state. It also examines CBFM accomplishments in pursuing upland development and analyzes current issues and concerns relevant to its implementation. Finally, it explores the prospects of CBFM in light of the major challenges it currently faces.

2. THE EVOLUTION OF COMMUNITY-BASED FOREST MANAGEMENT

It was the Spanish colonial government that first initiated the centralization of control and management of forests in the Philippines into the hands of the state. Even after the country gained independence, state-controlled forest management continued to help perpetuate the three-pronged problems of forest degradation—inequitable access to forest benefits in favor of the elite, poverty, and marginalization of forest-dependent communities. In an effort to arrest these problems, the notion of community-based forest management was born. This section traces the historical development of CBFM by examining four periods of forestry: the colonial period, the post-colonial logging years (1946–1970s), the emergence of community forestry, and its expansion and institutionalization as CBFM.

2.1 Forestry During the Colonial Period

Land ownership in the Philippines before the entry of the Spanish colonizers was generally communal. Forests were accessible to anybody; whoever cleared and cultivated a piece of land first would own that portion, the opposite of limited land ownership under colonial rule.

Through the creation in 1863 of the Inspection General de Montes (the Forest Service) the Spanish colonial government placed the control, administration, and regulation of forest resources with the state (Borlagdan et al. 2001). The utilization of forests became strictly regulated and granted only to a privileged few (Makil 1982). Vesting the country's land ownership under the Spanish crown, the Regalian doctrine gradually changed people's attitudes about the forest—away from the notion of stewardship and towards being more exploitative (Lynch 1982).

A subsequent series of royal decrees buried the issue of land ownership and prior claims of indigenous cultural communities (Lynch 1982). The Spanish land laws also led to the weakening of customary systems of land tenure. Moreover, the communal associations of Indios (Filipino elite) and indigenous people deprived the latter of their rights to their ancestral lands in favor of the colonial government and the local elite (Poffenberger and McGean 1993; Sanvictores 1997).

Despite the negative social impacts, the noted illicit claims as well as the laws and decrees of the Spanish colony were still not remarkable enough to overly exploit the country's natural resources. Since the colonizers had limited power to control the entire archipelago, many of the communal forests of the indigenous people were maintained under their traditional forest management systems. Also, their small population barely contributed to the pressure of extensive forest use (Borlagdan et al. 2001).

In 1898 the United States defeated the Spaniards and took control of the Philippine Islands. To maintain the state-controlled management of the forests, the U.S. administration adopted the Regalian Doctrine initiated by the former regime. American logging companies perforated the country and operations became mechanized (Boado 1988). This period marked the export of logs and timber to the world market and the introduction of "scientific forest management" in the country (Poffenberger 2000).

With the issuance of General Order No. 50 (14 April 1900), the Inspection General de Montes became the Forestry Bureau, which was vested with the power of designating land as private and public domains (Makil 1982). During this period, the American's insatiable demand for wood motivated them to formulate forest policies similar to the previous regime that benefited only a privileged few (Boado 1988), leading to severe forest destruction.

Forests were made increasingly more accessible through road construction. In addition, many mines were permitted to operate, which destroyed large areas of public forestland. At the same time, lowland forests were converted into agricultural plantations—attributed more to colonial trade interests than to the pre-war population growth of only 2.2 percent a year (Boado 1988; Borlagdan et al. 2001).

In response to the increasing areas of open grassland and degraded forest, the Americans instituted a forestry curriculum. In 1910 the Forest School was established at the University of the Philippines in Los Baños, formally adopting the new science of forest management (Borlagdan et al. 2001). The government's heightened concern paved the way to different reforestation activities, using several local and exotic species (Sajise 1998; Boado 1988).

After six years, Act 2567, otherwise known as the Administrative Code, was passed. It was the first policy on community forestry that assured the people's right of continuous supply of forest products for their personal use (Kirchhoff 1993). Still, the code proved ineffective and, until recently, forestry policies have consistently privileged the wealthy and politically more influential concessionaires under timber license agreements (TLAs) (Pulhin 2001).

The forest industry flourished despite the drawbacks in combating the socio-economic and political marginalization of the rural population, as well as the onslaught of the country's forest resources during the American regime. With the operation of large-scale commercial timber extraction and processing facilities and exports of logs and timber to the United States, the timber industry became a major contributor to the country's economy. Just before World War II broke out, the lumber industry ranked fourth in terms of production, second in employment, and third in monthly payments. It also provided the government with average annual revenues of about 2.5 million pesos (approximately U.S.$1.25 million) from forest charges (Boado 1988).

Amidst the development of the forest industry, however, the reality unfolded that there was a "steady loss of forest throughout the era of American rule" (Poffenberger 2000).

Not a spark of hope glimmered when the Japanese took over the Philippines in 1942. They seized every opportunity to exploit the forest resources to advance their own colonial interests. This resulted in more severe forest destruction and a devastated wood industry. Of the 163 sawmills operating before World War II, 147 (87 percent) were completely destroyed during the Japanese occupation (Boado 1988).

2.2 Forestry During the Post-Colonial Logging Years

The Regalian Doctrine initiated by the Spaniards and affirmed by the Americans was perpetuated in the post-colonial period (Sajise 1998; Boado 1988). The Constitution of the independent Philippine Republic, established 4 July 1946, vested the state with sole ownership of all timberlands (Boado 1988). As such, the government held the rights of allocating, classifying, regulating, and managing the forests and timberlands. This circumstance reinforced massive timber exploitation by the private sector and intensified agricultural expansion, such as the conversion of lowland forests to pasturelands and mining areas. Increased forest exploitation, therefore, marked the post-war period (Borlagdan et al. 2001).

To meet the high demand for tropical timber, particularly in Japan, Europe, and the United States, large-scale logging operations were expanded (Boado 1988). The logging business grew, became very profitable, and generated even more revenues for the government (Borlagdan et al. 2001). This was considered timely, as the government needed additional capital for its rehabilitation and to accelerate industrialization. The proliferation of timber licenses, however, again only benefited the more influential investors—usually those with political or military ties—representing a very tiny portion of the Filipino citizenry (Boado 1988; Borlagdan et al. 2001; Sajise 1998).

Early efforts to promote sustainable timber utilization led to the establishment in 1953 of a policy called the Philippine Selective Logging System (PSLS) (Borlagdan et al. 2001). But the population explosion and demand for foreign exchange earnings from timber exploitation contributed to a sustained onslaught on the country's forests. The administration of President Ramon Magsaysay, for instance, allowed the opening up and conversion of vast tracts of forestlands for resettlement and agricultural production in order to contain the growing population (Borlagdan et al. 2001).

Meanwhile, the era of President Ferdinand Marcos was recognized as the zenith of logging. Marcos used the TLAs to strengthen his political network, dispensing them as rewards to loyal cronies. Used as an arm for political patronage, the number of TLAs soared to as many as 400 during his regime (Vitug 2000), and the highest deforestation rate ever, amounting to as much as 300,000 hectares per year, was recorded (Pulhin 1996).

2.3 The Emergence of Community Forestry

Decades of forest exploitation stimulated the formulation of programs to bolster local participation in forest conservation. In the 1970s the Marcos administration adopted three major people-oriented programs: Forest Occupancy Management (FOM) in 1975, Family Approach to Reforestation (FAR) in 1976, and Communal Tree Farming (CTF) in 1978. These programs basically focused on the involvement of local people and communities in reforestation activities (Sajise 1998). In addition, the government introduced the Program for Forest Ecosystem Management (PROFEM) in 1978, obligating all citizens to plant one tree a month for five consecutive years (Sajise 1998). These government programs, however, failed to recognize the people as partners and instead treated them as mere labor providers in forest conservation and development (Boado 1988; Pulhin 1998).

Considering the volatile political situation in the 1970s, political analysts viewed these programs as a form of counter-insurgency measure to maintain political stability and order in the countryside (Porter and Ganapin 1988). Despite this, it was during this period that the forestry sector started to realize that the problem of deforestation was not solely rooted in technical limitations but was also socio-political in nature (Pulhin 1998).

In the 1980s the government introduced two major people-oriented forestry programs: the Integrated Social Forestry Program (ISFP) and the Community Forestry Program (CFP). Right after the lifting of martial law in 1982, the Marcos administration issued Letter of Instruction (LOI) No. 1260. Under this policy, the CTF, FOM, and FAR programs were consolidated into the Integrated Social Forestry (ISF) Program (Pulhin 1998; Sajise 1998). It granted a 25-year stewardship contract to qualified forest occupants, allowing them to settle and till the upland areas; in return they agreed to protect and reforest these lands. Priority was given to those who had occupied their lands before 31 December 1981. This agreement, however, barred the upland communities from commercial timber extraction—a privilege retained by the TLA holders.

In contrast, the Community Forestry Program (CFP) gave the participating upland communities the leeway to commercially use forest resources to improve their economic welfare. Started in 1989, the CFP extended its coverage to natural forests, allowing communities to sustainably utilize forest resources subject to appropriate social and technical preparation. With the implementation of the ISF and CFP, the contribution of local people was increasingly recognized, not simply as laborers in reforestation activities but as real partners in forest conservation and development. Community forestry was then seen as a catalyst for upland poverty alleviation, attainment of social justice, equitable resource distribution, and sustainability of forest resources (Pulhin 1998, 1999).

2.4 Expansion and Institutionalization of Community-based Forest Management

The period of the 1990s saw broadening of the coverage of community forest management to include other land-use types not originally covered before—productive residual forest, existing forest plantations, old growth forest, degraded watershed areas, protected areas, and practically all types of forest where there are indigenous cultural communities or forest settlers (Pulhin 1997; Borlagdan et al. 2001).

Also, during this time, increased technical and financial support was coming from international funding agencies, such as the Asian Development Bank, World Bank, GTZ,[1] Japan's Overseas Economic Cooperation Fund, United States Assistance for International Development (USAID), the Swedish and Italian governments, and other multilateral and bilateral donors (Vitug 2000; Pulhin 1998). Diverse types of tenure instruments also emerged during this period with the aim of providing tenure security to forest communities. Furthermore, different income-generating technologies were developed in an attempt to alleviate upland poverty while ensuring the sustainability of forest resources (Pulhin 1999).

The evolution of varied forms of institutional arrangements continued. Community forestry expanded from plain government-implemented projects in the 1970s facilitated by the formation of people's organizations (POs) to receiving support from other sectors, such as non-government organizations (NGOs), local government units (LGUs), academics, etc. (Pulhin 1998; Vitug 2000).

With the passage of the Local Government Code (LGC) on 10 October 1991, some aspects of forest management were devolved to the different units of local government, including the provincial, municipal, and *barangay*[2] levels (Sabban 1998). The code provided for the inclusion of cities and municipalities in the implementation of community-based forestry projects like ISF and other similar projects. Moreover, provinces, cities, and municipalities were encouraged to manage and control a communal forest with an area of (at most) 50 square kilometers, establish tree parks or greenbelts, and participate in the enforcement of forestry laws, pollution control laws, small-scale mining laws, and other laws on environmental protection (Sajise 1998).

A major milestone in the evolution of people-oriented forestry in the Philippines occurred in 1995 with the enactment of Executive Order No. 263, adopting community-based forest management as a national strategy for sustainable forestry and social equity. It unified and integrated the earlier community-oriented forestry programs and projects, and provided a comprehensive mechanism for their implementation, thereby institutionalizing CBFM in the country (Pulhin 1997; Borlagdan et al. 2001).

With the expansion and institutionalization of CBFM, erring licensees were cancelled and expired ones were not renewed. This shift in the government's forest management approach led to the restructuring of the once corporate-controlled timber industry. From 1971 to 1977, TLAs controlled one-third of the country's total land area

[1] GTZ, translated, is the acronym for the German Society for Technical Co-operation, a worldwide active federal enterprise for international cooperation.

[2] Barangay is the smallest political unit in the Philippines.

of 30 million hectares; but starting in the late 1980s, TLA areas declined to the present 1.036 million hectares (FMB 1998).

The 1997 Department of Environment and Natural Resources (DENR) Strategic Action Plan for CBFM envisioned that by 2008 the communities would manage a total of nine million hectares of the country's forestlands that are mostly under existing and potential open-access areas. This comprises about 58 percent of the country's total forestland area. Meanwhile, it was expected that only half a million hectares would be allocated for industrial plantations and other purposes (Pulhin 1998).

3. CATEGORIES OF COMMUNITY-BASED FOREST MANAGEMENT

A recent assessment of CBFM in the Philippines, initiated by the Ford Foundation Inc., came up with three categories, based on the major initiator or supporter: self-initiated, locally-assisted, and national programs and projects (Borlagdan et al. 2001). Below are brief discussions of each.

3.1 Self-initiated Community-based Forest Management

Self-initiated CBFM is associated with the indigenous peoples or indigenous cultural communities who, since time immemorial, have been the stewards of their forest resources. Their management of the local forest resources pre-dates any government-initiated community forestry projects.

Similar to other Asian countries, most indigenous people in the Philippines employed indigenous forest management practices in some period of their history, as suggested by anthropological studies, but few of these practices still exist today. Some of those which do still exist were recently documented, including (1) the *muyong* system of the Ifugao, (2) the *tayan* of the Bontoc, (3) the *saguday* of the Sagada; (4) the practices of the Ikalahans, and (5) the *gaop* system of the Higaonons (Borlagdan et al. 2001). Based on the documentation and field observations, it may be safe to assume that the few remaining protected natural forests in the country are mostly located in areas where indigenous forest management systems still persist. Despite this, there appears to be a lack of comprehensive study of these systems that can serve as a sound basis for the recent search for sustainable forest management (Bennagen 1996).

3.2 Locally-Supported Community-based Forest Management

Locally-supported CBFM pertains to local government and other partner-sectors, such as NGOs, people's organizations, and private corporations, which involve local communities in the establishment of their own community forestry projects. The options of the LGUs to implement their own programs were formally laid down through the enactment of the 1991 Local Government Code, which devolved some of the DENR functions to the LGUs. Some of the successful people-oriented forestry projects that are being implemented include the Barobob Watershed Project, initiated by the provincial government of Nueva Viscaya in Luzon, and the Bukidnon Environment Small-Scale Tree Farm Project by the Bukidnon provincial government in Mindanao (Pulhin 2001).

At the same time a number of NGO and PO initiatives gained recognition for their notable contributions in community forestry. Two of the well-documented cases are the Mag-uugmad Foundation in the Visayas and the Mindanao Baptist Rural Life Center in Mindanao. Both successfully responded to the needs of the areas concerned and contributed to the sustainable development and preservation of local forest resources.

Aside from LGUs, NGOs, and POs, there are also some concerned private corporations that tapped forest communities as partners in the development of forestry projects. Most of them deal with watershed management, particularly the protection and rehabilitation of watershed areas. Two of these corporations, which were mandated by the government to protect the different watersheds and forest reservations under their jurisdiction, are the National Power Corporation and the Philippine National Oil Company. They forged alliances with the local communities by engaging them in forest-related, mutually beneficial activities such as reforestation and forest protection.

In general, the implementation of the projects mentioned above transformed these sectors into being more responsive to the needs of the concerned communities. To some extent they have contributed to addressing rural poverty, while attempting to sustain local forest resources. "While their limited area coverage may be a constraint in terms of achieving broader impacts, their site specificity and ability to focus give them greater advantage to penetrate on their specific target groups" (Pulhin 2001).

3.3 National Programs and Projects

It is through the Philippine government, particularly the DENR, that most of the national programs and projects on community forestry have been realized. These include different people-oriented forestry programs and projects that were integrated and unified under the umbrella program known as the CBFM Program, such as the Integrated Social Forestry Program (ISFP), Upland Development Program (UDP), Forest Land Management Program (FLMP), Community Forestry Program (CFP), Low Income Upland Communities Project (LIUCP), Regional Resources Management Project (RRMP), Integrated Rainforest Management Project (IRMP), Forest Sector Project (FSP), Coastal Environment Program (CEP), and Recognition of Ancestral Domains/Claims (Pulhin 1998).

The concept of "people first and sustainable forestry will follow" serves as the guiding principle of the CBFM Program in performing its tasks as it attempts to address the issue of upland poverty (considered as a root of the country's deforestation problems) while responding to the exacerbating condition of the Philippines' natural forests. Other than forest rehabilitation and protection, and their accompanying livelihood activities, a major strategy is the issuance of the different types of tenure instruments to qualified and organized communities. These instruments include the Community-based Forest Management Agreement (CBFMA), the Certificate of

Stewardship Contract (CSC), and the Certificate of Ancestral Domain Claim (CADC-CBFMA). Eventually, the organized and empowered communities are given the rights and responsibilities to directly manage and benefit from forestland resources (Sabban 1998).

4. STATUS AND ACCOMPLISHMENTS

At present, national CBFM programs and projects cover around 5.3 to 5.7 million hectares of forestlands, with the participation of more than 355,000 families (Tesoro 1999; DENR 2002). This accounts for around 57 to 63 percent of the DENR's target under its CBFM Strategic Action Plan, to be completed by 2008. Among the noted accomplishments of CBFM relates to provision of resource access to local communities, forest rehabilitation and protection, provision of livelihood opportunities, and enhancing the political capacity of communities.

4.1 Resource Access Provision

Until recently, forest occupants, including indigenous people, were treated as "squatters" or "encroachers" in forestlands, even if they may have occupied these areas since time immemorial. This, to a large extent, has been corrected under the CBFM program and its forerunners over the last three decades.

Since 1983 six different types of tenure instruments have been developed. Such instruments recognize the vested rights of forest occupants, both migrants and indigenous people, and allow them to have access and enjoy the benefits from forest resources. Under the recent CBFM program, the different tenure instruments were rationalized by providing only four types instead of the original six. The objective was to minimize confusion among the participating communities and facilitate the administrative efficiency of the DENR, the sole government agency responsible for their issuance.

As of September 2000, a total of 442,971 tenure instruments, covering around 5.332 million hectares of public forests and forestlands, have been issued to individuals and communities living within or adjacent to forestlands (Borlagdan et al. 2001).

4.2 Forest Rehabilitation and Protection

Most of the households and communities involved in CBFM are engaged in forest rehabilitation activities. While no comprehensive national assessment has been done so far on the area reforested through the program, it is believed that it has significantly contributed to the overall reforestation accomplishment of the government, recorded at 549,807 hectares from 1987 to 1998 (FMB 1998). In addition, CBFM program participants have likewise contributed to the establishment of plantations in areas outside the forests, such as in their own private areas and other alienable and disposable lands, amounting to 8,223 hectares as of 1999 (Tesoro 1999).

CBFM communities also served as the first line of defense in the protection and management of forest resources. Accounts of local communities stopping illegal logging and other forest violations—in three instances, to the point of losing their lives—are common in almost all CBFM sites. It was established that the government is able to save at least 127 million pesos (approximately $2.53 million based on 2003 exchange rate) annually by allowing the communities to manage and protect CBFMP sites, instead of hiring additional forest guards to protect and secure these areas (Tesoro 1999).

4.3 Community Livelihood Opportunities

Various livelihood schemes have been incorporated into the different CBFM projects to provide livelihood opportunities to the participants. A common approach is to contract out the different site development activities to a PO, such as reforestation, agroforestry, assisted natural regeneration, and timber stand improvement. Part of the proceeds from these activities are set aside for livelihood projects, which are jointly identified with the communities concerned (Pulhin 1999).

In support of the above activities, credit and/or marketing cooperatives have also been organized in some communities through the project's community organizer. Cooperative members are trained on various livelihood activities, including simple bookkeeping and financial management, for more effective functioning of the organization.

Other than project funds, sources of capital for livelihood activities may come from limited utilization of forest products such as timber and rattan. This applies particularly to those projects which have been given a resource-use permit to utilize forest products to augment participants' incomes. The scheme is based on the principle of "borrowing from nature to finance community and forest development." The idea is to plow back part of the sales income from legally harvested products from natural forests or existing plantations to finance livelihood projects, reforestation, and related forestry activities to ensure forest sustainability.

4.4 Building Political Capacity

Studies indicate that CBFM projects have, to a certain extent, provided some space towards building the capacity of forest communities to influence decision-making relevant to forest development and management (Pulhin and Gata 2001; Contreras 2000). With the assistance of NGO staff and other professionals, members of people's organizations are given the appropriate training to analyze their problems and to plan, implement, monitor, and evaluate their own plans towards the improvement of their socio-economic and environmental situation. In the process, local people are organized and mobilized towards collective action that builds their confidence and political capacity to influence decision-making on matters relevant to forest management. But considering the historical and political context of the Philippines forestry sector, characterized by state-cum-elite domination and control, building the political capacity of the marginalized upland communities takes a long time and is not easy to achieve.

5. ISSUES AND CONCERNS

Despite the noted accomplishments, a number of issues confront CBFM that may imperil the achievement of its objectives of poverty alleviation, social equity promotion, and forest resource sustainability. Some of these issues are briefly discussed below.

5.1 Identification and Depiction of the "Communities" in CBFM

There is a need to delineate the "communities" who are to benefit from the implementation of CBFM projects and programs. Delineation will certainly resolve the inequity of access to forest benefits in favor of the local elite. Moreover, the different interest groups whose sources of livelihood are mainly dependent on the local forest resources reserve the right to enjoy the benefits from CBFM in return for their participation towards sustainable forest management.

5.2 Operationalizing the Concept of Sustainable Forestry

Sustainability of forest resources is the very essence of forest management. A central concept in forestry science, sustained-yield forestry deals with the management of forest resources in perpetuity. It involves the application of "scientifically-based rules for balancing harvest with growth" (Lee 1984) to provide a continuous supply of forest resources through time. The concept of sustained-yield forestry was originally developed to maintain stable timber supply. It was later broadened to include other outputs from the forests (Clawson and Sedjo 1984). In practice, however, the application of the concept has been confined to commercial timber, particularly in developing countries like the Philippines (Colchester 1994).

Obviously, community forestry requires a broader approach to sustainability than currently provided under the foresters' concept of sustained-yield forestry (Pulhin 2001). At the moment there appears to be no consensus yet in the forestry sector as to the broader meaning of sustainable forestry and how it should be achieved. "Until wide agreement has been reached on what constitutes sustainable forest management (and how it should be realized via CBFM), pledges to achieve it will need to be taken with a grain of salt" (Poffenberger and Stone 1996 [statement in parenthesis added]).

5.3 Providing Sustainable Livelihood and Equitable Sharing of Forest Benefits

Local communities will continue to participate in CBFM projects only as long as they receive a steady stream of social, economic, and ecological benefits at levels that exceed the value of their input (Vergara 1997). As Byron (1996) argued, "There were [sic] some naiveté in the expectation that communities would take collective action for managing forests, retaining existing remnants, or in reforesting through plantations or assisted natural regeneration with little prospect of benefit for themselves." At present, however, the experiences in many CBFM areas have little to show in terms of clear trends that a stream of benefits can be assured from participating in these projects. In some CBFM projects, delayed budget releases or, at worst, unpaid reforestation wages, have rendered the participating communities economically worse off. Inability to provide sustainable livelihood to participating communities through a combination of forest and non forest-based activities remains to be one of the greatest obstacles that hinders the success of most CBFM projects.

It should be added, however, that a sustained flow of socio-economic, environmental, and other benefits from community forestry is not a guarantee that the welfare of the participants will automatically improve. As mentioned earlier, benefits from community forestry (and other rural development initiatives in general) are usually captured by the elite sector of the population. Unless benefits are shared equitably, community forestry will only contribute to the reproduction of rural poverty and lead to division and disharmony among those affected.

5.4 Bureaucratic Roadblocks to Effective Implementation

The CBFM approach demands a paradigm shift within the DENR organization, from its traditional, regulatory-oriented forest land management strategy towards one that is developmental and oriented to people and service. Such a shift should be appreciated and internalized by the whole DENR constituency, from the central down to the field personnel, in order to effectively perform the multiple responsibilities of promoting social justice, social equity, and better socio-economic conditions in the uplands, together with the sustainability of the country's forest resources. There are, however, indications that most DENR personnel, including those involved in the implementation of CBFM projects, have not yet appreciated or internalized their new roles and responsibilities in catalyzing sustainable forest management. Moreover, there is a lack of field personnel equipped with the multiple competencies necessary to facilitate a more effective and efficient implementation of CBFM. Inability to further simplify some policies of CBFM also serves as a bureaucratic roadblock to effective implementation (Dolom 1998).

5.5 Competing Interests of the Different Sectors and Actors

The policy and practice of CBFM in the Philippines has continued to prosper, despite some reservation and resistance from the other sectors of society, particularly the wood industry. Yet the argument that private corporate forestry also plays a central role in the country's economic development cannot be totally eclipsed by the community forestry paradigm. Indeed, as one of the country's main advocates of corporate forestry alluded, it has "its wealth of experience and capability in technical and managerial matters, and its reservoirs of infrastructure, including market access and perhaps financial resources" (Sanvictores 1997). These features provide private corporate forestry with a comparative advantage over community forestry, especially if the country is to pursue its march towards industrialization. Questions on the complementarities between upland communities and the private corporate sector under the CBFM approach, and under what conditions such complementarities will work, need more detailed assessment based on actual experience on the ground.

The multiple actors in community forestry, such as the POs, NGOs, and LGUs, including those from the various layers of the DENR bureaucratic hierarchy, also raise parallel concerns. Experience in CBFM projects has established that each actor would have varying interests and motivations for their involvement that may be contradictory or conflicting, sometimes breeding misunderstanding and disharmony among these actors. Evidently, an appropriate institutional mechanism has to be developed that would allow dialogue, negotiation, and conflict resolution, and more importantly, partnership, among various interest groups for the common good of the communities.

5.6 Ensuring Continuous Financial Support

Most of the programs and projects unified under the CBFMP are foreign-funded, in the form of either a loan or grant. The targeted nine million hectares of forestland to be placed under community management will definitely require a huge amount of financial resources, which may not be supported under the current DENR budget. Considering that the current model of community forestry is highly dependent on external funding, even the former DENR Community Forestry Program Office expressed some fear that "the enormous financial and technical assistance given to foreign-funded community-based projects make(s) them non-replicable and thus not sustainable" (National CFP Coordinating Office 1996). This is especially true, considering the present budget deficit of the government and the continuing budget cuts in the forestry sector.

6. THE PROSPECT OF COMMUNITY-BASED FOREST MANAGEMENT

Considering its noble objectives and potential, and recognizing the initial gains of the different initiatives discussed above, CBFM provides a glimmer of hope in the shadowy horizon of Philippine forest management. Its viability remains crucial in the next five to ten years. During this period, it is hoped that the targeted nine million hectares of forestlands will be in the hands of the local communities who are expected to manage them sustainably.

Equally crucial is the challenge facing CBFM to prove that it can effectively address the problems of social inequity, upland poverty, and forest depletion. The success of CBFM and its "moving into the twenty-first century" depends, however, on its ability to address three major challenges: sustaining innovative policy initiatives, promoting a more responsive practice (and consequently improving outcomes), and responding to changing contexts and needs.

6.1 Sustaining Innovative Policy Initiatives

In light of the continuing destruction of natural forests in the Philippines, one of the major challenges facing CBFM is to create an enabling policy environment where grassroots community forestry initiatives will flourish. The DENR has to revitalize these types of local initiatives other than what it previously emphasized—that is, the interventionist and highly programmatic type of community forestry. Parallel to this, policies will have to focus more on harnessing indigenous knowledge and the role of local institutions (both formal and informal) in forest management. Also, a great challenge is the creation of an enabling policy environment that will facilitate the taking over of forest management by the local community, instead of reinforcing government territorial jurisdiction and control.

6.2 Promoting Responsive Practices and Improving Outcomes

If CBFM is to live up to the expectations of its advocates, there is a need to demonstrate that it is indeed a viable and better alternative to the TLAs in terms of promoting social justice, equity, and sustainability in managing the country's forest resources. The challenge is to advance more responsive practices in synergy with the local communities to attain existing objectives and move beyond current boundaries and strategies. Among the specific ways to achieve this include: privileging the forest dependent groups so as to benefit from CBFM initiatives, developing CBFM criteria and indicators with the local communities to operationalize sustainable forestry, building the knowledge base on CBFM as the basis for policy and decision-making, transforming the forestry bureaucracy from being a regulatory to a developmental agency, promoting sustainable livelihood and building political capacity of CBFM participants, and forging meaningful partnerships with the LGUs and the private sector (Pulhin 2001).

6.3 Responding to Changing Contexts and Priorities

The greatest challenge to CBFM in the twenty-first century, presumably, is its ability to respond appropriately and timely to the changing contexts and priorities of forest management (Pulhin 1997). As local communities are increasingly incorporated into the market economy, the mode of production systems, technologies, and social institutions needed will certainly shape and change the direction and form of what is now called community-based forest management. Considering the changing macro and micro socio-political, economic, and ecological forces that will shape the future of CBFM, the degree of local people's participation in the decision-making process will determine whether satisfying their interests will be the main agenda of future projects and programs. Ensuring that the direction of forestry management will be fashioned towards this agenda is a major challenge facing CBFM in the Philippines.

7. CONCLUSIONS

The last 30 years saw the incorporation of the concept of community forestry from the periphery to being the central forestry policy agenda of the Philippine government. It is now considered as the core national strategy in pursuing sustainable upland development. Considering the radical shift in policy orientation, from a highly inequitable private corporate type of forest management to a community-based approach, as well as some initial gains from its implementation, there is reason to be optimistic about the future of CBFM. This optimism, however, should be balanced with a caution: Unless the above-mentioned issues and concerns are appropriately addressed, the initial gains of community forestry may very well "go down the drain." Challenges confronting CBFM include sustaining innovative policy initiatives, advancing best practices, improving outcomes, and responding to changing contexts and priorities. It is hoped that these challenges, if effectively addressed, can advance community-based forest management in the twenty-first century.

REFERENCES

Bennagen, P. L. 1996. Consulting with spirits, working with nature, sharing with others: An overview of indigenous resource management. In *Consulting with spirits, working with nature, sharing with others: Indigenous resource management in the Philippines,* ed. P. L. Bennagen and M. L. Lucas-Fernan, 1–22. Quezon City: Sentro Para sa Ganap na Pamayanan.

Boado, E. L. 1988. Incentive policies and forest use in the Philippines. In *Public policies and the misuse of forest resources*, ed. R. Repetto and M. Gillis, 165–204. New York: Cambridge University Press.

Borlagdan, S. B., E. S. Guiang and J. M. Pulhin. 2001. Community-based forest management in the Philippines: A preliminary assessment. Project report. Quezon City: Institute of Philippine Culture, Ateneo de Manila University.

Byron, N. 1996. Research needs for community forestry. Proceedings of the international conference, Community Forestry as a Strategy for Sustainable Forest Management, 24–26 May, at Manila, Philippines, 411–423.

Clawson, M. and R. Sedjo. 1984. History of sustained-yield concept and its implications to developing countries. In *History of sustained-yield forestry: A symposium,* 3–15. Portland: Forestry History Society.

Colchester, M. 1994. Sustaining the forests: The community-based approach in South and South-East Asia. *Development and Change* 25(1): 69–100

Contreras, A. P. 2000. Creating space for local forest management: The case of the Philippines. Paper presented during a policy forum of the Project Creating Space for Local Forest Management, 27 October, at Angelo King International Center, De La Salle University, Manila, Philippines.

Department of Environment and Natural Resources (DENR). 1990. *The Philippine master plan for forestry development.* Quezon City: DENR.

———. 2002. The Philippine forestry sector. Paper presented during the National-level Civil Society Forum on the Tropical Forest Conservation Act, 5 September, at the PhilDRRA Office, Loyola Heights, Quezon City, Philippines.

Dolom, B. L. 1998. Research and development issues in the Sierra Madre: The case of community-based forest management. In *Sustainable mountain ecosystems: Philippine's challenge for the 21st century*, ed. J. M. Pulhin, L. L. Rebugio and F. B. Pulhin, 137–144. Laguna, Philippines: College of Forestry and Natural Resources, University of the Philippines Los Baños.

Forest Management Bureau (FMB). 1998. *Philippine forestry statistics*. Quezon City: Department of Environment and Natural Resources.

Garrity, D. P., D. M. Kummer and E. S. Guiang. 1993. *The upland ecosystems in the Philippines: Alternatives for sustainable farming and forestry.* Washington, D.C.: National Academy Press.

Kirchhoff, J. F. 1993. Community participation in forest management. Paper presented at the Conference Workshop on Natural Resource Policies in the Philippines: Bottlenecks and Solutions to Implementation, 9 December, at the Asian Institute of Management, Makati, Philippines.

Kummer, D. M. 1992. *Deforestation in the post-war Philippines*. Quezon City: Ateneo de Manila University Press.

Lee, R. 1984. Sustained-yield and social order. In *History of sustained-yield forestry: A symposium*, ed. H. Streen, 90–100. Durham, NC: Forestry History Society.

Liu D. S., L. R. Iverson and S. Brow. 1993. Rates and patterns of deforestation in the Philippines: Application of geographic information systems analysis. *Forest Ecology and Management* 57:1–16.

Lynch, O. J. 1982. *Tribal land law: A mechanism for upland participatory development.* Manila: Integrated Research Center, De La Salle University.

———. 1987. The legal bases of Philippine colonial sovereignty: An inquiry. *Philippine Law Journal* 62:279–316.

Makil, P. Q. 1982. *Towards social-forestry oriented policy: The Philippine experience*. Quezon City: Institute of Philippine Culture, Ateneo de Manila University.

Poffenberger, M. 2000. *Communities and forest management in Southeast Asia*. Berkeley: University of California.

Poffenberger, M. and B. McGean. 1993. Upland Philippine communities: Guardians of the final forest frontiers. Research Network Report No. 4. Berkeley: Center for Southeast Asia Studies.
Poffenberger, M. and R. D. Stone. 1996. Hidden faces in the forest: A twenty-first century challenge for tropical Asia. *SAIS Review* (Winter-Spring 1996).
Porter, G. and D. J. Ganapin. 1988. Resources, population and the Philippines' future: A case study. World resource paper no. 4. Washington, D.C.: World Resources Institute.
Pulhin, J. M. n.d. Community forestry country profile: Philippines. International Network of Forests and Communities (INFC): Victoria, Canada. <http://forestsandcommunities.org/Country_Profiles/Philippines.html>.
———. 1996. Community forestry: Paradoxes and perspectives in development practice. Ph.D. dissertation. Canberra: The Australian National University.
———. 1998. Community forestry in the Philippines: Trends, issues and challenges. In proceedings of the international seminar, Community Forestry at a Crossroads: Reflections and Future Direction in the Development of Community Forest, 17–19 July, at Bangkok, Thailand, ed. M. Victor, C. Lang and J. Bornemeier, 201–215. Bangkok: Regional Community Forestry Training Center.
———. 1999. Community forestry in the Philippines: Approaches and emerging lessons for sustainable forest management. In proceedings of the international seminar, Sustainable Forest Management, 31 August–2 September, at Pokhara, Nepal, ed. P. Mathema, I. C. Dutta, M. K. Balla and S. N. Adhikary, 103–116. Pokhara, Nepal: Institute of Forestry, Tribhuban University.
———. 2001. Community forestry for sustainable upland development: The case of the Philippines. Paper presented at the Global Change and Sustainable Development in Southeast Asia: A Regional Science-Policy Conference, 17–19 February, at Chiang Mai, Thailand.
Pulhin J. M. and M. L. L. P. Gata. 2001. Creating space for local forest management: The Philippines' Bicol national park. Paper presented at the Regional Conference on Sustaining Upland Development in Southeast Asia: Issues, Tools & Institutions for Local Natural Resource Management, 28–30 May, at the ACCEED Conference Center, Makati, Metro Manila, Philippines.
Sabban, M. V. M. 1998. Community forestry and decentralization policies: Reflections on experiences from the Philippines. In the proceedings of the international seminar, Community Forestry at a Crossroads: Reflections and Future Direction in the Development of Community Forest, 17–19 July, at Bangkok, Thailand, ed. M. Victor, C. Lang and J. Bornemeier, 175–180. Bangkok: Regional Community Forestry Training Center.
Sajise, P. E. 1998. Forest policy in the Philippines: A winding trail towards participatory sustainable development. In *A step toward forest conservation strategy (1): Current status on forests in the Asia-Pacific region, Interim Report 1998*. Tokyo: Institute for Global Environmental Strategies (IGES).
Sanvictores, E. 1997. Philippine forestry in the 21st century: Challenges and prospects. In *Proceedings of the Society of Filipino Foresters: 49th anniversary and national convention*, 16–19. Manila: Society of Filipino Foresters.
Tesoro, F. O. 1999. Community resource management as a strategy for sustainable economic development. DAI-NRMP Document No. 099-056-POL-014. Quezon City: Development Alternatives Inc./Natural Resources Management Program.
Vergara, N. T. 1997. Is community forestry management implementable? Paper presented at the 1997 National Convention and Workshop of the Society of Filipino Foresters, June 5–7, at Imus Sports Complex, Imus, Cavite, Philippines.

Vitug, M. D. 2000. Forest policy and national politics. In *Forest policy and politics in the Philippines: The dynamics of participatory conservation*, ed. P. Utting, 11–39. Quezon City: Ateneo de Manila University Press.

Chapter 10

FOREST POLICY TRENDS IN VIETNAM

Do Dinh Sam and Le Quang Trung
Forest Science Institute of Vietnam

1. INTRODUCTION

Vietnam's forestry sector has undergone large changes over the last decade under the country's *Doi Moi* reforms. These have made significant progress, not only in food production and changes in the make-up of flora in the mountain areas but have also resulted in a recovery in forest resources, damaged by a combination of war, industrial forestry, and poverty over the last few decades. In recent years, the country's forest area has rapidly increased in both quantity and quality because of the institution of new forestry policies. According to the Ministry of Agriculture and Rural Development, by the end of 1999 the total forest area in Vietnam was 10.9 million hectares, or about 33.2 percent of the total natural area, including 9.4 million hectares of natural forest and 1.5 million hectares of production forest (MARD 2000).

The new policies have caused a transition in the forestry sector, from being centrally planned and subsidized (by co-operatives and the state) to one that is multi-sectoral, market-oriented, and participatory. Among these policies is a reformulation of various types of forest ownership for more effective management, protection, and use of the country's forests. Forest land is now classified under three categories: special-use forest (national park, natural conservation area, historical area, etc.), protection forest (watershed, sandy, sea wave, etc.), and production forest. At present there are 5.35 million hectares of protection forest, 1.53 million hectares of special-use forest, and 4.04 million hectares of production forest. By 2010, according to government plans, 650,000 hectares of protection forest, 480,000 hectares of special-use forest, and 3.96 million hectares of production forest will be planted, and each of these three categories will be managed differently. The forestry policies are continually being reviewed and amended, with a focus on increasing the participation of rural communities in forest management through the "rights of forest owners," benefit-sharing, and management policies. As well, logging practices are moving from using natural forests for wood to

M. Inoue and H. Isozaki (eds.), *People and Forest—Policy and Local Reality in Southeast Asia, the Russian Far East and Japan, 157–167.*

using mainly planted forests. This chapter aims to shed some light on forest policy trends in Vietnam.

2. DEVELOPMENT OBJECTIVES AND TRENDS IN VIETNAM'S FORESTRY SECTOR (2001–2010)

2.1 Development Objectives in the Forestry Sector by the Year 2010

To implement the resolution of the country's National Assembly, Vietnam's prime minister issued an official decision, detailing the following forestry development objectives to be accomplished by 2010 with the following goals (Vietnam Government 1998):

- Establish five million hectares of new forest, while protecting existing forests, in order to increase forest cover to 43 percent of the national territory; protect the environment; decrease the severity of natural disasters; increase water availability; preserve gene resources; and protect biodiversity.
- Provide material for construction as well as raw materials for the production of paper, wood-based panels, non-wood products, and fuelwood, both for local consumption and export; develop the forest products processing industry; and make forestry an important economic sector, contributing to improvement in the socio-economic situation in mountain areas.
- Use open land and bare hills efficiently; create employment opportunities; contribute to hunger elimination and poverty reduction; support sedentary cultivation; increase the incomes of rural mountain people; create stable social conditions; and strengthen national defense and security, especially in border areas.

According to the government's forest development strategy and other governmental decisions, specific targets of the forest development plan to be achieved by the year 2010 are given below.

Table 1: Forest development targets to be achieved in Vietnam by the year 2010.

Forest Type	Area (hectares)
Protection forest	6.0 million
Special-use forest	2.0 million
Production forest	8.0 million
Total forest area	16.0 million

2.2. Trends in Forestry Policy

Forestry policies have been elaborated via various papers issued by the Vietnam state (the National Assembly, the government, and ministries). Since 1990, about 150 important policies related to forestry have been issued in the following forms:

- The Forest Law, issued by the National Assembly
- Resolutions by government
- Decisions by the prime minister
- Regulations and circulars
- Ministerial directives

The following are important and core forestry policies:

- The Law on Protection and Development of Forests (reviewed and amended)
- The Land Law (1993), reviewed and amended in 1998–2000
- Government Resolution 01/CP (1995), on the allocation and contracting of land for agriculture, forestry, and aquaculture production to state enterprises
- Government Resolution 02/CP, on the allocation of forestland to various economic sectors for management and use for long-term and sustainable forestry development
- Prime Ministerial Decision 661/QD-TTg (1998), on the objectives, tasks, policies, and organizations for the establishment of five million hectares of new forest
- Government Decree No. 163/1999/ND-CP, concerning allocation and lease of forestland to organizations, households, and individuals for long-term forestry purposes
- Ministry of Agriculture and Rural Development Circular No.56/1999/TT/BNN-KL (1999), guiding the development of regulations on forest protection and development for villages, hamlets, and communities
- Prime Ministerial Decision No. 187/1999/QD-TTg (1999), on the renovation of the organization and management mechanism of state forest enterprises
- Prime Ministerial Decision No. 08/2001/QD-TTg (2001), which issued regulations on management rules for special-use forest, protection forest, and production forest
- Prime Ministerial Decision No. 178/2001/QD-TTg (2001), on the rights and obligations of households and individuals that are allocated and contracted forest and forest land for benefit-sharing
- Other policies on development investment credits, taxes, and extensions related to forestry activities

Based on the forestry policies issued, the major trends in forestry policy are now focused on three main objectives:

- Developing forestry in the direction of social forestry, with the participation of multi-sectoral economic entities.
- Oversee a transition in the forestry sector from simply harvesting natural resources to sustainably managing and utilizing forests.
- Develop forestry in the direction of a market-oriented economy and increase its contribution to the country's economy, create employment, and improve the lives of people in rural communities.

2.2.1 Social Forestry Development

This change in orientation was prompted by three main causes:

- Failures in the operation of the centrally planned and subsidized economy.
- Ineffective forestry activities of state units (co-operatives and state enterprises).
- Unsuccessful efforts at forest development, caused by poverty and lack of attractive policies for local participation.

At the beginning of the Doi Moi process, the National Assembly passed the Land Law and the Forest Protection Law. The government also later passed policies related to forests and forest use, such as Resolution 02/CP and Government Decree 163/1999/ND-CP. These policies determined the following:

- Land ownership belongs to the people. Households, individuals, and organizations are allocated land and water surface rights for long-term use and management for agriculture, forestry, and aquaculture development purposes. They have the right to exchange, transfer, rent, inherit, or mortgage the land allocated to them, including the right to use it as capital for joint ventures with domestic and foreign organizations and individuals in order to boost production.
- Agencies, households, and individuals do not have to pay land-use fees on allocated forest land (for households and individuals, limited to not over 30 hectares for a duration of 50 years). Upon expiry, they may extend this period, as long as the land has been used correctly for the purposes allocated. The government also contracts out land for forest production to organizations, households, and individuals, with the duration depending on each type of forest (for protection forests and special-use forest, it is 50 years; for production forest, it depends on business rotations).
- The benefit-sharing policy should be formulated and amended to increase the benefits to households and individuals involved in forest land protection, management, and reforestation. Two prime ministerial decisions were issued to flesh-out the benefit-sharing regulations and the obligations of households and individuals allocated or contracted forest and forest land (including natural forests, production forests, and barren lands) for reforestation and maintenance

for all three categories of protection, special-use, and production forests (Vietnam Government 2001a, 2001b).
- The renovation of state forest enterprises should be addressed in order to implement efficient production and business practices, and also to increase the function of their services to farmers (reducing forest area managed to make it suit to their operating capacity).

Other policies were issued on credit extension services, which also created good opportunities for rural inhabitants to participate in the management, protection, and maintenance of forests, as well as tree-planting.

According to a Ministry of Agriculture and Rural Development report, nearly 8.8 million hectares of forest land (50 percent of forest land) have been allocated to various targets, of which 2.61 million hectares (29.7 percent of the area allocated) have been allocated to over 450,000 households. Besides this, there are about 50,000 households that have received a total of over 1.86 million hectares of forest and forest land on contract from state forest enterprises, protection forest and special-use forest management boards, foreign companies or joint venture companies, etc. After the allocation of forest land, cooperative relationships in forest production have arisen in almost all localities in the country. Many households took the initiative and invested capital and labor to organize production activities, such as tree-planting, protecting and tending trees, and practicing agroforestry. Due to national cropping systems and the suitable structure of planted crops, tens of thousand of households have transformed bare land and denuded hills into wealthy and luxuriant farms, upgraded land after slash-and-burn cultivation to establish forest gardens, and created tree farms with forest trees species and agricultural crops of fruit trees of high economic value. Below are some examples of what has been accomplished.

- In Yen Bai province, 9,500 farms combining forestry and agriculture have been established, representing 11.9 percent of the total farming households in the province. The average land area of a household is five to ten hectares, but large-scale farms have also been created. An example is Mr. Do Thap, who has invested in planting 30.4 hectares of forest, planted 20,000 *Cinnamomum cassia* trees, and contracted for forest protection services and tending forest re-growth on hundreds of hectares.
- In Lao Cai province, there are 1,500 farms producing annual revenues of 15 to 20 million Vietnam dong (VND) each (U.S.$1,200 to $1,600).[1] Every farm has an average of three hectares of fruit trees or industrial crops.
- In Thanh Hoa province, over 13,000 households are engaged in forest gardens and forest farm production. The area of each farm is about five to ten hectares, with annual incomes of 5 to 20 million VND per farm.
- Many households that were allocated land that included bare land, denuded hillsides, or depleted natural forests were at first only able to produce enough firewood and timber for their own use, but by practicing agroforestry

[1] One U.S. dollar is equivalent to 12,000 to 13,000 VND.

production they have made important contributions in protecting biodiversity and the environment, soil improvement, and improving crop productivity.

In short, forestry activities under the direction of social forestry have been developed by creating rights, obligations, and benefit-sharing involving multi-sectoral bodies in forest management, protection, and development. This change of forest policy has contributed to Vietnam's recent successes in forest development.

2.2.2 Sustainable Management and Utilization of Forests—Improving Forest Functions of Production, Environmental Protection, and Biodiversity Conservation

Vietnam's abundant and diverse forest resources, with many valuable kinds of timber and forest products, have been seriously degraded in both area and quality. Forest loss has impacted not just the wood and non-wood supply to meet national economic demand but also environmental protection capacity and biodiversity. Most natural forests in Vietnam were heavily exploited and now produce much less wood than they did historically. Only about a half million hectares of forest are classified as "rich," with more than 100 cubic meters of timber per hectare. Another 1.7 million hectares are classed as "medium" stocked, while some 2.8 million are "poor," with timber volumes below 80 cubic meters per hectare. Fifty percent of the man-made forest is made up of young plantations or newly planted ones (less than ten years old), because plantation activities have received more attention only recently. At the same time, many of the endemic species of flora and fauna found only in Vietnam are in danger.

In order to deal with these serious circumstances, the Law on Forest Protection and Development, issued in 1991, identified three categories of forest: (1) protection forest (watershed, sandy forest, and forests in coastal areas), (2) special-use forest (national parks, natural conservation areas, historical and cultural parks, etc.), and (3) production forest. Each type of forest is managed differently.

Great attention has been paid to natural forests for protection and sustainable management through a number of activities, including those listed below:

- Creation of the Five Million Hectares Afforestation Project (Prime Ministerial Decision 661/QD-TTg).
- Reduction of annual timber harvests from natural forests from one million to 300,000 cubic meters a year.
- Issuing the benefit-sharing policy ensured the sustainable management of forests by increasing the participation of rural inhabitants in forest management, in accordance with their rights and obligations (Prime Ministerial Decision 08/2001/QD-TTg).
- Application of the participatory approach in land-use planning.

2.2.3 Market-oriented Forestry

The forestry sector is managing about 16 million hectares of forest land, but its annual benefit is very small. In addition, of the over 30 million of people presently living in forested areas, up to 30 percent are still living in hunger and poverty, with monthly incomes below 15 kilograms of rice per person. Thus the majority of the households are forced to pay more attention to their immediate needs; they invest money and labor in production activities that bring quick returns. This explains why they have not yet paid much attention to investing in forest production, with households putting only 30 percent of the forest land allocated to them to use. This highlights the fact that forest development in Vietnam must occur together with a poverty alleviation strategy. In supporting poverty alleviation, the forest development activities not only protect the environment and biodiversity but also create employment and income, contributing to the improvement of living standards for the rural populations in mountainous areas.

To this end, the prime minister issued Decision No.661/QD-TTg (1998), which laid out objectives, tasks, policies, and organization to create large areas of new forest. It included the following:

- The establishment of five million hectares of new forests, of which two million hectares is economic forest.
- Objectives of using open land and bare hills efficiently, creating employment opportunities, contributing to hunger elimination and poverty reduction, and increasing incomes for rural mountain people.
- Goals of providing material for construction as well as raw material for production of paper, wood-based panels, non-wood products, and fuelwood for both local consumption and export.
- Objectives of creating a raw material supply area for the development of a wood processing industry to help make forestry an important economic sector.

Policies mentioned in Decision 661 related to forestry include the following:

- As an investment policy for forest protection, the government will continue to provide funds for forest protection contracts of up to 50,000 dong per hectare over five years. It will also contribute a maximum of one million dong (about $80) per hectare, distributed over six years, for natural regeneration or for planting and tending trees over three years. It will also provide up to 2.5 million dong per hectare for planting and tending forests with valuable tree species that have production cycles of over 30 years.
- Policy on credit with priority interest.
- Scientific research and forestry extension will be supported to contribute to increasing the yield of production forests and development of the wood processing industry.
- In developing the market for forest products, all forest products from plantations (i.e., bamboo and other non-wood forest products) are free to be

sold in the market. Processing and the export of forest products from plantations, round wood included, will be encouraged.

3. TRENDS IN FORESTRY POLICY RELATED TO RURAL PARTICIPATION

In terms of the role of beneficiaries, the participation of individuals, households, and village communities in planning, protecting, managing, and planting forests is increased via a series of policies, whose main concepts are described below.

3.1 Creating a Legal Basis for People to Become Owners of Forests and Forest Land

The wide variety of owners of forests and forest land must be clearly identified. These include state forest enterprises, management boards of special-use and protection forests, local authorizations (where land allocation has not occurred) by provincial people's committees, district people's committees, or communes, social organizations (schools, cooperatives, army institutions, etc.), village communities, households, and individuals.

The role of households and individuals as forest owners is detailed in the Land Law, the Law on Forest Protection and Development, and in government resolutions.[2] One of the changes in policy is the category of forest land that can be allocated to households and individuals. In the past only barren land and plantations could be allocated to them. While now, the categories of forest land open to them for management, protection, and development include special-use forest (less than 1,000 hectares), protection forest (less than 5,000 hectares, or scattered plots), and natural forest considered to be production forest. This means that households, individuals and communes are owners under these forest categories.

3.2 Increasing the Participation of Households, Individuals, and Village Communities in Forest Management

Besides now being able to own forest land, households, individuals, and village communities can participate in the management of special-use, protection, and production forests (including land with and without forest cover) under contract with the forest owners (state forest enterprises, management boards of special-use and protection forests). Resolution 01/CP (1995) and Prime Ministerial Decision

---

[2] I.e., No. 02/CP; 163/1999/ND-CP, concerning the allocation and leasing of forestland, and Decision 08/2001/QD-TTg by the prime minister, which issued regulations on the management of natural special-use, protection, and production forests.

08/2001/QD-TTg issued regulations on the management of natural special-use, protection, and production forests. Decision No.178 detailed the rights of benefit-sharing and the obligations of households and individuals who are allocated and leased forest land and forests.

Attention has been paid to community forest management recently. The Ministry of Agriculture and Rural Development issued Circular 56/1999/TT/BNN-KL, guiding the development of rules for the protection and development of forests within village communities. It includes rules on forest protection and the mobilization of internal resources to tend, maintain, and develop forests (watershed, historical, landscape, and religion forests) owned by village or hamlet communities. Developing rules can be done in either a community meeting or a meeting of household representatives, where they discuss, comment, and agree on implementation. According to the forest protection department within MARD, by June 2001, 1,203 communes in 146 districts of 24 provinces and cities had formulated community forest management plans encompassing a total forest area of 2.35 million hectares.

3.3 Benefit-sharing Policy

The benefit-sharing policy, considered important for encouraging rural people to manage and develop the forests, has been reviewed and amended to increase the benefits to the people. In 2001, the prime minister issued two important decisions on the management of natural forests and benefit-sharing (Vietnam Government 2001a, 2001b). The following describes the amendments made to both:

- *Rights of households who sign contracts for and invest in protection forests*—Households are entitled to payment for protecting, regenerating, and planting forests, in accordance with the contracts signed with the forest management boards. Households have the right to collect fuelwood and non-wood forest products under the forest canopy (20 percent in timber forests; 30 percent in bamboo forests) and 85 to 90 percent of harvesting products, after taxes are paid. In addition, they are entitled to all agricultural and forest products when the forest is ready for harvest, if the households themselves carried all the costs for regeneration and planting on land without forest cover. In addition, they can use 20 percent of allocated land for agricultural production.
- *Special-use forests*—The forest owners (households or individuals) are allowed to conduct harvesting, research, and ecological tourism activities, in accordance with the laws and regulations.
- *Rights of households and individuals allocated natural production forest to manage*—They have the right to collect dead trees, trees damaged by fire or other natural calamities or damaged during the process of applying silviculture technologies, and they can harvest forest products to meet their own family's needs. They are also allowed to submit for approval their wood requirements to construct a house for newly established families, not to exceed ten cubic meters per family. They can conduct official harvesting, when allowed, and are entitled to 100 percent of products from poor regeneration forest or 70 to 80

percent of products in the case of regenerated forest (after shifting cultivation), and 2 percent per year of forest that has 100 cubic meters of growth per hectare annually. In addition, they are entitled to 95 percent of forest products if from a bamboo forest (after taxes are paid).

- *Plantations established by state funds and allocated to households and individuals*—They are entitled to 75 to 85 percent of total forest products (after taxes are paid). If they themselves carried all the costs for establishing plantations, they are able to choose the species, technical norms, and planting and harvesting techniques. The products harvested from plantation forests can be sold freely in the market.
- In short, the benefit-sharing policy deals with watershed forests, special-use forests, and natural production forests in detail, whereas they were not mentioned in earlier policies.

4. CONCLUSIONS

Through changes in policy Vietnam has achieved positive results in forestry, with forest cover increasing to 33 percent of the national area. As well, efficiency in forest regeneration has been improved, along with the tending of natural protection forests, and the growth of plantations to provide raw material for the production of paper and wood panels, etc. And, at last, over the last few years a national park system has been formulated and consolidated.

Vietnam's forestry policy shows the government's determination to protect and develop forest resources through the participation of households, individuals, and other non-state organizations. The changes in policy have created three directions in forest development under: (1) promoting social forestry, with multi-sectoral participation; (2) changing the focus from harvesting natural resources to managing and utilizing forests sustainably; and (3) increasing the forest sector's contribution to the country's economy, while creating employment and improving the lives of people in rural communities. The participation of households, individuals, and village and hamlet communities plays an important part in implementing the forest development objectives. The nation has approved and enforced land laws; forest protection and development laws; resolutions on forest land allocation, leasing, and the allocation and contracting of forests; and regulations detailing the rights and obligations of households and individuals on allocated and/or contracted forest land and forests. This process has established different forest owners and ownership styles in Vietnam, and launched major renovations of state forest enterprises and forest management boards to better achieve their main functions of guiding, providing technical services, and being a bridge between the state and rural people.

Despite clear trends in forestry policies, still being reviewed and amended, challenges remain in the process of implementation in the face of an increasing demand for wood and other forest products from the natural forests. At the same time, the allocation and contracting of forest land is still operating at a low level, as is the efficiency in the use of barren and poor lands (land without forest cover), while the renovation of the state forest enterprises process has not yet produced satisfactory results. There are still obstacles and shortcomings in land-use planning, typically in the identification of the three forest categories (especially protection forest), as well as inappropriate buffer zones around national parks, low participation in micro-level land-use planning, limited multi-purpose use of forest products, and limited investments in forestry.

On the other hand, the forestry development program in Vietnam has received considerable interest from various donors. The 5 Million Hectares Partnership and the Forest Sector Support Programme (FSSP) were created with the participation of nearly 20 international donors and organizations to support the development of Vietnam's forestry sector.

In summary, there is great potential and opportunity for forest development in Vietnam. But in order to implement the country's forest development objectives, forestry policies still require improvements so as to meet the nation's requirements for environmental and biodiversity protection at the same time as promoting economic development.

REFERENCES

Dang, Nguyen Van. 2001. *Forestry in Vietnam 1945–2000.* Hanoi: Agricultural Publishing House.

Phuong, Pham Xuan. 2001. *Introduction to forestry policy.* Hanoi: MARD.

Sam, Do Dinh and Nguyen Hoang Nghia. 2001. *Research on natural forest.* Hanoi: Statistical Publishing House.

Trung, Le Quang. 2000. *Research on establishing policy to attract people participating in pinus merkusii forest for five million ha afforestation programme.* Hanoi: Forest Science Institute of Vietnam.

Tuynh, Vu Huu. 1999. *Key forestry policies for forestland allocation and land use in the mountainous area of Vietnam.* Hanoi: MARD.

Vietnam Government. 1998. Prime Ministerial Decision No. 661/QD-TTg, 29 July 1998. Objective, tasks, policies, and organization for the establishment of five million hectares of new forest.

———. 1999. Decree No. 163/1999/ND-CP, 16 November 1999. Allocation and lease of forestland to organizations, households, and individuals for long-term forestry purposes.

———. 2001a. Prime Ministerial Decision No. 08/2001/QD-TTg, 11 January 2001. Regulations on management of natural special-use, protection, and production forest.

———. 2001b. Prime Ministerial Decision No. 178/2001/QD-TTg, 12 November 2001. Regulations on the benefit right, responsibilities of households/individuals who are allocated, leased, and contracted forestland.

Vietnam Ministry of Agriculture and Rural Development (MARD). 2000. *Forest resources statistics 1999.* Hanoi: MARD.

Chapter 11

INFORMATIONAL BASIS OF POLICY JUDGMENTS: THE CASE OF THE ROYAL FOREST DEPARTMENT OF THAILAND

Jin Sato
Graduate School of Frontier Sciences, The University of Tokyo

1. INTRODUCTION

1.1 Objective

This report presents a preliminary analysis of policy formation in the area of forest management in Thailand. Despite its remarkable record of industrialization and economic growth, the country still has a large agricultural population that is in daily interaction with the natural environment. With the majority of people living in the countryside, policies on land use have a significant impact, not only on the conservation of biodiversity but also on the welfare of local people. Because of both an increasing scarcity of land and the emphasis on forest protection for the conservation of biodiversity, forest land has become a "politicized space" over which local people, government, and private industry strive for control (Sato 2000). Any initiative regarding land use must, therefore, be planned within this context; otherwise, the policies adopted will be neither socially nor ecologically sustainable.

It is often claimed that sound policies must rely on accurate information. In this context, many social scientists tend to think that more information is better than less. But this claim has to be examined carefully, since we often see reports that are never read, with papers piled up in corners of the library that are written just for the sake of writing them. We need to examine how information is connected to policy processes. In other words, some form of "strategy" needs to be built into the research framework to make it relevant, not only in connection with science and logic, but in its utilization and impact on policy.

M. Inoue and H. Isozaki (eds.), *People and Forest—Policy and Local Reality in Southeast Asia, the Russian Far East and Japan, 169–185.*

This study examines the role of information provision in such policy decisions. Work in the social sciences on policy formation in developing countries tends to rely on official documents and records, such as laws, cabinet resolutions, ministerial announcements, and so on. But these policy documents are in fact worded vaguely, leaving much room for differing interpretations. We know very little of the patterns, if any, of interpretation of ambiguous information and its translation into policy outcomes. The aim of this study is, therefore, to determine, by systematic analysis, the degree of freedom that exists in policy formulation at various levels of governmental authority (i.e., local, district, provincial, and regional levels), and to investigate the manner in which information is transferred from one level to another. In the process we may identify the informational bases of policy generation, and this may contribute to a more strategic utilization of natural sciences research.

This paper is primarily a case study of the Royal Forestry Department (RFD) of Thailand.[1] Although Thailand was known as the "Kingdom of Forests," it had, in fact, lost much of its natural forest assets by the late 1980s. Figure 1 illustrates the decreasing forest cover since the early 1960s and the increasing efforts of the government to expand protected areas.

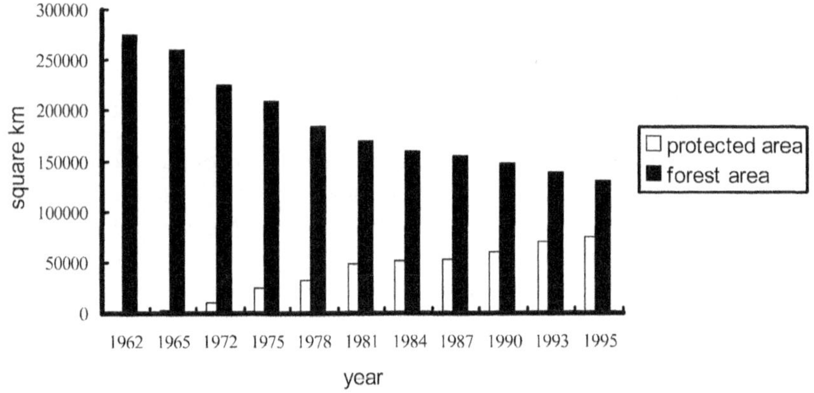

Figure 1: Decrease in forest cover and expansion of protected areas.
Source: Royal Forestry Department (1997).

One may wonder how the RFD has been able to financially sustain itself, or even strengthen itself, despite the dramatic loss of forests under their responsibility. The RFD

[1] Thailand went through a major restructuring of its government in October 2002, and the Royal Forest Department was split into two groups—one responsible for conservation and the other responsible for production. The former groups were absorbed into the newly established Ministry of Natural Resources and Environment. The answer to whether the new government structure has completely changed the decision-making processes described in this paper or not requires further investigation.

was able to more than double its budget and its staff during the "loss" period. One way to explain this mystery is to redefine the central mandate of the forestry department.

Despite the dramatic loss of forests, the RFD still has direct control over almost half the land area of Thailand, as "forest lands," and remains powerful, despite occasional challenges by non-government organizations (NGOs) and local people. This is because control of land itself is as important as the control of forests; and the loss of forests can also serve as a lever for increased funding. Of the territory under RFD control, forests with relatively rich biodiversity (i.e., national parks and wildlife sanctuaries), representing about 15 percent of total land area, are directly under the control of the RFD (RFD 1997). It is therefore clear that RFD operations have significant implications, not only for biodiversity but also for the welfare of farmers living in those forest areas.

Specifically, this study aims to (1) identify policy-relevant and policy-irrelevant information, with a focus on the work of frontline forestry officers who operate in the field; (2) identify "policy-relevant information" in relation to other factors that have an impact on policy formulation, such as politics and the bureaucratic culture, in various situations of decision-making; and (3) suggest an agenda for future research which will have immediate policy implications. Finally, I attempt to answer the apparently obvious, yet seldom investigated question: Where does policy come from?

1.2 Literature Review

There are broadly two strands of scholarly tradition from which this study can be seen to depart. First is the study of bureaucracy and administrative behavior in government (Simon 1958; Kaufman 1960; Allison 1971; Wilson 1989). This group of work focuses on decision-making structures in public administration and the conditions under which certain courses invite certain outcomes. This study relies particularly on two sources. One is the classic work by Kaufman (1960), *The Forest Ranger*. Kaufman examines how policy decisions, made at the top level of government, overcome centrifugal tendencies towards organizational fragmentation, and are translated into coherent action by forest rangers in the field. Kaufman's work is especially useful, because, although dealing with the Forestry Department of the United States of America, it employs an anthropological approach to the study of bureaucratic behavior. The other is a more recent work by Feldman (1989), *Order Without Design: Information Production and Policy Making*. In the context of the Energy Department of the U.S. government, Feldman asks why bureaucratic analysts persist in producing documents and information that are seldom used by policymakers. These studies indicate the complexity of organizational behavior, which defies easy analogy with the processes of individual rationality.

The second strand of scholarship on which this study relies is specialist work on Thailand. Siffin (1966) and Riggs (1966) published pioneering studies of the Thai bureaucracy, and some more recent contributions examined are by Yoshida (1985), Tamada (1992), and Hashimoto (1998). This work illuminates the historical development of the bureaucratic polity in Thailand and its structural problems, such as corruption. However, the question "Where does policy come from?" is not really addressed.

In addition, there are documents and reports concerned with forest policies in Thailand that are not necessarily academic, yet contain some useful information. Despite the relatively abundant published material on forest law and administration, there are few systematic analyses. Very few Thai scholars are working on issues related to land and forests from the viewpoint of the social sciences.[2] Why few academics work on the issue of land tenure and forest policies is an interesting question in itself, and deserves explanation. One possible reason for the paucity of study may be that the nature of the subject falls between the cracks of traditional disciplines. The faculty of forestry has not concerned itself with the social aspects of forest management; policy-related subjects are not considered "science," and are therefore felt to be inappropriate contenders for university resources. Second, it is a politically hot subject, which makes it difficult to collect sufficient empirical data. This has left the NGOs as the principle agents of critique of forest and land policy.

1.3 Research Methods

The primary method of this investigation was a series of interviews with RFD officials in Bangkok and in selected regions. Former RFD officials and university academics were interviewed as well. Three provinces were selected for case study, Uthaithani, Nakorn Sawan, and Ubonrachatani, because of their comparatively rich forests and the many conservation projects that have taken place there.

The initial research was conducted in the Lansak and Hoi Kot districts of the province of Uthaithani, interviewing frontline forestry officials who work in the field in order to investigate the way in which these officers interpret policy. The interviews were conducted in the Thai language without the use of interpreters, with one research assistant who was able to help me record the interviews. Formats for making reports and the minutes of monthly meetings were obtained whenever possible.

---

[2] This does not mean that there are no studies on the subject. Sayamon (1995), for example, looked at the historical development of the RFD, the structure of power in land administration, and various master's theses that are relevant to the subject.

2. BACKGROUND OF FORESTRY POLICIES OF THAILAND

2.1 Introduction to the Royal Forestry Department

Before reporting the core matter of this study, it will be helpful to outline the basic features of forest policies in Thailand.[3] In Thailand, forest policies are implemented under the direction of the Royal Forestry Department (RFD), established in 1896 by the British. The mission of the RFD at that time was to regulate and control the logging businesses, which had been dominated by feudal chiefs, in the northern region of Thailand. Centralized control by the RFD made it easy for the British and the Siamese governments to profit from valuable timber such as teak. The major task of the RFD in the early days, then, was to negotiate with the feudal chiefs in order to gain control of forest resources in the north.

Control over valuable timber soon made the RFD realize the importance of controlling land. Although the idea of demarcating "forest reserves" was present from the first years of the RFD's establishment, the actual policies did not materialize as the Forest Conservation and Protection Law until 1938. This law for the first time made clear the government's intention that certain "areas" were government property, which should not be invaded. This policy was further strengthened by the designation of "permanent forest" in 1961 and the National Forest Reserve Act of 1964. Both acts are still current. But conflict with local people who lived inside the boundaries of the designated areas became intense, and land disputes between people and government increased dramatically as the population grew.

A well-publicized conflict was the years-long street demonstration by the "Assembly of the Poor" (Baker 2000). In March 1996, more than 10,000 farmers from all over the country rallied on the street in front of the parliament. Among the demands the assembly made, the most vital was concerned with rights to forest and land use. There were numerous violent conflicts between the RFD and local people over the right to stay on public land, land on which people had often resided prior to its designation as permanent forest. Farmers seldom had official documentation to prove their residence, however, and therefore faced the threat of eviction from their homeland.

Although 42 percent of land in Thailand is at present designated as forest reserve, most of that area is already occupied by farmers. There is farming even in areas where protection activities are stricter, such as national parks and wildlife sanctuaries. How to deal with people inside RFD territory is one of the most pressing issues, not only for the RFD, but also for the government of Thailand.

---

[3] One should also note that forest policy is one of the least studied areas, with no dedicated academic staff even in the forestry school at Kasetsart University, the only institution offering higher degrees in forestry.

2.2 Levels of "Policy"

"Policies" in this discussion are interpreted as "ways of exerting power, of getting people to do what they otherwise might not do" (Stone 1997). Therefore, policy, by definition, is not neutral; it makes a claim to rearrange power and authority over certain resources, and always results in an unequal distribution of costs and benefits. Official government policies often take the form of "documents" that must be interpreted and then translated into action. At the highest policy level, there is the Constitution, and below it there are several royal decrees and laws that define legal and illegal activities related to forests. The multiple layers of policies are not always consistent with one another, thus causing confusion.[4]

At present, six royal decrees (*prarachabanyat*) govern forest administration. Cabinet resolutions are numerous and can be revoked when a new cabinet takes over from the old one. Although the resolutions have no legally binding power, they often carry an authority that is equivalent to law. A well-known one concerning forestry was the Wang Nam Kiaw cabinet resolution of 1997, which permitted farmers to live inside protected areas, but only when they could prove prior residence. Because the method of proof was not clearly defined, the Chuan administration revoked this resolution in the following year.

The formation of cabinet resolutions normally follows decisions of the National Forestry Policy Committee (*Kanakamakaan Nayobaai Paamai Hengchaat*), held bi-monthly, and composed of representatives of forest-related departments in the government.[5] This committee was formed in 1985 when the government issued the National Forestry Policy that continues to serve as the basis for forestry development and conservation. It states that 40 percent of the country's forests should be conserved and that 25 percent will be strict conservation forests, while the remaining 15 percent will serve as forests for economic production. The object of the bi-monthly meetings of the committee is, therefore, to discuss the detailed steps toward this goal and to overcome obstacles. The agenda of the committee is normally provided by the director general (DG) of the RFD, and the secretariat is provided by the RFD's planning division.[6]

The DG of the RFD is also the source of less formal, but often more influential, policies. The present DG, for example, published a booklet for officials in the department entitled *Policies 101 Points* immediately after his appointment as DG in 1998. The duration of the appointments of DGs has varied from a few months to three

---

[4] Some experts critical of forest policies in Thailand claim that cabinet resolutions are intentionally framed to be broad and vague so that enough room remains for the power élites to exercise influence in their interests.

[5] The committee is chaired by the deputy minister of agriculture and the secretariat is the RFD (Planning Division). The committee is composed of representatives of departments and academics. The departments include the Department of Land, the Department of Land Development, the Agricultural Land Reform Office, and the Department of Local Administration.

[6] In addition, the annual National Forestry Meeting primarily focuses on research, rather than policy.

years. Upon completing his appointment, a DG often proceeds to the rank of permanent secretary-general of the Ministry of Agriculture, depending on age and political popularity with the government. Because the DG holds enormous power, especially with regard to promotions, informal policies originating with the DG play an influential role inside the RFD.[7]

Sub-legal policies such as these, geared to the operational level, together with their frequent ambiguity, can serve to resolve inherent contradictions among the higher-level policies. They provide the "working principles" for frontline officials. Before moving on to the analysis of information flow, let us step back a little to take in a wider picture of land administration in Thailand.

2.3 Classification of Forest Lands

One important dimension that makes the study of policy in Thailand so difficult is the complexity of land and forest classification. The roots of this complexity can be traced to the definition of "forests" in Thai law. According to the Forest Law of 1941, still in effect today: "Forests are pieces of land that are yet to be occupied by individuals according to land law." Since only about 20 percent of farmers have official property rights, this allowed the RFD to claim the majority of the land as under its control. Actual forest (tree) cover has nothing to do with the definition.

Following the logging ban in 1989, the RFD initiated a zoning project for all forests inside the forest reserves to identify potential areas for plantation, degraded forests that could be given to landless farmers (economic zone [E]), land suitable for agriculture (A), and land that should be strictly under state protection (conservation [C]).

Policies on zoning of forest therefore have significant implications for land use and the welfare of local people in rural areas, and this is the central reason why, when we look at forest policies, we must examine other land-use policies at the same time.

Among the categories of forest land, wildlife sanctuaries, national parks, and forest reserves each have their own legislation, other areas are protected by cabinet resolutions or ministerial and departmental regulations. This not only gives national parks and wildlife sanctuaries higher status, it also privileges the RFD who have *direct* control over these lands with little interference from provincial governments.

Figure 2 illustrates the anatomy of forest land classification in Thailand. In 1961, the area of permanent forest (the outer boundary, 50 percent of total land) was demarcated based on the policies established by the Land Classification Committee in 1957. To accelerate legalization of this area as forest, the National Forest Reserve Act was promulgated in 1964. Because of the time lag between the two, however, the area covered by this law had become much smaller than in the original plan. It should be noted also that some forest was left outside the permanent forest boundary. This forest category includes private forests and forest on public land under the control of the Land Department. There is also an area of dual authority (depicted by the gray arrows in the figure) where wildlife sanctuaries and national parks overlap with forest reserves.

[7] The DG's policy is clearly stated in the annual report of the RFD.

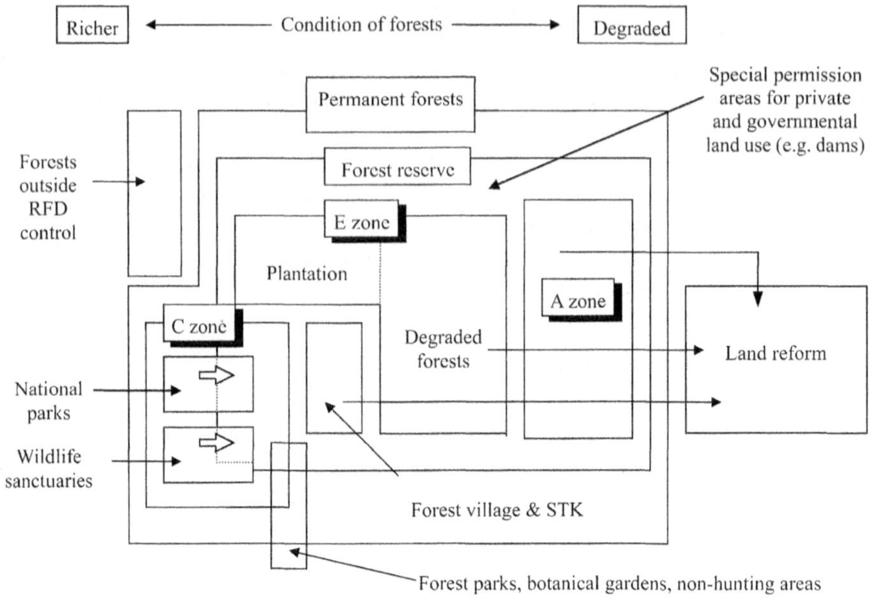

Figure 2: Anatomy of forest land classification in Thailand.

The basic strategy of the RFD is to secure the maximum area of land under the strictest regulations (the laws on wildlife sanctuaries and national parks); but if it finds a substantial number of villagers in such areas, it will downgrade the status of the land to forest reserve. The final lines of the RFD's defense of forest are the C-zones, which will eventually be protected by closure to human settlement. By contrast, parts of the E-zones are to be privatized and released for commercial tree planting.[8] The rapid growth of C-zones since the late 1980s was intended to protect such territory ahead of demographic decentralization, which might give farmers stronger claims to encroach on reserved state lands.

3. INFORMATION

3.1 Types of Information: District Officer Level

Behind any policy decision, there must be an informational base. Information may alert policymakers to a problem to be solved, it may help to identify the right means of tackling the problem, and it may provide feedback as to the success of a policy. But not

[8] Some NGOs claim that the government's hidden agenda in privatizing the E zone and expanding the land reform area is to allow already wealthy sections of society to legally grab "unused" land (IUCN 1996).

all the information available is utilized in policymaking. Some information is collected as a routine procedure without it having any potential for direct influence on policy formulation. The initial task of this study is, therefore, to classify available information as relevant and non-relevant to policymaking.

The primary agents of information collection at the frontline of the RFD are the district forestry officers (*Paamai Ampeu*). They are ground-level officers who work closely with villagers and forests that lie outside the protected areas. One or two officers are often responsible for 15 to 30 villages. A district normally consists of five to ten sub-districts (*thambon*). There are 769 districts in Thailand, but there are only 540 forestry offices (*paamia ampheu*), since not all districts have forest lands within their administrative boundaries (Hashimoto 1998).

By regulation, the official duties and responsibilities of the district forest officer are as follows:

1. Give permission to outsiders for the conduct of research on forestry products, in keeping with forest legislation.
2. Encourage and develop the conservation of forest resources and forest animals.
3. Collaborate with other government agencies on given assignments.

The frontline officers, in practice, occupy the most important position in regard to access to information, since they are located in the field close to people and the forest, thus exposed to the changes in the field. They often have informal contacts with local people and gather information about their forest activities. There are primarily four ways in which frontline officials collect village-level information: (1) informal information gathering from daily patrols; (2) villagers requesting dispute mediation at the district office; (3) informal reporting by village leaders (school teachers, monks, etc.); and (4) formal data collection as determined by orders from the top (often relating to forest land demarcation and property claims).

Personnel transfer policy is another element in the success of information accumulation. Since the time when communist insurgency was a major national concern, frequent transfer of personnel (normally every four to five years) became customary to avoid co-optation between local officers and farmers, but mobility prevents field officers from acquiring intimate knowledge of their areas. Because forest management work by its nature requires consistent observation of slow-growing forest over many years, personnel policy can be critical. A policy of frequent transfer means that frontline officers are hampered in both forest observation and, perhaps more importantly, in developing a sound rapport with the villagers who can provide them with local information.

3.2 Hierarchy of Reporting and Levels of Discretion

In the monthly reports to the provincial forest office, the district forest officer supplies information regarding the quantitative *achievement* of (1) public extension service activities (e.g., distribution of leaflets, organization of seminars and training workshop, etc.), (2) organization of study tour and visits, (3) distribution of information other than the above, (4) distribution of seedlings, and (5) tree plantation. In addition to providing information in these pre-established categories, the report pro forma contains a section for listing difficulties and complaints that arise. This section is often used to notify upper-level offices of inadequate budget and equipment, but these complaints are seldom responded to.

In addition to the ordinary monthly report, district forest officers furnish occasional "special reports," which usually deal with farmers' encroachment on public land. These special reports are usually addressed to the district chief (*naai ampeu*), and thence to the provincial governor. They are mostly written *in response* to previous orders from the top, and are seldom drafted on the initiative of the officials themselves.

Although the district officers are positioned at the lowest level in the hierarchy, they are the initial contact point for the public in matters relating to policy, and may exert considerable practical influence through granting "permission" to use forest reserves. They are able to grant "permission" for such important matters as factory establishment, mining projects, and plantation projects.

But despite their intimate knowledge of villagers and forest conditions, the rich information that frontline officers have is seldom utilized to inform policy. There is no format for conveying their ideas, and not all incidents they handle are recorded in reports. District officers are there to implement policies initiated at the top; they should not come up with new ideas.

Use of the speed-of-response requirement, measured by the classification of "Urgent" stamped on the tops of letters requesting information, is an indication of their position. There are four categories of urgency: most urgent (*duan thiisut*), very urgent (*duan maak*), urgent (*duan*), and normal letters, which are not stamped. The urgency level does not necessarily correspond to the importance of the matter, but nevertheless is a measure of RFD prioritization of information. In general, the information classified as "most urgent" relates to land encroachment, forest fires, illegal logging and hunting, and tenurial disputes.[9] It is significant that the urgency ranking of letters is determined by the top officials of departments and is directed "downstream," and that frontline officers do not have the authority to classify letters directed "upstream." As a result, matters arising locally rarely climb the ladder through the official channels of reporting.

The formal data provided by district officers is collected at the provincial level, where the provincial forestry officer summarizes it for reporting to the Bangkok central office. Reports on the use of budget are made on a seasonal basis. The level of discretion exercised at the provincial level is quite low compared to that at the district officer level. This is because the heads of provincial forestry offices must accommodate the policies of both the provincial governor and the DG of the RFD in Bangkok, and sometimes must deal with conflicts between the two.[10]

---

[9] There are also categories of letters, based on their "secrecy." "Classified" letters are often those related to sensitive international issues around the border, personal penalties, and promotion related matters (Virawat, 20 December 2001, personal interview).

[10] The policies of the governor of the province often tend to show concern for the welfare needs of the people in the province, whereas the policies of the DG of the RFD more often prioritize conservation above public welfare. Tourism is often supported by governors, yet rejected by the RFD. There are examples of the governor rejecting development projects within the forest land,

Bureaucratic decisions follow the command line, as depicted in Figure 3. Normally, the monthly report follows the arrows in the figure. Note that two arrows extend from the district forest officer. Information related to land and property must be addressed to the district chief in the Ministry of the Interior command line, as well as to the RFD.

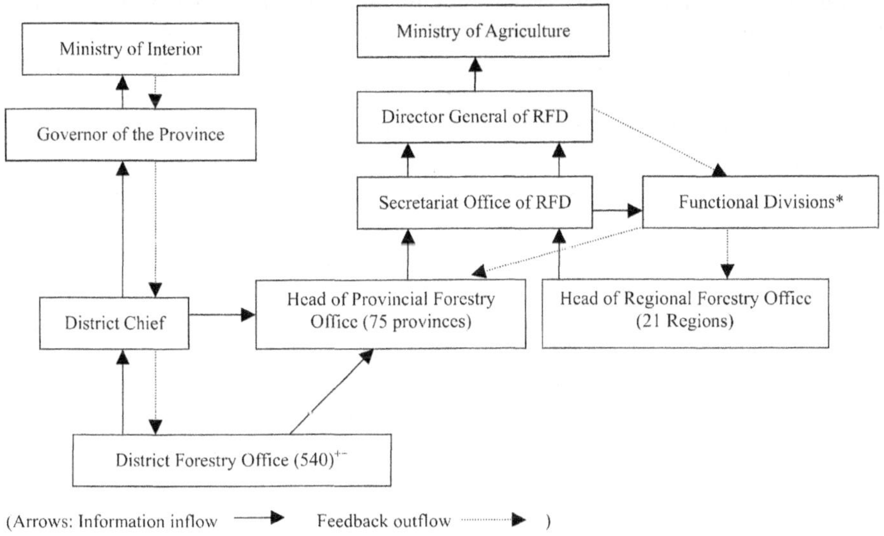

Figure 3: Command line: Ministry of the Interior and the RFD.

* "Functional divisions" are offices and divisions within the RFD.
** Monthly salary of district forestry officers is paid by the RFD.

Despite the impression given by this diagram, most information collected in the field by the frontline forestry officers is *not* utilized in policymaking. Forestry officers implement policies formulated at the top, but the information they supply seldom goes to *formulate* policies.

3.3 Feedback Mechanisms

If policies are *not* based on this flow of information from the field, what then is the basis of policy formulation? The answer is that policies *are* based on information from the field, but on information collected via different channels and for different reasons. The ordinary data collected by the frontline officers are not used directly in policymaking. Their reports are merely records of activities, bureaucratically designed to prove that officers are fulfilling their daily obligations.

however, such as the KEMO project in Kanchanaburii province. This was a mining project which the RFD had already approved, and which the governor rejected in response to the increasing pressure of public protest against it.

Feedback derived from routine reporting is limited. The case study in the province of Uthaithani showed that only problems relating to protected areas (i.e., national parks and wildlife sanctuaries), such as large-scale forest fires and legal matters, received policy feedback from Bangkok. In other cases, the reporting is one-way and there is no response or evaluation. Matters which require the signature of the DG provide an indication of information prioritization within the RFD. Only matters concerning changes in law, general policies, and promotions or transfers must be signed personally by the DG, whereas other issues can be authorized by the deputy DGs. The specifics of this division of labor change with the preferences of the DG in place.

Information used in policymaking is primarily collected by special order from provincial forestry offices and from Bangkok, triggered by a media report of an incident related to forest lands and resources. Information from the media reaches the top decision-makers much faster than via the ordinary channels, and orders to verify such information at the field level proceed faster than usual.

This is especially so when scandalous events, involving high-ranking people (e.g., politicians), are disclosed by the media. It is very difficult for the RFD to take no action if there is public pressure to do so. Of course, the RFD can pretend it is dealing with a problem according to public expectations, when in fact it is not. When the media reported the construction of a resort inside the Taplan National Park in Prachin Buri province, the RFD reportedly reacted by implementing a new demarcation to separate the resort from the national park, so as to clear the legal hurdles (*Bangkok Post* 2000).

Media reporting plays an important role in alerting high-ranking officials and politicians, who are normally insensitive to information from the field, as to issues of policy. It was reported to us that farmers, aware of the insensitivity of the bureaucracy, strategically use the media to inform and pressure top officials, and to place them in a position in which they had to act. The exact nature of farmers' use of the media, however, requires further investigation.

4. DECISION SITUATIONS

4.1 Three Types of Situations

In relation to policymaking, Fukui (1974) classified decision situations into three categories: normal, urgent, and critical. The distinctions between the categories are based on the criteria of "predictability" and "time available to decide." If there is sufficient predictability and time for decision, the situation is classified as "normal." If either predictability or time available is constrained, the situation is classified as "urgent." If both factors are constrained, then the decision situation is "critical."

From the bureaucratic decision-making point of view, as the situation moves from the normal to the critical, it will be more difficult to apply the procedures detailed in the operational manual or the standard operating procedure, which normally work well. The significance of Fukui's classification of decision situations is that government operations are often structured to follow a standard operating procedure, which directs the manner in which decisions should be made. It is when the manual is not applicable that bureaucratic failures tend to manifest themselves.

This typology can be applied to the present study, but for forestry policy there are extra complexities. Influential decisions, affecting large areas and many people, must be made in Bangkok, which is located far from the forests and from local people. Furthermore, in the RFD a manual for decision-making does not, in fact, exist. The set of forestry-related laws provides the only guidance, but the laws are not coherent and contain many contradictions. It therefore falls to the frontline officers to interpret them in a coherent way at the ground level. For this reason, the organizational *culture* within which these officers work is a crucial guide in their interpretation of policy and policy judgments.

4.2 Examples of Decision Situations

Several events in the recent history of forest policy in Thailand can illustrate the nature of "critical" and "urgent" decisions. The most recent incident I wish to take as an example is the unusually speedy decision of the government to ban commercial logging in January 1989, which represented a "critical" decision situation. The ban was triggered by the tragic loss of 300 lives in a flooding disaster in the southern part of Thailand (Project for Ecological Recovery 1992). The flooding was reportedly caused by excessive logging in the area.[11] Decision-making was "critical" because the tragic incident was unpredictable and the accumulated pressure of public opinion against the logging business allowed little time for consideration. The unusually high speed of decision, by cabinet resolution, makes manifest the seriousness of the issue. The speed of decision can be explained by several factors. First, frustration and protest had accumulated against the growth of destructive logging operations in various parts of Thailand by the year 1988. Second, an especially high rate of deforestation in the 1970s and 1980s had almost exhausted the timber supply potential, so that a change of governmental attitude toward logging was imminent. Third, the already mentioned flooding in Nakorn Sithamraat province occurred at the "right time" to conjoin all three forces, thus determining a swift decision. It is also worth mentioning that the increasing cost of replantation at the logging site was becoming a serious burden for the logging companies. The ban took away their resources, yet it also freed them from paying the recovery cost.

An example of an urgent decision-making situation was the series of street demonstrations by the "Assembly of the Poor," mentioned earlier. Because it attracted nation-wide public attention, it could not be ignored by politicians. Nevertheless, the government had some time to negotiate. Several months were spent in negotiation over the demands, with new cabinets often reneging on agreements made by the previous administration. During the period of negotiation and tentative resolution, investigations were ordered to obtain facts from the field, especially with regard to property rights and the critical issue of proof of residence prior to nationalization. During negotiation,

[11] There was, in fact, a debate as to whether or not the flooding was caused by excessive logging.

information from the field became relevant to the formation of policy. Finally in September 2001, the Assembly won a series of compromises from the Taksin government and dissolved after four years of struggle. This does not mean final resolution. Cabinet resolutions have, historically, been dishonored by new cabinets. A legal act, such as a community forestry bill, is necessary to guarantee the long-term rights of farmers living in forest areas.

Finally, the ten-year debate over the community forestry bill can be taken as an example of "normal" procedure in bureaucratic decision-making. The community forestry bill will legalize the customary uses of state forests practiced over a long period, if registered properly. The main debate had been over whether to allow local people to register community forests *inside* protected areas, and the inclusion of the possibility of logging naturally regenerated forests (Makarabhirom 2000). The chief factor delaying enactment has been the RFD's resistance to allowing villagers access to forests inside the protected areas. The environmentalist NGOs joined forces with the RFD to support the conservative version of the draft bill, which restricted villagers' rights to establish community forests inside protected areas.

Because dramatic events like the flooding incident were not involved, nor were there high stakes involved from a political viewpoint, there was no social or political pressure to "speed up" the process, and this allowed the bureaucratic culture of the RFD to play a central role in shaping the community forestry policy (Table 1). Despite the evidence accumulated by NGOs and academics of sound forest management by villagers, mainstream RFD officials were immovable. Information from the field, therefore, played a minimal role in this policy formulation.

Table 1: Relative speed and influence of factors involved in decisions.

Decision Situation	Politics	Bureaucratic Culture	Information from Field
Critical	●	○	□
Urgent	○	□	○
Normal	□	●	□

Key: ● influential and determines the direction, ○ influential but not directional, □ not influential.

Although the empirical support provided for the above analysis is not yet comprehensive, it is fairly safe to conclude that "information from the field" has a marginal impact on the development of policies, except in emergency cases, where specific data from the field is demanded from the top, on an ad hoc basis. Information from the field is not "formative" of policies but only "supportive" of the policies that are already being formed. Data collected by local officers is such as to have the potential to inform policy in many ways, especially in regard to the forest and villager interrelationship, but is unlikely to be made use of with the present design of information flow. It should be noted again, however, that frontline officers are the gatekeepers when people apply to make use of state land. Thus, although district officers may not be able to influence national policy, they can shape local policy, important areas of which lie within their discretion.

So where does policy come from? It comes from a combination of factors determined by the nature of the problem, the political environment, the bureaucratic culture, pressure from the mass media, and information from the field. It would be a mistake to suggest that one factor or actor is exclusively responsible for the ultimate character of policy decisions.

5. CONCLUSION AND IMPLICATIONS

This discussion has argued that (1) district forest officers (*paamai ampeu*), though placed lowest in the organizational hierarchy, possess extensive discretion in policy applications at the field level, (2) information from field officers, however, has a marginal impact on policy formulation, and (3) the utilization of field information in policy development is a function of political context, which is strongly influenced by the mass media. Although this discussion has primarily focused on frontline officers, policy discretion at the provincial level is another area to be examined. Since not all information will be reported to Bangkok, there is likely to be some opportunity for autonomy in the interpretation of policy at this level as well. The key to this exercise of autonomy is, again, an ambiguous relationship with the command line of the Ministry of the Interior and the provincial governor.

If non-stipulated discretion plays a large role both at the field level and the level of the DG in policy decisions, then policy analysis should shift its focus from published documents to the organizational culture within which decision-makers work. The organizational culture of RFD bureaucrats is firmly rooted in the Forestry Faculty at Kasetsart University, which supplies the large majority of officials at the RFD (90 percent of executives are graduates). This concentration of influence is quite unusual even in Thailand, where competition between universities (such as that between Chulalongkorn and Tammasart Universities) is common. The culture of a "pure-blooded" department is extremely difficult to change, as was demonstrated when the present DG, who is a graduate of the Faculty of Fishery, was resisted by many high-ranking officials. With the increasing speed of environmental change, this intractable element of the organizational culture is becoming a problem, i.e., the problem of organizational inability to inform policies based on field data and the inability to adapt policies to situational variations.

Explanation of the causes of biodiversity loss under the administration of the RFD will necessarily have two perspectives: as a physical system and as a product of socio-economic forces. Although both these perspectives are required for an understanding of conservation of biodiversity, only the first has usually been considered seriously. I have tried to demonstrate that entirely non-local factors, often categorized as "socio-economic," are important determinants of the local physical interaction between people and forests. This view is supported by evidence that field-based, bottom-up information seldom influences policy directions. Although this is difficult to prove empirically, our observational experience is that prescriptions based *exclusively* on physical local factors will not satisfactorily solve any problem.

To reiterate a point made at the beginning of this chapter, the interaction of forestry and land policies is so far a surprisingly under-studied subject, despite its societal importance. There are therefore numerous promising areas for future research. The most important agenda, which I wish to initiate in the near future, is the identification of the triggering mechanisms of bureaucratic learning in the context of natural resource management. Simply put, how do bureaucrats learn and change? This type of research should inform those, in turn, who are striving to find out the most effective way for local people to participate in the decision-making process of how natural resources should be used.

REFERENCES

Allison, G. 1971. *Essence of decision.* Harvard University Press.

Baker, C. 2000. Thailand's assembly of the poor: Background, drama, reaction. *South East Asia Research* 8(1): 5–29.

Bangkok Post. 2000. Who will nip the forestry bullies? *Bangkok Post*, December 7.

Feldman, M. 1989. *Order without design: Information production and policy making.* Palo Alto: Stanford Univ. Pr.

Fukui, H. 1974. Negotiation over the reversion of Okinawa: Policy making process in the Japanese government (in Japanese). *International Politics* 52.

Hashimoto, S. 1998. Thailand (in Japanese). In *Local governments in Asia,* ed. A. Morita. University of Tokyo Press.

Kaufman, H. 1960. *The forest ranger: A study in administrative behavior.* Baltimore: Johns Hopkins University Press.

Makarabhirom, P. 2000. The evolution of the policy making process: will there ever be a community forestry bill? *Asia-Pacific Community Forestry* 13(2): 58–62.

Project for Ecological Recovery. 1992. *The future of people and forests in Thailand after the logging ban.* Bangkok: Project for Ecological Recovery.

Riggs, F. 1966. *Thailand: A modernization of a bureaucratic polity.* Honolulu: East-West Center Press.

Royal Forest Department 1997. *Forest Statistics of Thailand.* Bangkok: Royal Forest Department.

Sato, J. 2000. People in between: Conversion and conservation of forest lands in Thailand. *Development and Change* 31:155–77.

Sayamon, K. 1995. *Panhaa tangkotmai lae nayobaai khong kaanchai amnaatrat thi kiyokap kaan thiidinnai kheet paa* (Problems of law and policy in the use of power inside forest land). Master's thesis submitted to the Faculty of Law, Chulalongkorn University (Bangkok, Thailand).

Simon, H. 1958. *Administrative behavior.* New York: The Macmillan Company.

Stone, D. 1997. *Policy paradox: The art of political decision making.* New York: Norton.

Tamada, H. 1992. Modernization of bureaucracy: A case of Thailand (in Japanese). In *Politics of Southeast Asia*, ed. T. Yano. Tokyo: Koubundo.

Wilson, J. 1989. *Bureaucracy: what government agencies do and why they do it.* New York: Basic Books.

Yoshida, M. 1985. Bureaucracy in Thailand. In *Bureaucracy and economic development in developing countries,* ed. M. Ohuchi and E. Shimoyama. Research series No. 328. Tokyo: Institute of Developing Economies.

IUCN (The World Conservation Union). 1996. *A Review of the Thai Forestry Sector Master Plan*. Geneva: IUCN.

THAI GOVERNMENT DOCUMENTS

Monthly Report addressed to the Uthaithani Provincial Forester (August 2001) (in Thai).

Monthly Report addressed to the Governor of Uthaithani (August 2001) (in Thai).

Chapter 12

THE DEVELOPMENT OF FOREST POLICY IN THE RUSSIAN FEDERATION—WITH A FOCUS ON KHABAROVSK KRAI

Alexander Sheingauz
Russian Academy of Science, Far Eastern Branch, Economic Research Institute

Hiroaki Kakizawa
Hokkaido University, Graduate School of Agriculture

1. INTRODUCTION

Russian forests are of global importance, both for timber resources and for their ecological value. Russia holds the largest forest resource in the world, consisting of one-fifth of global forest stocks. It is often claimed that Russian forests are the last "wood basket" and have great potential in the world timber market. Russian forests also hold great significance for biodiversity conservation and have been recently recognized as an important sink for greenhouse gases.

The Russian Far East region's share of rich forest resources is about 42 percent of Russia's total forest area. As the region is adjacent to China, where timber demand has rapidly increased, its role as timber supplier in Northeast Asia will likely continue to grow. The Far East also contains important boreal forest ecosystems, especially the southern taiga, famous for its unique features and significant proportion of virgin forests.

But because of political, social, and economic confusion, forest management has become unstable and effectively insufficient, and it is often pointed out that forest conservation is in a critical condition. To consider future directions of sustainable forest management, it is necessary to analyze the development and current state of forest policy and the logic behind recent changes.

The former U.S.S.R. had a forest legislation system conforming to the totalitarian state, but the new political and economic structure of the Russian Federation demanded

M. Inoue and H. Isozaki (eds.), *People and Forest—Policy and Local Reality in Southeast Asia, the Russian Far East and Japan, 187–200.*

a new forest legislative framework. During the period of *perestroika* and reforms (after 1985), the main forest act was changed twice (in 1993 and 1997). Many other forest acts were developed or transformed, and local forest acts arose as a new phenomenon. But, according to expert opinion, the current overall federal legislative framework is inefficient. Some local governments, including that of the Khabarovsk krai, tried to develop a local framework to improve the situation. Efremov et al. (1999) argued that although the local legislative framework of Khabarovsk krai had become more developed than the federal framework, conflict over resource management issues between the federal and local governments has caused uncertainty. This problem has become serious, as Russia's president, Vladimir Putin, has begun to establish a strong central government in order to restore political and economic order.

In this chapter, the development of forest policy in the Russian Federation, the logic behind the changes, and its influence on the forest sector will be analyzed. Analysis of the local framework will focus on Khabarovsk krai, because it has one of the most advanced and well-organized forest policies in the federation and it is also the center of forestry and the forest industry in the Russian Far East. The period covered by this study stretches from 1985, the year *perestroika* started, to the end of 2000. The period studied will be divided into four sub-periods: before the collapse of the U.S.S.R., when some institutional changes began (1985 to the end of 1991); from 1992 to 1996, when the initial institutional transformation from a totalitarian system to a market economy was implemented; from 1997 to 1999, when the institutional framework was developed under the decentralized regime; and after Putin became president in 2000, when administrative reforms were implemented.

2. FOREST POLICY BEFORE THE U.S.S.R.'S COLLAPSE

The first forest law that covered the forest management of the entire former Soviet Union (U.S.S.R.) was the Basic Forest Law of the U.S.S.R. and Union Republics of 1977. Forest codes in each republic came after this. These laws did not entail any fundamental changes to the management practices that had evolved over time, as a consequence of normative actions that defined the hierarchical system (Carlsson et al. 1998; Sheingauz et al. 1995). Formulation of the fundamental direction of forest policy and management was attributed to the Supreme Soviet, while utilization, regeneration, control, and protection of forests was in the hands of the U.S.S.R. State Committee of Forestry and its subordinated organizations,[1] the local branches of committees and *leskhozes*.[2]

After *perestroika* began, the former U.S.S.R. introduced several resolutions to adopt the market economy and environmental protection. For example, in 1988 the Resolution of the Communist Party Central Committee and the U.S.S.R. government (the usual

[1] In terms of the use of forest resources, the Ministry of Forest Industry and the Pulp and Paper Industry held large influence in the decision-making process.

[2] *Leskhoz* is the Russian name for fundamental forest management organizations at the field level.

form of decisions at that time) introduced the forest lease system, analogous to the tree farm license system of the province of British Columbia, Canada. Another, the resolution devoted to Nature Protection in 1989, required environmental controls on forest utilization. But these resolutions contradicted each other and could not provide a clear legislative system. In the case of the forest lease system, its introduction caused serious confusion at the field level and was cancelled in 1991 (Kakizawa 1997). Nonetheless, they were important, as they formed the platform of the new forest legislation system of the Russian Federation (Sheingauz et al. 1995).

Another important point following *perestroika* was that many local governments started to introduce their own forest legislation. With the failure of economic reform, the political power of the U.S.S.R. had weakened, and local governments sought to expand their political independence. Khabarovsk krai was one of the leading local governments in this respect, and it introduced the Temporary Rules of Forest Use in Khabarovsk krai in 1992 (Sheingauz et al. 1995). It introduced a forest lease system based on competition, and was generally credited with forming the basis for elements of the Fundamental Forest Law of the Russian Federation of 1993 (Mabel 2000).

3. INITIAL DEVELOPMENT OF THE INSTITUTIONAL FRAMEWORK OF THE FOREST SECTOR

3.1 Federal Level

The U.S.S.R. collapsed in 1991 and the Russian Federation became an independent state. The State Committee of Forestry was abolished at the end of 1991, and the Forest Service was organized to replace it in 1992.[3] The new constitution was enacted in 1993, and it assigned the issues related to natural resources to joint jurisdiction between federal and local governments.[4] This article clearly stated that local government is not subordinated to the federal government, but has certain authority in the field of forest policy. There is no clear definition of "joint jurisdiction," however, which made the roles of federal and local governments ambiguous.

As mentioned above, when local governments started introducing their own forest legislative systems, the federal government was urgently pressed to introduce new federal forest laws. In 1993, the Fundamental Forest Law of the Russian Federation (hereafter, the Fundamental Law) was enacted. It was largely a mixture of old U.S.S.R.

---

[3] Between 1991 and the end of 1992, the Committee of Ecology and Natural Resources was responsible for forestry issues.

[4] In this chapter, local government means the political entities which make up the Russian Federation. These entities include republics, krai, states, and autonomous states. Khabarovsk krai is one of these entities. A municipality is a self-governing body under the local government.

laws along with new rules concerning the market economy (Sheingauz 2000). The hierarchical structure defined by the Fundamental Law was basically the same as under the old U.S.S.R. system; that is, the federal government set the framework of forest policy, local governments formulate concrete forest policy, and municipalities assign the rights to use forests and collect user fees. Municipalities possess relatively large powers concerning the control of forest use.

One of the main features of the Fundamental Law was the introduction of the lease system. The right to use forests was granted by two types of lease: the long-term lease (from one to 50 years) and the short-term lease (less than one year). The Fundamental Law stipulated that the assignment of leases should be carried out through competition, in line with the market economy; but it also left loopholes allowing lease agreements through direct negotiation between the forest user and the authorities. As the authority to distribute licenses was attributed to municipalities, implementation of the lease system will be discussed in next section.

The Fundamental Law also had drawbacks, such as the fact that some provisions were unclear and contradicted other laws (Sheingauz et al. 1995), but the most important one was that it did not clarify ownership of the forests. Many of the local governments, including Khabarovsk krai, claimed that they held the rights to their forests, and the controversy continued between federal and local governments (Kakizawa 1997).

3.2 Krai Level

The Russian Federation was politically unstable from the beginning of its establishment, because republics like Tatarstan, Chechen, etc., insisted on one degree or another of independence from Russia. President Yeltsin, also in an unstable political position, faced struggles with conservative groups. For these reasons, he tried to get support from local governments to strengthen his political position by promising more authority to them (Hakamada 2000). More than 30 local governments, including that of Khabarovsk krai, concluded an agreement with the federal government that gave more authority to local government (Nakai 2000). Under these circumstances, many of them developed their own forest policies.

In 1994, the Khabarovsk krai administration established the Krai Commission on Forest Use (hereafter, the Commission), and it became the main organization to make the final decision on lease agreements. The Commission was chaired by the deputy governor and consisted of representatives of the krai administration, the krai branch of the Forest Service, specialists, and so on. Lease conditions were set by the Commission, including the obligation for regeneration, and it examined not only the tender price but also the applicant's ability to carry out forestry operations. As mentioned before, the fundamental authority to assign forest use was attributed to municipalities by the Fundamental Law, but it also gave authority to local governments to assign long-term leases. The Khabarovsk krai administration used this clause fully, and the Commission actually handled all the long-term leases, which made up the majority of forest use. The aspiration to arrange the majority of forest use as long-term leases (25 years and more) is unique to Khabarovsk krai among other Russian local governments until now.

The krai administration also issued a decree on leasing and payments for forest use in 1995, in which it mandated that 25 percent of forest user fees should be retained for the krai budget, instead of the whole fee going to municipalities as directed by the Fundamental Law. Through this reallocation of user fees, the krai administration secured financial resources to invest in forest regeneration and restoration, which were mandated by the Fundamental Law.

4. DEVELOPMENT OF FOREST POLICY UNDER THE DECENTRALIZATION REGIME

4.1 Federal Level

The Fundamental Law of 1993, which was incomplete and included contradictions with existing legislation, was replaced by the Forest Code of the Russian Federation of 1997 (hereafter, the Federal Code).

Compared with the Fundamental Law, the Federal Code had certain distinct characteristics. First, it defined all forests as being owned by the federal government. But because of strong claims from local governments to the rights to their own forests, the code also allowed that parts of forests could be transferred under agreement between the two levels of government (Kakizawa 1997). Second, the Federal Code gave broad authority to the local governments, including the authority to assign leases and the right to keep 60 percent of forest user fees. This meant that while local governments were given wide authority on forest policy and management, municipalities were deprived of almost all of it, a reversal of that stipulated by the Fundamental Law. Third, the Federal Code defined more sophisticated conditions for forest leases, and emphasized that the leases should be granted through competition.

Although the Federal Code was an improvement compared to the Fundamental Law, it still had many defects, starting with the definition of terms and concepts, and including the fact that the competence and authority of each organization were not defined clearly. The monopoly of the forests by the federal government also contradicted the constitution, which allowed various types of property rights for natural resources. It also lacked coordination with other natural resources legislation, and ecological provisions were weak (Sheingauz 1999).

4.2 Krai Level

Under the Federal Code, which gave strong authority to local governments, the Khabarovsk krai administration became more active in developing its own policy framework and increasing its autonomy.

First of all, it sued the federal government through the Russian Constitutional Court, claiming that the Federal Code violated the authority of local governments in the ownership of their own forests. The Republic of Komi also brought the same kind of lawsuit against the federal government (Mabel 2000; Sheingauz 1999). Although both cases were defeated, these lawsuits indicated that local governments were becoming more independent from the federal government and claiming more autonomy in forest policy.

Second, in 1999 the krai administration enacted the Forest Code of Khabarovsk krai (hereafter, the Local Code). The Local Code was formulated by specialists and representatives from forest-related authorities in Khabarovsk krai, supported by the U.S. Agency for International Development. The characteristics of the Local Code were as follows (Sheingauz 1999):

1. it was developed through public participation;
2. it took local characteristics into account and integrated existing local institutions, including provisions of the Commission on Forests;
3. it was more ecologically oriented than the Federal Code;
4. it contained more organized definitions of forest use and placed greater stress on the competitive base for the allocation of leases; and
5. it stipulated that one-third of revenues from user fees to the krai should be distributed to municipalities to avoid municipal financial crises.

This was the first comprehensive local forest law in Russia, and many local governments with rich forest resources have followed Khabarovsk krai by introducing their own local forest legislation.[5]

Third, the Krai Commission on Forest Use became more active. First of all, the krai administration adopted the policy of canceling lease agreements when the lessees failed to pay user fees and/or repeatedly violated forest practices and environmental regulations. The commission cancelled more than 100 lease agreements between 1997 and 2000 based on this policy. The cancelled leases were redistributed to new forest users through competitive tenders. The krai administration further issued orders in 1999, which mandated that the lessee apportion a certain amount of harvested timber to be processed in the municipalities, so as to contribute to the local economy. The Commission on Forest Use began to add this requirement to newly organized bidding, and also pressured existing forest users to introduce processing facilities by threatening to cancel their leases.

The organizational structure of the forest sector in Khabarovsk krai as of April 2000 (Figure 1) had the Forest Service and its subordinated organization managing the forests. The State Committee of Environmental Protection and its subordinate organizations were responsible for monitoring the status of forest management and use. The Forest Service set the foundation of forest policy and management under the supervision of the federal government, and the krai administration developed their own forest policy within its delegated power. The krai administration established its own Department of Natural Resources and Extracting Industries, to create strategies for the

[5] New provisions of the Local Code are yet to be implemented, because Putin's reform confused the federal/local relationship, and federal forest management organizations began to resist the local forest policy under the centralized regime.

use of forest resources, and the Commission on Forest Use, to allocate forest leases to users. The krai branch of the Forest Service was also under double supervision by the local government and the Federal Forest Service.

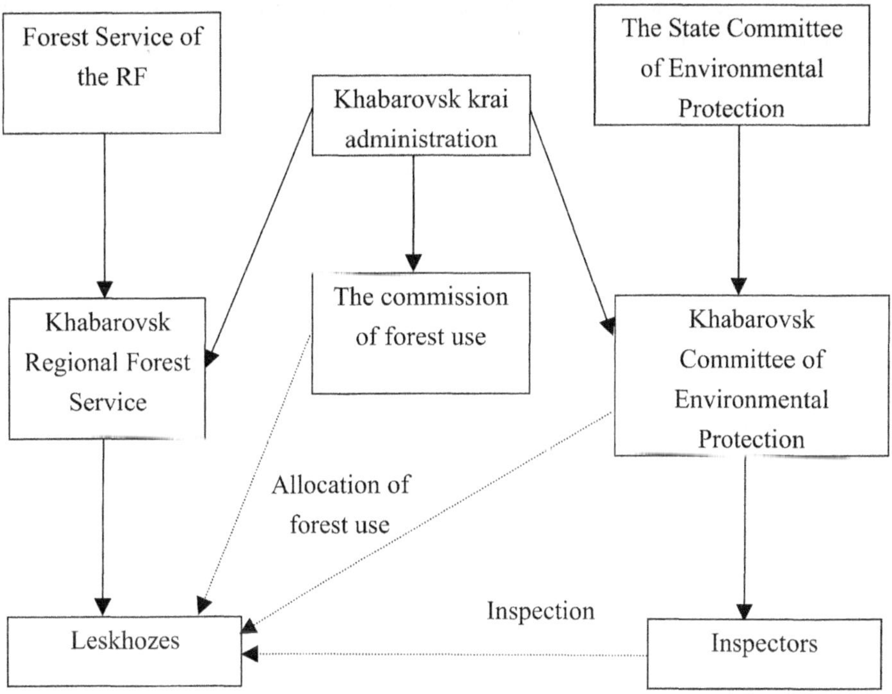

Figure 1: The organizational structure of the forest sector in Khabarovsk krai as of April 2000.

5. THE CENTRALIZED POLICY FRAMEWORK OF THE PUTIN ADMINISTRATION AND ITS EFFECTS ON LOCAL FOREST POLICY

5.1 Federal Level

The framework of decentralization changed under President Putin, who was elected in early 2000. The Putin administration considered that the major reasons for the chaotic situation in Russia were the weakness of authority of the president and corruption in the administrative system, and attempted to restore social and economic order through strong federal authority and dictatorship of the law. Thus, the focus of reform was to establish a strong, reliable federal institution and speed up the liberalization of economic policy. In this direction, the Putin administration drew up a series of policies to gain stronger political influence for the federal government. First, seven federal regional offices (*okrugs*) were established, and a representative of the president was appointed to each office to control the local governments. One of their main tasks was to match local laws to federal laws. Second, a law was approved by the State Duma (lower house of the Russian parliament) that gave the president the authority to dismiss the governor of a local government if the governor conducted illegal activities. This gave the president a strong tool to control local governments. Third, the Law of Restructuring the Federation Council (the upper house) was approved, and the system which made the governor and the chairperson of the local assembly automatically members of the upper house was cancelled. At the same time, some fundamental economic acts (i.e., Tax Code, Land Code) were put into force.

The effects of centralization have appeared in the field of natural resource management. Another policy change which struck directly at local forest policy was the abolishment of the Forest Service. Immediately after Putin was inaugurated as president, he issued a decree on administrative reform to reduce the number of administrative units and personnel, in order to improve the Federation's financial situation. The decree ordered the abolishment of the Federal Forest Service and the State Committee of Environmental Protection, and the Ministry of Natural Resources was given authority over forests and environmental issues. Because the Ministry of Natural Resources was originally established to control mineral resources and was biased towards resource development, environmental groups strongly opposed the reform (Revedov 2000). But Putin maintained the decree and the Forest Service was abolished on 15 August 2000. About two-thirds of the employees who worked at the headquarters of the Forest Service were hired on by the Ministry of Natural Resources, but many of those dismissed were specialists with higher qualifications.

For the control of natural resources as well as environmental policy and management, the Ministry of Natural Resources organized seven regional branches, mirroring the *okrug* offices, with the intention of developing more federal commitment to local forest and environmental management. The Far Eastern branch was established to control the entire Far East region, including Khabarovsk krai, and located the headquarters in Khabarovsk City. All local organizations of the Forest Service and local committees of environmental protection were abolished, and forest and environmental divisions were established under the Far Eastern branch. Even by the middle of 2002, the direction of restructuring the *leskhozes* was unclear. The local environment committees had already undergone serious downsizing, with almost 70 percent of the staff dismissed. All the environmental inspectors who worked at the municipal level were also dismissed, resulting in the abolishment of federal environmental inspections at the municipal level.

Such restructuring has had serious effects on forest management and environmental protection. As the Ministry of Natural Resources is biased in favor of resource development, the environmental aspect of forest management is slighted and the reduction of staff has weakened the forest management system. Forest fire control was especially weakened, owing to the decrease in financing. Abolishment of federal environmental inspection at the municipal level has also had serious consequences, because these had played a very important role in monitoring illegal and improper logging operations and management activities.

5.2 Krai Level

Reorganization of the natural resource management system has had the greatest impact on local governments where the headquarters of the new federal okrug*s* were placed, as in the case of Khabarovsk krai. In these local governments, the local organizations and committees of federal organs were absorbed by the Far Eastern branch of the Ministry of Natural Resources. This destroyed the management system to much more of a degree than in other local governments where local committees of natural resources were organized (Figure 2).

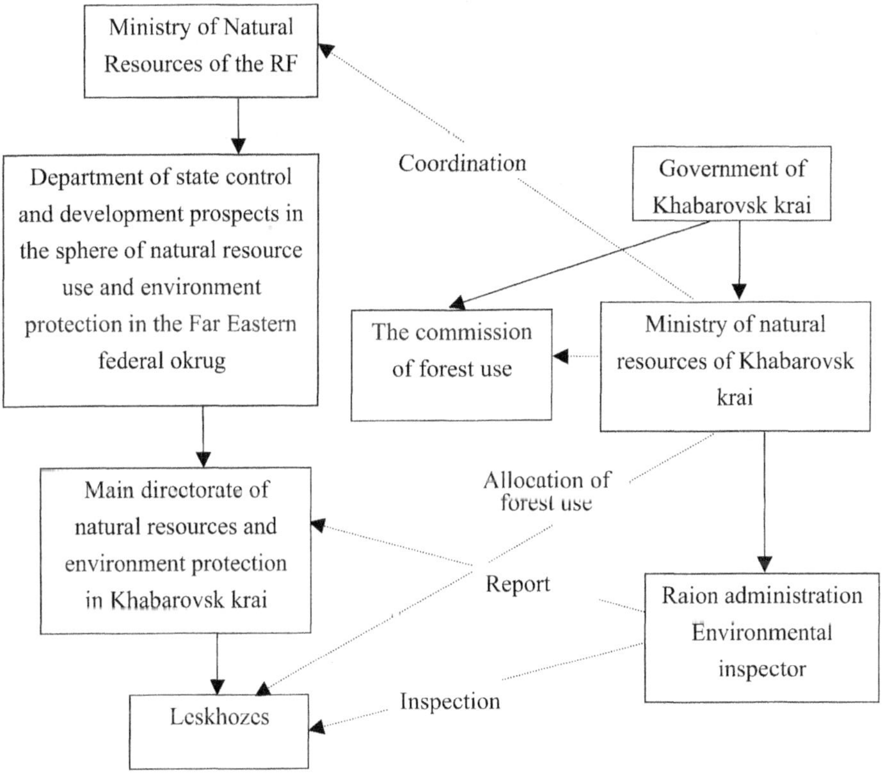

Figure 2: The organizational structure of the forest sector after Putin's reform as of April 2003.

Matching local laws to the federal ones by the okrug also made the situation worse. The Council of Khabarovsk krai was forced to exclude many specific provisions from its local forest code. The federal government even questioned the legal basis of the Commission on Forest Use, because federal legislation did not clearly give any authority to the local government to establish such a commission. The federal government started to examine the framework of local forest management and found it necessary to reconsider policy.

To counter the destruction of natural resource management at the federal level, the krai administration tried to reinforce it on its own. Two ministries were established, the Ministry of Natural Resources and the Ministry of Forest Industries, in the transformation of the krai's administration into the krai government.[6] In 2001, the Principal Orientation of the Forest Sector Development for Middle- and Long-term Outlook in the Khabarovsk krai was drafted and officially adopted by the governor (Sheingauz 2001). It includes a core chapter on forest policy and proclaims the transition to sustainable forest management as a main goal. Unfortunately, the krai authorities suffered from a lack of funds to implement these measures.

Because of this situation, the krai authorities have tried to attract not only forest sector investors but also different international programs, mostly linked with forest conservation. As of the middle of 2002, the following notable programs were working in the krai territory: the World Bank's Sustainable Forestry Pilot Project, the U.S.-Russian Forestry Resources and Technologies (FOREST) Project, the GEF (Global Environment Facility) project, the Protected Areas Network for Sikhote-Alin Mountain Forest Ecosystems Conservation in Khabarovsk krai (Russian Far East), and the WWF's (World Wide Fund for Nature) project, Ecoregion-2000.

But municipal authorities remain without power to exert real influence upon forest management in their territories.

6. DISCUSSION AND CONCLUSION

6.1 The Current State of Forest Policy in Russia

The following aspects have made forest policy in Russia a hot issue: an increase in forest mismanagement with simultaneous growth of illegal cutting and forest fires; Russia's striving to integrate with the world market economy, especially by joining the World Trade Organization and other international economic associations; insufficient contribution of the forest sector to the state budget; and increasing Russian and international public demands to improve the ecological situation by conserving forests and biodiversity.

Under these circumstances, redevelopment of the national forest policy has become an urgent issue. One can observe active discussion of this topic in specialized Russian magazines and newspapers, conferences, and workshops over the last two years

[6] In 2002 the Khabarovsk Krai Administration changed its name to the Khabarovsk Krai Government and turned its departments into ministries, with the intention of strengthening its political identity. Both of the new ministries of natural resources and the forest industry expanded their functions in comparison with their former entities. For example, the Ministry of Natural Resources established new divisions for nature protection and indigenous people, and also tried to reestablish its environmental inspection system.

(Moiseev 2001). According to a questionnaire survey of Russian forest experts conducted by the U.S.-Russian FOREST project in May 2002, 96 percent of respondents gave the first rank (out of eighteen possible ranks) to the development of official national forest policy, and 83 percent ranked development of official local forest policies as second.

This new management situation has given rise to a discrepancy between the Forest Code and current tasks, and the necessity for intensive correction of the code has become widely called for (Kukuev 2002). Many specialists even consider that it is time to develop an entirely new forest code, and that the new code should aim to fix forest policy as one of its main charters. In reality, renovation and development of forest legislation remains the main task in forest policy formulation; and since the biggest downfall of Russian forest management is the simple lack of law implementation, it must be supported by strong enforcement of acts already in place.

6.2 Outcome of Khabarovsk Krai Policy

The Khabarovsk krai is one of the few local governments in Russia that has maintained a consistent forest policy framework.

The first reason for this is that the forest sector is one of the main industries in the krai. It plays not only an economic but also a large social role, because many towns and villages depend completely on the activities of forest enterprises. The second reason is that the forest industry is one of the major export industries in Khabarovsk krai. This creates a strong incentive for the krai administration to develop forest policy to establish a strong forest sector that contributes to improving the economic situation and acquiring foreign currency (Mabel 2000). To accomplish these targets, the krai administration developed a competition-based lease system, with the intention of encouraging competitive forest industry enterprises. The third reason is that there are already good human resources in place to develop an institutional framework for the forest sector in the krai. The city of Khabarovsk has the biggest concentration of forest managers, designers, researchers, and other specialists within a vast territory east of Lake Baikal. Furthermore, there was foreign technical and financial support to develop forest institutions. These resources made it possible to formulate an advanced policy framework.

As a result, the Khabarovsk krai administration succeeded in formulating a progressive institutional framework for the forest sector. The forest lease system which has been developed not only assigns the right of forest use to users based on competition, but it also serves as a strong policy tool to control forest users. Through these controls, the krai administration has succeeded in dramatically increasing the forest user fees collected as well as reducing illegal logging. The krai administration also established a local code, which set concrete guidelines for forest institutions in the krai.

This locally dominated policy framework has a weak foundation, however. It was manifested with the shrinking authority of local governments as President Putin developed policies to establish strong federal authority. Another weakness is the limited human and financial resources of the krai administration. In 1999, employees in the krai forest sector made up 15.3 percent of those engaged in total krai industries, representing 3.3 percent of the krai economy. Revenues from the forest sector to the krai budget were 413.1 million rubles (U.S.$17 million), i.e., 7.5 percent of total budget revenues (Sheingauz 2001).

Even though the local government developed a progressive forest policy framework and a lease assigning system, the weakened federal forest management organization has become a serious obstacle to establishing sustainable forest management. Based on federal ownership of forests, the federal government has the authority to organize forest management as a federal administrative unit and has the funds for them, which made the federal government the key player in the forest institution.

The development of an institutional framework of the forest sector of Khabarovsk krai indicates that locally-based policy development has advantages for elaborating forest policy that reflects local conditions and contributes to sustainable forest use.

6.3 Future Direction of Forest Policy

It is natural for different groups to try to include different concepts that they support in the national forest policy (Moiseev 2001). On the other hand, the wide differentiation of natural and socio-economic conditions among the regions of Russia requires different approaches and solutions. This is why state national forest policy can exist only as very general formulas, such as the following: implement only rational use of resources, especially cuttings; ensure complete and complex use of forest utilities; halt illegal cutting; implement certification of forest use; and make a transition to sustainable management. In this sense, development of local forest policy frameworks is critically important.

The future direction of regional forest policy in the Russian Far East and Transbaikal (325.4 million hectares of forest, or 45.3 percent of Russia's forests) is outlined now in official documents, such as the federally targeted program, Social-Economic Development of the Far East and Transbaikal for 1996–2005 and till 2010, adopted after revision by the Russian government on 19 March 2002. It designated the followings tasks:

1. stop the degradation of forest resources and improve forest resources;
2. create reliable forest fire control systems;
3. provide reliable information on forest resources;
4. ensure the comprehensive utilization of forest resources and implementation of modern harvest technologies;
5. develop wood-processing industries to provide comprehensive utilization of timber;
6. optimize business sizes, especially with support of small business;
7. develop forest infrastructure, especially roads and communications facilities; and
8. provide state institutional support of the forest sector.

The officially adopted forest policy of Khabarovsk krai to 2010, mentioned in the previous section, consists of the following positions:

1. transition to sustainable development of the forest sector as the main goal;
2. an end to the waste of forest resources, first of all timber;
3. development of timber and non-timber products processing;
4. change of machinery and technologies according to social, ecological, and economic demands;
5. solution of the forest fire problem;
6. constant conservation of forest coverage;
7. timely reforestation;
8. provide reliable information on forest resources;
9. solve contradictions between the normative and legislative bases;
10. improve the institutional base of the forest sector's management and production;
11. radically change managers' methods and psychology, and include the involvement of all social groups in forest management;
12. improve the forest-use fee system;
13. develop forest infrastructure, especially roads and communications facilities;
14. utilize non-timber forest products and intangible forest benefits to diversify the forest sector's output;
15. increase the effectiveness of resource consumption and labor productivity within the forest sector to world standards;
16. increase and improve social uses of forests, especially recreational use;
17. encourage a wide range of forest cooperation with foreign countries in economic and ecological spheres, in combination with an increase of inland forest sector activity; and
18. develop forest research and design for all constituents of the forest sector.

Thus, one can see the total re-orientation of forest policy, and increasingly detailed elaboration and concrete definition, in its movement from the federal to the regional level, and then to the local government level. But realization of the outlined policy will demand huge efforts and resources, and it seems that some of the policy goals may not be achievable until farther into the future, after 2010.

REFERENCES

Carlsson L., M.-O. Olsson, T. Malmloef, B. Lehmbruch and O. Mashkina. 1998. Initial analyses of the institutional framework of the Russian forest sector, IIASA interim report (IR-98-027). Laxenburg (Austria): International Institute for Applied Systems Analysis

Efremov, D. F., L. Carlsson, M.-O. Olsson and A. S. Sheingauz. 1999. Institutional change and transformation in the forest sector of Khabarovsk krai, IIASA interim report (IR-99-068). Laxenburg (Austria): International Institute for Applied Systems Analysis.

Hakamada, S. 2000. *Russia under Putin—The dictatorship of the law* (in Japanese). Tokyo: NTT Publication.

Kakizawa, H. 1997. Forest policy reform and the restructuring of the forest industry in Russia. *Journal Forest Research* 2:25–30.

Kukuev, Yu. A. 2002. Interview on a new Forest Code (in Russian). *Izvestiya* (newspaper), June 6.

Mabel, M. 2000. *The flexible domestic state: Institutional transformation and political economic control in the Khabarovsk krai forest sector.* Laxemburg: International Institute for Applied Systems Analysis.

Moiseev, N. A., ed. 2001. National forest policy of Russia: Proceedings of the International Scientific-Practical Conference (in Russian), 11–12 October 2001. Moscow: Publishing House of Moscow State Forest University.

Nakai, T. 2000. Local system reform under the Putin administration (in Japanese). *Economic Bulletin of ROTOBO* 1165:1–15.

Revedov, A. 2000. The flip side of new strategy of Russia (in Japanese). *Global Net* 121:24–25.

Sheingauz A. 2000. Forest policy in the Russian Far East: Current status in pace of economic reform. In *The Third IGES International Workshop on Forest Conservation Strategies for the Asia and Pacific Region* (7–9 September 1999), 134–157. Tokyo: Institute for Global Environmental Strategies.

———, ed. 2001. *Forest sector of Khabarovsk krai: Principal orientation of development* (in Russian). Khabarovsk: Publ. House RIOTIP.

Sheingauz, A., S. Nilsson and A. Shvidenko. 1995. Russian forest legislation, IIASA working paper (WP-95-45). Laxenburg (Austria): International Institute for Applied Systems Analysis.

Chapter 13

PUBLIC PARTICIPATION IN JAPAN'S FOREST PLANNING SYSTEM

Takashi Fujiwara
Forestry and Forest Products Research Institute

1. INTRODUCTION

For the sustainable management of forests, it is necessary to define appropriate principles and criteria and then implement them at the management level. In many countries, progress has been made in forest policy, and new strategies have been introduced in recent years. It has been pointed out, however, that "there is not yet any strong evidence that the strategies are being acted upon" (Poore and Chiew 2000). Most developing countries cannot solve the problems of limited management resources and high costs originating from having scattered sites, complicated by the diverse interests of multiple parties in forest management.

Public participation methods reflect the local conditions, carry a lower cost than other approaches, avoid disputes, and are the key to unlocking this situation, even in developed countries (Inoue 2001).

In this chapter, the reality of public participation in the process of forming forest plans in Japan is analyzed, including a historical examination. The intent of this analysis is to contribute to future policy development in Japan as well as to international efforts to realize sustainable forest management.

M. Inoue and H. Isozaki (eds.), *People and Forest—Policy and Local Reality in Southeast Asia, the Russian Far East and Japan, 201–212.*

2. FOREST PLANNING SYSTEM AND LEGISLATION IN JAPAN

2.1 New Forest Legislation in Japan

With a population of 120 million and a national land area of 37 million hectares, Japan is a leading importer of wood. It relies on foreign forests for 80 percent of its wood consumption, accounting for 40 percent of the global trade volume of industrial round wood. Meanwhile, two-thirds of Japan's national lands are covered by forests, with the potential of supplying more than half of the domestic wood demand. As forests also play an important role in the conservation of national lands and headwaters and as places for cultural education, recreation, and other purposes, their sustainable management is an important task for Japan.

In June 2001, the Basic Law on Forestry of Japan was revised for the first time in 37 years and renamed the Basic Law on Forest and Forestry. The Basic Law on Forestry was enacted in 1964 to strengthen the productivity of forestry, improve the status of forestry workers, and eliminate the productivity gap between forestry and other industries, based on a vigorous demand for wood and strong interest in forestry management among forest owners in the aftermath of the Second World War. Various measures under the Basic Law on Forestry helped to increase forest resources as well as wood supply capabilities. But the share of domestic industrial wood declined from 70 percent in 1964 to 20 percent in 2000, because of the failure of the domestic timber industry to increase production sufficiently to compete with imported wood.[1] As a result, individual forest owners faced extremely tough management conditions. During this period, public interest in and expectations of the functions of forests expanded to various fields, including disaster prevention, headwater conservation, biological diversity, and global warming mitigation (Figure 1). In an international context, forest management was put on the agenda of global environmental issues, and "sustainable forest management" became an internationally shared common target. It was in response to the changing expectations of the public in regard to forests and the increasing international trend towards sustainable forest management in recent years that the Basic Law on Forestry of 1964 was revised and became the Basic Law on Forest and Forestry in 2001 (Forestry Agency 2001).

While the former Basic Law stipulated increased production and the stable development of forestry as policy targets, Chapter 1 of the new Basic Law stipulates "fulfillment of the multifunctional roles of forests" (Article 2) as the basic premise for forest and forestry policies, and cites "the conservation of land water resources and the natural environment, the protection of public health, the prevention of global warming, the provision of forest products" as examples of the multi-functionality of forests. It then declares: "Recognizing that sustaining the various functions of forests is

[1] It is important to note that the wood imported included that which did not include costs of sustainable forest management in the price.

indispensable for maintaining stability in the quality of life and the national economy, forests shall be properly managed and conserved for the future."

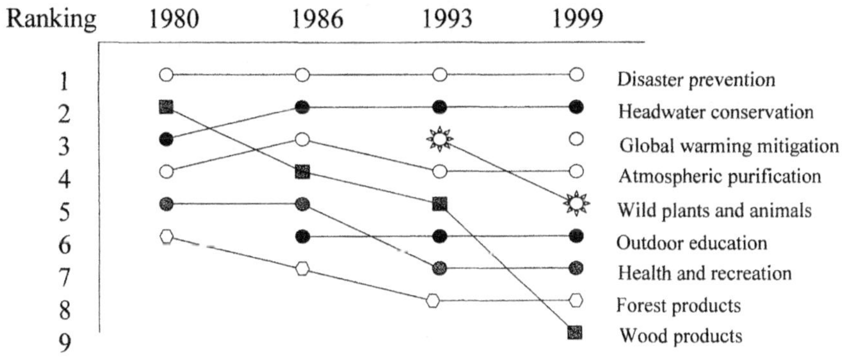

Figure 1: Trend of public expectations with regard to forests.
Source: Public opinion poll on forests and human life (Cabinet office 1999).

A technically ambitious approach was adopted in that forests are classified into three functional categories in a forest plan in order to apply the policy targets to actual forestry practice. The inclusion of the expanded use of wood as a policy target should prove highly significant to facilitate public recognition of wood as an eco-material, in addition to the promotion of forestry in general.

In order to pursue these targets under the serious economic circumstances of forestry in Japan, some issues must be properly addressed, including financial assistance for proper management, the participation of and improvement of public awareness, including urban dwellers, and to enable financial assistance and the proper management of a forest plan to support such financial assistance and public participation, etc.

2.2 Planning System and Forest Law in Japan

According to Article 11 of the Basic Law, the government formulates the Basic Plan for Forest and Forestry for the comprehensive and systematic implementation of related measures. The Forest Law also stipulates other forest planning programs to realize the above-mentioned targets in the field of forestry. The forest planning system is outlined in Figure 2 (Yamamoto 1999).

Japan's forest planning program consists of official plans at four levels: the nationwide forest plan, formulated by the Minister of Agriculture, Forestry and Fisheries; the regional forest plan, formulated by the relevant prefectural governor; the municipal forest plan, formulated by each head of the more than 3,000 municipal districts; and the forest management plan, formulated by each private forest owner.

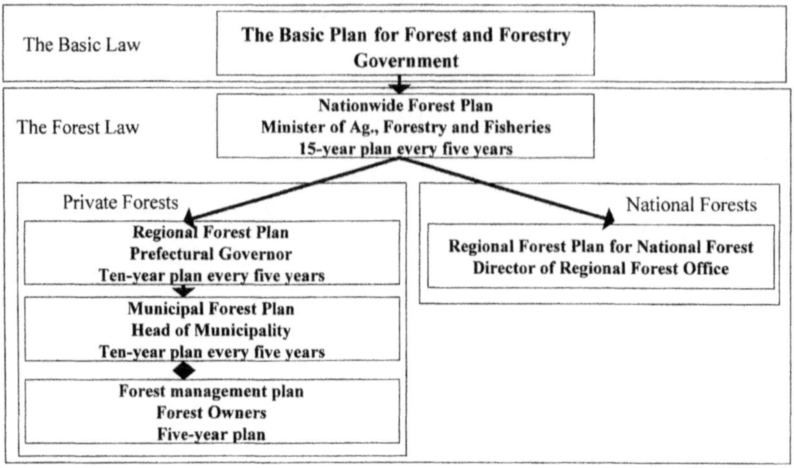

Figure 2: Forest planning system in Japan.

A slightly different planning system applies to national forests, which are managed by the government on its own initiative (Figure 3). Apart from the Forest Law, the Law Concerning Management of National Forests stipulates a different planning system for the management of national forests. The Minister of Agriculture, Forestry and Fisheries is required to formulate the Basic Management Plan for Entire National Forests, and the directors of the Regional Forest Offices are required to formulate regional management plans for 155 planning areas nationwide

3. FOREST PLANNING SYSTEM AND PUBLIC PARTICIPATION

The manner of public participation in the process of formulating different forest plans is explained here.

3.1 Forest Planning System for Public Participation and Non-National Forests

3.1.1 Official Plans and Outline of Public Participation

3.1.1.1 National Plan

There are two types of national-level forest plans: the Basic Plan for Forest and Forestry and the Nationwide Forest Plan. The formulation procedure for the former was set forth in Article 11, Paragraph 5 of the Basic Law, which states: "The Government shall seek opinions from the Forestry Policy Council in establishing the basic plan." Article 11, Paragraph 6 of the Basic Law stipulates that "The Government shall, without delay, submit to the Diet and publish the basic plan after it has been established." In regard to the procedure relating to the Nationwide Forest Plan, the Forest Law demands consultation with the heads of related administrative bodies, including the Minister of the Environment, and the hearing of the opinions of the Forest Policy Council as well as prefectural governors. These are the existing statutory provisions regarding public relations for the national level Basic Plan and the Nationwide Forest Plan. As seen above, there are no mandatory statutory provisions to make forest plans directly reflect public opinions, but the forestry agency invites public opinions by publicly announcing the draft Basic Plan and the Nationwide Forest Plan as part of the formulation procedure in the spirit of the New Basic Law introduced in 2001.

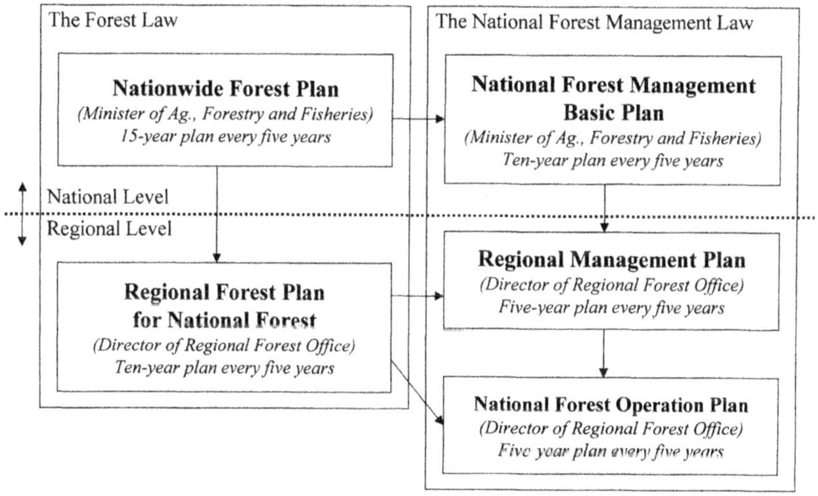

Figure 3: Planning system for Japan's national forests.

3.1.1.2 Regional Plan

In regard to regional plans formulated by prefectural governors, Article 6, Paragraph 1 of the Forest Law stipulates that when such a plan is to be formulated, the intention to do so must be publicly announced, accompanied by exhibition of the draft plan for inspection for 30 days. Anyone with an opinion on the draft regional forest plan is entitled to submit opinions backed by their reasons to the relevant prefectural governor in writing (Article 6, Paragraph 2 of the Forest Law). While the prefectural governor is required to listen to the opinions of the prefectural forest policy council and the heads of related municipalities (Article 6, Paragraph 3 of the Forest Law), he must submit a summary of the opinions on the draft plan to them (Article 6, Paragraph 4 of the Forest Law). Moreover, when a prefectural governor has formulated a regional forest plan, he must publicly announce the plan without delay, together with a summary of the opinions submitted to him and the processed results of such petitions (Article 6, Paragraph 5 of the Forest Law) (Figure 4).

The district forest plans formulated by the heads of municipalities basically require the same process, i.e., public announcement and exhibition of the draft plan to invite public opinions (Article 10-5, Paragraph 5 of the Forest Law).

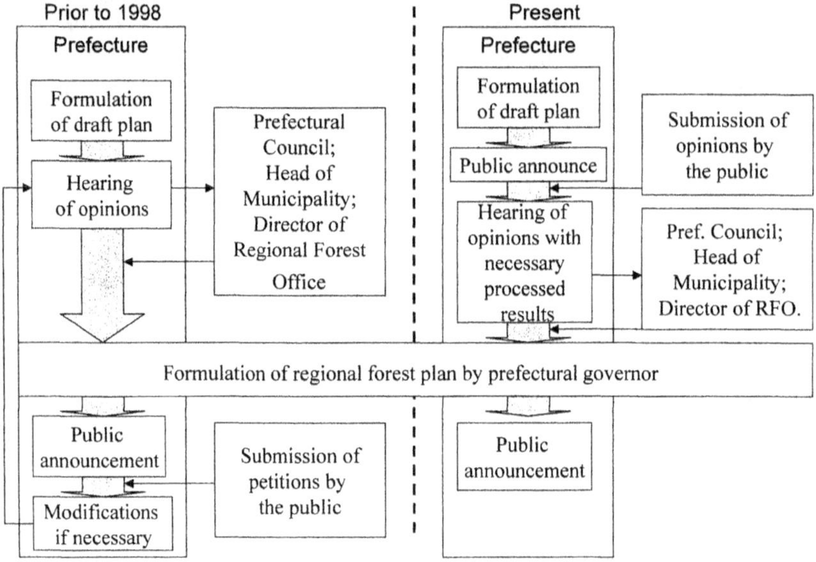

Figure 4: History and outline of regional forest plan formulation procedure.

3.1.2 Historical Review of Public Participation in Regional Forest Plans

The provision mentioned in 3.1.1 regarding public participation in regional forest plans was introduced following the revision of the Forest Law in 1998. Unlike the revised law, which stipulates the hearing of public opinions prior to the finalization of a plan, the mechanism to reflect public opinions on a plan was described up to 1998 as the "*ex post facto* procedural system" dating from the time when the Regional Forest Planning System was first introduced in 1962. To be more precise, the old Forest Law stipulated that "should persons have opinions with respect to regional forest plans announced according to the provisions of Article 5, Paragraph 6, they shall submit a petition with reasons to the prefectural governor within 30 days of the date of announcement" and that "prefectural governors must process the petitions of opinions stipulated in the previous paragraph in good faith and notify the petitioners of the results. If such results require modification of the regional forest plan, the prefectural governor shall make the necessary modifications" (Article 7 of the old Forest Law) (Figure 4). These provisions of the old Forest Law were intended to make regional forest plans reflect wide-ranging opinions and, since its enactment in 1962, these provisions are pointed out as constituting "a liberal procedure compared to other similar laws" (Society of the Study of Agriculture and Forestry Legislation 2002). The system of allowing the *ex post facto* submission of opinions, however, had two shortcomings: (1) a lack of public awareness that a plan could be modified based on the submission of opinions; and (2) a lack of incentives on the part of the administration for plan modification, because of the complicated process of the *ex post facto* modification of the plan. As a result, it was said that no petition was lodged regarding regional forest plans.

As the background for the revision of the *ex post facto* system to a system with the prior invitation of opinions, two factors were pointed out. One was the growing regulatory character of regional forest plans following the introduction of the development permit system, as well as the system to order changes of notified cutting in 1974, and the designation of forests in need of thinning in municipal forest plans in 1983 (Table 1). Another was the necessity to further promote the participation of local public bodies and local residents in forest improvement so that adequate forest improvement could proceed in the midst of stagnant forestry production activities.

3.2 Management Plan for National Forests and Public Participation

The public participation method in the formulation process of forest plans obtained its present form following the revision of the Forest Law in 1998, which was directly prompted by the fundamental review of the system governing national forests. In short, taking the opportunity of the reform of the national forest management system, the underlying spirit of reform was extended to the planning system for non-national forests. The background of the reform of the national forest management system and various forms of public participation are explained next.

3.2.1 Fundamental Reform of National Forests and Background

National forests currently cover an area of 7.9 million hectares, accounting for 31 percent of the forest area, or approximately one-fifth of Japan's total national land. When the national forest management system was set up in 1947, after the Second World War, a self-supporting accounting system was introduced to finance all expenditures with self-raised income through the sale of wood and other means. In the face of the rapidly growing demand for wood in the 1950s for post-war reconstruction, management emphasis was placed on enhancing wood productivity, and the balance basically remained in the black up to the mid-1970s. But the financial situation thereafter deteriorated due to sluggish wood prices under the pressure posed by imported wood, limited forest resources, and increasing fixed costs, resulting in an accumulated debt of ¥3.8 trillion (equivalent to U.S.$29 billion) in 1998. The reform in 1998 to rectify the situation consisted of three main pillars: (1) introduction of a new accounting system which did not rely on self-supporting finance while disposing of the accumulated debt, (2) change of the management stance to emphasize the function of benefiting the public, and (3) strict rationalization of the organizational structure and staff members. The basic idea was to manage "national forests as common property of 'the public' through 'public participation' for 'the public' to make national forests 'forests of the public' in both name and reality" (Forest Policy Council 1997). Public participation in various forest planning processes became crucially important to embody this idea.

3.2.2 Reflection of Public Opinions on Various Plans

3.2.2.1 Process to Establish a Basic Management Plan

The planning details for national forest management are set forth in the Law Concerning Management of National Forests (hereafter referred to as "the Law" in this section) as mentioned earlier. Article 5 of the Law stipulates that when the Basic Management Plan for Entire National Forests is to be formulated, the purport of the plan must be publicly announced in advance, together with exhibition of the draft plan for public inspection for 30 days. Article 5 of the Law also stipulates that anyone with an opinion on the plan can lodge a petition in writing, accompanied by reasons, before the end of this 30-day period, that a summary of the lodged petitions must be submitted when the opinion of the Forest Policy Council on the draft plan is sought, and that public announcement of the established plan must be accompanied by a summary of the lodged petitions and the processed results (Article 5, Paragraph 2 through 4 of the Law). The processes involved are basically the same as those for the formulation of regional forest plans for non-national forests. The difference is that while there is no mandatory provision for the hearing of public opinions in the formulation process of the Nationwide Forest Plan for non-national forests, as described in 3.1, the revision of the planning system for national forests in 1998 introduced the mandatory requirement for the hearing of public opinions.

3.2.2.2 Process to Establish Regional Forest Plans

A similar procedure to hear public opinions is stipulated in the formulation process for the regional management plans for national forests and the forest management implementation plans to be formulated by directors of regional forest office. As these regional plans are closely related to the administrative practices of local public bodies, they must be formulated after obtaining the opinions of prefectural governors, heads of municipalities, and academics after the processed results of public opinions have been forwarded to them (Article 6 of the Law).

3.2.3 Historical Review of Public Participation in National Forest Management

The immediate reason for the policy revision regarding public participation in forest management in Japan was the need to solve the financial strain faced by national forests, as mentioned earlier. This clearly suggests that the question of national forest management stimulated and developed tense debates among policymakers on the issues of public participation in a process of forest policy formulation in Japan's post-war history. An epoch-making debate on the management of national forests occurred on the Shiretoko Peninsula in eastern Hokkaido in the second half of the 1980s.

In 1987, the Kitami Regional Forest Office conducted partial selective cutting of national forests in Shiretoko, based on the relevant regional forest plan, which commenced in 1986. Objections were lodged against this on the grounds that cutting would destroy the habitat of the Blakiston's Fish Owl (*Ketupa blakistoni)*, designated a precious indigenous species, and that the procedure to form consent in the plan formulation process was questionable, sparking a nationwide debate in the mass media. This situation prompted the Forestry Agency to establish the Committee to Examine Forestry and the Protection of Nature, the members of which included mass media personnel, academics, and those involved in the protection of nature; the committee continued its discussions for one year. In the end, it made several recommendations, including that the existing principles for national forest protection lacked the idea of establishing a buffer zone which would cushion the negative impacts around a core conservation site, and that it was necessary to examine a mechanism to reflect the opinions of the general public as well as local residents at the formulation stage of a regional management plan. Following these recommendations, the hearing of the opinions of academics prior to the formulation of a plan was stipulated as part of the revision of the Management Rules for National Forest Land in 1991.

Learning from this and other incidents to deal with controversial projects, an approach has since been introduced to thoroughly discuss draft plans by all interested parties to solve local disputes.[2] What will be required of national forest management in the future is to secure active public support for the said management and to firmly establish a system to improve and consolidate the management through the introduction of public wisdom, rather than promoting a conflict-solving type of public participation. There are already signs of positive progress as a result of the efforts to create forests for the people, and further progress is anticipated (Committee of Forest of 21st Century 2001).

---

[2] In 1992, dealing with a petition against the development plan of recreational facilities in the Mt. Ontake area in Gifu Prefecture, a series of roundtable meetings, consisting of representatives of environmental groups, heads of related municipalities, and staff members of related administration bodies, was held over the period of six months, and consensus was reached before the start of the project.

Table 1: History of major revisions of forest law and forest planning system in Japan.

Year	Major revisions	Background
1951	Creation of the forest planning system; introduction of the compulsory planting system and the cutting permit system	Compulsory regulations to deal with forest devastation caused by indiscriminate cutting during WWII
1962	Fundamental revision of the forest planning system (introduction of a two-tier system consisting of the Nationwide Forest Plan and regional forest plans); abolition of both the cutting permit system and the compulsory planting system	Progress of reforestation efforts allowed to change from legal compulsion to inducement based on guidance and subsidy
1968	Creation of the forest management plan approval system (voluntary plans formulated by individual owners may be approved by the prefectural governor); extension of the forest plan period (five-year extension for nationwide and regional forest plans)	Increased wood demand under high economic growth; introduction of the inducement system based on the voluntary planning system
1974	Introduction of the forest land development control system (a permit by the prefectural governor is required for the conversion of forest land of one hectare or more); strengthening of this system by requiring notification of cutting; introduction of the forest management plan approval system for groups of owners	Pollution, the indiscriminate development of golf courses and housing sites and concern in regard to the protection of nature became social problems; emphasis on a balance between the economic function and public benefit function of forests
1983	Introduction of (i) the municipal forest improvement plan system, (ii) the system to recommend forests in need of thinning by the heads of municipalities and (iii) the profit-sharing silviculture system	Emergence of delayed thinning, etc., due to the sluggish demand for wood
1991	Unification of regional planning areas for national and non-national forests; introduction of arbitration system by prefectural governors for forests needing thinning	Further aggravation of delayed thinning, etc.; emergence of the question of reform for national forests
1998	Revision of forest management plan approval system (thinning criteria become part of the approval conditions; authority for approval is transferred to heads of municipalities); change of planning procedures for regional forest plans and district forest improvement plans (introduction of system to invite opinions prior to plan finalisation)	Increasing social concern in regard to the problem of thinning; review of the role played by administrative bodies in the form of the transfer of authority to local public bodies and others; fundamental reform of national forest operation
2001	Extension of planning bodies for forest management plans to those who are not forest owners; strengthening of planning system to classify subject forests for management based on different functions to benefit the public	Fundamental reform of the forest administration following the enactment of the Forest and Forestry Basic Law

Source: The author.

4. CONCLUSIONS

As explained earlier, the need for public participation in forest management can be understood from a practical point of view to prevent "a huge management cost arising from disputes, as it is more effective to prevent social disputes regarding the method of forest utilization." It is important to recognize that this approach is more essentially part of the "adaptive approach," which assumes the impossibility of gaining perfect knowledge of the ecological functions of forests, given the complexity and dynamic nature of forest ecology, which involves trial and error.[10]

In the midst of drastic changes in the expectations placed on forests, from just wood production to multi-functionality, the mechanism for public participation in plan formulation and the management of national forests in Japan has been developed to fulfill two objectives: (1) adjustment of interests and (2) public assistance for forest management. Japan is characterized by a concentration of people involved in forest management as a profession in the national and prefectural forest administration sector, with few working in the private sector, reflecting the situation of an extremely small number of large private forest owners. Accordingly, a situation where constructive dialogue regularly takes place regarding the plans formulated by local public bodies is lacking. Nevertheless, the public expectations of forest functions are becoming more diverse, and it is true to say that the number of people positively involved in forests is increasing. It is hoped that the forest administration will actively promote dialogue with the public on all aspects of forest management, based on the idea of adaptive management, which accepts that some of the planned matters are on trial, while possessing a well-constructed database which is open to public access.

REFERENCES

Cabinet office. 1999. Public opinion poll on forests and human life (in Japanese), Cabinet office of Japan.

Committee of Forest of 21st Century of Nagoya Regional Forest Office. 2001. Forestation activities from dialogue with citizens (in Japanese). Report from committee meeting, chaired by Shiniti Yamamoto, Professor, Nagoya University.

Forestry Agency. 2001. *Annual report on trends of forestry 2000.* Forestry Agency.

Forest Policy Council. 1997. The direction of policies on forestry and the fundamental review of the system governing national forests (in Japanese). In *The fundamental review of the system governing national forests (2).* Forest Policy Council.

Inoue, Makoto. 2001. Policy recommendations for participatory forest management. In *IGES policy recommendations on forest conservation in the Asia-Pacific.* Institute for Global Environmental Strategies (IGES).

Poore, Duncan and Thang Hooi Chiew. 2000. *Review of progress towards the year 2000 objective.* Yokohama: International Tropical Timber Organization (ITTO).

---

[10] "The Principle of Ecosystem Approach," adopted at COP 5 on the Convention of Biological Diversity, stipulates "Decentralization of Management " in Principle 2 and "Involvement of All Sectors of Society" in Principle 12.

Society of the Study of Agriculture and Forestry Legislation. 2002. The explanatory of laws of agriculture and forestry (in Japanese), 773. In *Private forest, vol. 1.* Taisei Shuppan.

Yamamoto, Hirokazu. 1999. The forest planning system of Japan. In *Global concerns for forest resource utilization*, ed. Atsushi Yoshimoto and Kiyoshi Yukutake. Kluwer Academic Publishers.

PART 3

LOCAL REALITIES IN FOREST USE AND MANAGEMENT

Chapter 14

FOREST VALUES AND LIVELIHOOD UNCERTAINTY IN TWO INDIGENOUS COMMUNITIES OF INDONESIAN BORNEO

Martinus Nanang
Institute for Global Environmental Strategies

1. SCOPE AND SETTING

1.1 Scope

Forests are important for all living beings. For humans, their importance is traceable from the economic, social, cultural, political, ecological, and even spiritual points of view. It is usually recognized that the forest is important to the livelihoods of the local people. Their lives are centered around the forest and their cultures develop from their relationship with it.

This chapter deals with the issue of the importance placed on the forest by the people living in its vicinity, and shedding light on the value of the forest to them. It can be assumed that these values can be a driving force for action, either positive or negative; but the issue of values is not the only factor that sparks action. This study analyzes the uncertainty in the livelihoods of local people as a dominant factor that drives their actions. Uncertainty elicits a response, and the way people respond brings about consequences. These consequences are analyzed and their implications for policy are discussed. The focus is on how people adjust to new situations and the implications of these adjustments on the forest itself.

M. Inoue and H. Isozaki (eds.), *People and Forest—Policy and Local Reality in Southeast Asia, the Russian Far East and Japan, 215–229.*

1.2 Setting

This chapter is the result of participatory research conducted in two communities of indigenous Dayak ethnic groups in East Kalimantan (Indonesian Borneo).[1] The two groups are the Tonyooi (also called Tunjung) of Muara Jawa village and the Benuaq of Tanjung Jaan village (Figure 1). The ethnic groups are culturally similar in terms of cosmology, customary laws, rites of passage, and healing rituals.

Figure 1: Map of East Kalimantan showing the location of the study.

Muara Jawa is a large village of 1,600 people, dominated by the Tonyooi ethnic group. The main source of livelihood is swidden cultivation, but it is slowly expanding to perennial crop cultivation. Economically, the village has a large number of low-income households (Figure 2). It is near a large amount of critically degraded lands and marshlands, but most of the surrounding area is covered by secondary forest, with only a small part being logged-over primary forest. Water is both an important resource and

---

[1] The author expresses appreciation to the participatory action research (PAR) team members: V. Sukapti, Samuel, and Ary Yasir (Muara Jawa), Rujehan and Amir Riyantone (Tanjung Jaan), and the local members of the PAR teams.

a source of disaster. Every year, floods cover the lowlands and damage people's properties. A small river passing through the village is the main source of potable water, but it is endangered by the degradation of the forests upstream. Private logging companies have been operating in the village area since 1970, and this is considered a factor that limits the local people's access to forestland.

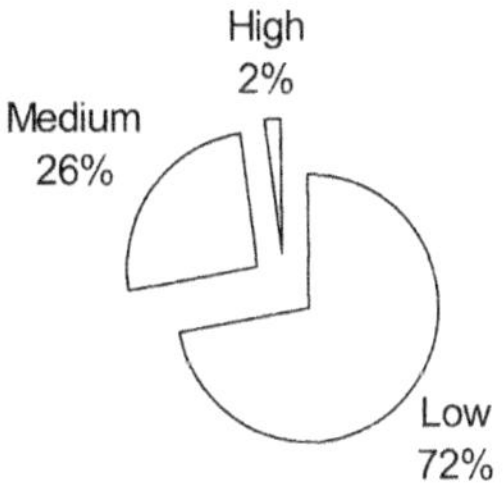

Figure 2: Wealth ranking of Muara Jawa community.

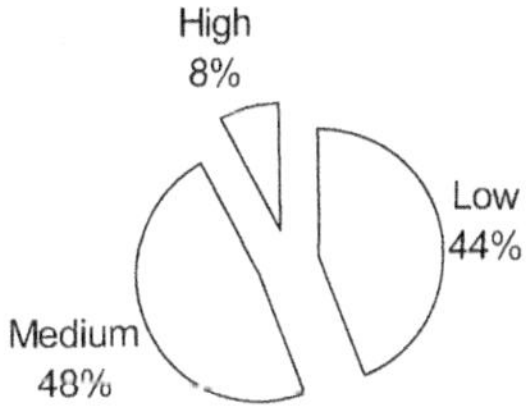

Figure 3: Wealth ranking of Tanjung Jaan community.

Sources: Field study and wealth ranking interviews (2001/02).

The Benuaq village of Tanjung Jaan is located on the shores of the largest lake in East Kalimantan, Lake Jempang, but the villagers are not dependent on fishing. Instead, their lives have been traditionally land-based, practicing swidden agriculture and perennial crop cultivation in a small area covered by secondary forest. The primary forest was completely destroyed by fires in 1997 and 1998. Now, water is the main problem for the community. The lake provided a way to the outside world for hundreds of years, but it has dried up annually in recent years due to the El Niño phenomenon. Besides these adversities, the community has been engaged in a conflict with an oil palm company over land, its main resource.

The relative well being of the people of Tanjung Jaan is represented in Figure 3. Compared with Muara Jawa, the people of Tanjung Jaan are better off, with better housing, more stable income, better education for their children, and ownership of motor vehicles and color television, among other indicators.

2. METHODS

Data for this chapter were collected during three field studies conducted from August 2001 to February 2002 in the two villages, in which participatory action research (PAR) methods were applied. During the fieldwork, several participatory techniques were used to gather firsthand data, including focus groups with women, men, and elders, village sketching, and individual interviews with key informants. To ensure the validity of the information the principle of triangulation, or cross-checking, of data has been applied, as well. The process of data gathering follows the dynamics of the community. Therefore, data on the same issue could provide different information.

For instance, in Muara Jawa, the information on forest collection was provided collectively by both men and women, but in Tanjung Jaan, the men and women provided it separately. This difference does not hamper the outcome of this study, because the purpose was not to make a comparison between the two gender categories.

3. FOREST VALUES

As this study deals with the issue of values, it is necessary to define what we mean by "values." Values are a basic element of culture, and culture is basically ideological. According to Lenski and Lenski (1982, as cited in Nanang 1997) ideology has three basic elements: beliefs, values, and norms. Beliefs are concerned with what people think the world is like, including their relationship with the physical environment. Values determine what behavior is considered desirable, moral, and just—working as a guiding principle (Kempton 1999). Human behavior is prescribed or proscribed in a particular situation through norms. Gentner and Stevens (1983; cited in Kempton 1999) assert that every individual has a "mental model," a simplified representation of the world that allows one to interpret observations, generate novel inferences, and solve problems. When a mental model is shared within a group, community, or society it is called a "cultural model," and people in the same culture often construct the same model. Thus, in spite of differences (in terms of gender, age, class, status, ethnicity, etc.), we can identify a common identity or shared characteristics among members of a culture.

By "forest value" we mean (refraining from any moral notions) how the desirable functions of the forest are valued, which may become a guiding principle in people's interactions with it. As cultural ecologists posit that culture develops from human interaction with the environment (Moore 1997), we can assume that forest values may differ between communities with different environments.

3.1 Perceived Importance of Forest Resources

The process of examining people's views on forest functions in Muara Jawa was done with a group of men and women together, and then men and women separately. This chapter presents the results of the separate processes. The technique of pair-wise ranking was applied in the process. Using this technique, group members were asked to identify forest functions and assign a number to each. The functions and numbers were put on vertical and horizontal axes and then compared with each other. The frequency of appearance of the number indicates the value of its respective function.

The exploration indicates that people are aware of the basic functions of the forest, which are not mutually exclusive. Therefore, individual functions listed in the tables below simply show the main tendencies. Most of the functions (livelihood, prevention of erosion and flooding, protection of water resources, protection from heat, and provision of construction materials, fuelwood, medicine, and food) are derived from their own experience of interaction with the forest. The men's and women's ranking of the forest functions of regulating climate, absorbing air pollution, providing recreation, and protecting flora and fauna have little to do with direct experience, but rather are based on learned knowledge.

Table 1: Summary of the pair-wise ranking of forest values in Muara Jawa.

Men			Women		
Points	**Rank**	**Forest function**	**Points**	**Rank**	**Forest Function**
14	I	Livelihood	12	I	Food supply
12	II	Fuelwood	11	II	Source of traditional medicine, fuelwood
11	III	Preventing floods, preventing soil erosion	10	III	Livelihood, protecting water resources, protection from heat
10	IV	Source of traditional medicine	9	IV	Preventing floods, providing construction material
9	V	Construction materials, absorbing air pollution	7	V	Protecting flora and fauna
8	VI	Protection from heat	6	VI	Preventing soil erosion
7	VII	Regulating climate/temperature	4	VII	Absorbing air pollution
6	VIII	Protecting water resource	2	VIII	Regulating climate/temperature, source of materials for traditional rituals
5	IX	Protecting flora and fauna	0	IX	Recreation
3	X	Materials for traditional rituals			
1	XI	Recreation			

Source: Field study and focus group interviews (2001/02).

Although there are differences between the views of the men and women, it is obvious that they agree on the economic importance of the forest (livelihood, food supply, fuelwood, construction materials). These functions received high scores in their judgment, even though the ranks are different. Actually, the main source of livelihood is still swidden cultivation, but it is common for the indigenous population of Borneo to have diverse sources of income. Forests provide more diversity of income source. During food crises, the forest often offers a buffer income for the people through collection of wood, rattan, fruit, and wild vegetables.

Both men and women also agreed on the importance of the forest as provider of materials for traditional medicine. Ecological functions (preventing erosion, preventing flood, regulating climate, protecting water resources, absorbing air pollution, protecting flora and fauna) also received similar ratings in their views, but were less important compared with the economic functions. Annual flooding and the large extent of critically degraded land are direct experiences that may have shaped their views of the importance of preventing floods and erosion. The heat of the sun leads them to appraise the importance of trees for shade.

The forest is less important from the religious and recreational points of view. Traditional religious rituals (rites of passage, healing rituals) require a lot of materials from the forest, but such rituals are rarely performed now because people have turned to modern medicine and herbal medicine. Moreover, such rituals are considered taboo by people who have converted to Christianity and Islam. The recreational function is considered much less important. For local people there is almost no desire to go into the forest simply to relax and be refreshed.

In Tanjung Jaan, the exploration of the people's views on the value of the forest was also done using pair-wise ranking. Fifteen men and five women were involved in the process. Table 2 details the men's and women's views separately.

Similarities between the men's and women's views can be seen in various aspects. First, for both men and women, ecological functions (regulating water source, preventing critical land degradation, preventing floods, preventing landslides, regulating climate, research) are ranked the highest. The importance of protecting soil fertility is prominent in the fact that the land in the area is composed of much sand and is thus less fertile. These responses may have been influenced by the activities of environmental NGOs, which have emphasized the ecological importance of the forest. Second, for the people, the economic function of the forest (construction materials, fuelwood, hunting sites, fruit, and spices) seems to be less important. In the process of discussion to identify forest functions, this appeared very late. This fact indicates that livelihood is not dominant in their spontaneous thinking about the forest. It is understandable for two reasons: one is that, economically, the people have turned to rely more on pineapple cultivation than on other sources of income; the other is because the territory has limited forestland which is not very rich in resources. The next section shows that the economic importance of the forest is mostly related to secondary forest. Third, men and women also agreed that the religious function is much less desirable and that the recreational function is nearly undesirable. Traditional rituals have been virtually abandoned in this Christian community. Although Catholicism (the majority) does not impose strict prohibition on performing the rituals, Protestantism teaches that any form of traditional religious ritual and artifacts is not right and against the true belief.

With respect to the forest function related to health (source of medicine) men and women differ sharply. For men it is less desirable, but for women it is very desirable. The reason for this difference was not explored.

By comparing the forest functions as perceived by the men and women of Muara Jawa and Tanjung Jaan, we can draw three observations. First, they disagree on the desirability of economic and ecological functions (in Muara Jawa, economic functions are the most important, while in Tanjung Jaan the ecological functions are the most desirable). Second, both communities agree that religious and recreational functions are less or nearly undesirable. Third, there is disagreement on the forest functions related to health as a source of medicine and protection from heat, even between the group of men and women in Tanjung Jaan (the reason is unexplained). Table 3 displays the comparison of the two communities, including gender divisions.

Table 2: Summary of the ranking of forest values as perceived by men and women of Tanjung Jaan.

Men			Women		
Points	Rank	Forest function	Points	Rank	Forest Function
13	I	Regulating water sources	14	I	Preventing floods, a place for swidden farming, protecting soil from being critical/preventing landslides, reservation for traditional medicine, vegetable source, water source
12	II	Preventing the land from becoming critical, preventing floods			
10	III	Regulating climate/air temperature			
9	IV	Place for research, livelihood			
8	V	Preventing landslides	13	II	Construction materials, firewood
6	VI	Construction materials, source of traditional medicine	10	III	Habitat for fauna and fish
			6	IV	Providing fruits, hunting sites, providing spices
4	VII	Firewood			
3	VIII	Hunting, source of material for traditional rituals	2	V	Source of rattan
			1	VI	Recreation, ritual materials
1	IX	Protecting fauna			
0	X	Recreation			

Source: Field study and focus group interviews (2001/02).

It is also noted that the views reflect their actual interactions with the forest, particularly in terms of making a living. Some have their roots in learned knowledge, either from school or other sources.

3.2 The Real Importance of Forests for the Economy

The exploration on the importance of the forest to the communities was done by ranking the collection, use, and sale of forest products as perceived by them. The ranking activities were done with focus groups using the "corn quantification technique" to help them estimate relative ranks over a one-year period before the forest fires occurred. Group members were asked to draw a products-use matrix. Various products were listed along the vertical axis, and three columns were drawn for products collection, use, and sale. Then they marked the columns with corn seeds to indicate the estimate of products they have collected, used, and sold within a one-year period. The maximum markers on a box were limited to ten seeds, representing the highest rank, and one seed marking the lowest. After completing the collection column, the informants were asked to make estimates of the use and sale of items in the collections, and the importance of the products for their household's economy.

In Muara Jawa, identification of forest products collection was made together by a group of men and women. The result of the discussion is presented in Table 4. The table shows varieties of forest products, both timber and non-timber, in several categories. It is important to note that only timber is a product exclusive to primary forest. Other products come from either fallowed forest alone or from both secondary and primary forest.

Table 3: Forest values in Muara Jawa and in Tanjung Jaan compared.

Forest Function		**Rank** (1=highest; 10=lowest)			
Category	**Item**	**Muara Jawa**		**Tanjung Jaan**	
		♂	♀	♂	♀
Economy	• Construction/house material	5	4	6	2
	• Fuelwood	2	2	7	2
	• Livelihood	1	3	4	5
	• Food supply (fruit, game, vegetable, spices)	4	1		4
	• Swidden farming	8			1
Ecology	• Preventing soil erosion	3	6	2	1
	• Preventing flood	3	4	2	1
	• Regulating climate/temperature	7	8	3	1
	• Protecting water resource	8	3	1	1
	• Absorbing air pollution	5	7		
	• Protecting flora and fauna (biodiversity)	9	5	9	4
	• Research			4	
Health	• Source of traditional medicine	8	2	6	1
	• Protection from heat	6	6		
Religious	• Material for traditional ritual	10	8	8	6
Leisure	• Recreation	11	9	10	6

Source: Field study and focus group interviews (2001/02).

Timber, rattan, honey, game, and resin have since become scarce. Other products are still abundant, but unfortunately, most of the scarcer products are highly valued by the people and collected as much as their availability permits.

According to the people's account, only timber, rattan, palm sugar, rubber, honey, and fruits are highly valued by the market. The rest have limited markets or none at all. This means that most of the products are collected for the people's own domestic (household) use or consumption. Thus, the importance of these products is not mainly for gaining cash income, but for domestic use.

In Tanjung Jaan, analysis of forest product utilization includes the collection and actual use of the products, including for earning money (sales). Men and women have differences in their use and collection of forest products. Table 5 indicates the similarities and differences. It shows that for men, wood for construction, fish, bamboo, medicinal plants, *doyo* (*Curculigo* spp.), firewood, and rattan are the most important products, but the collection of rattan and wood for construction is not so high. Most are collected for their own use. Only fish, deer, turtles, monitor lizards, snakes, and *doyo* are collected to generate cash. For women, only a few products have commercial importance. These are *doyo*, *keniih* (*Garcinia cowa*), rattan fruits, and *kajang* or *pandan* (*Pandanus tectorius*). Most of them are products from the secondary forest. It can be concluded that for the people the forest is very important, not just for cash income, but more importantly for household use.

Table 4: Ranking of forest products potentials, collection, and importance as perceived by the people of Muara Jawa (1=highest; 10=lowest).

Forest products	**Collected**	**Used**	**Sold**	**Importance to household income**
Timber	4	3	7	10
Rattan	3	2	8	10
Resin	5	1	9	2
Honey	1	1	9	4
Medicinal plants	5	5	1	6
Sugar palm	10	10	7	10
Game animals (incl. birds)	5	5	7	4
Latex (rubber)	5	5	9	2
Bamboo	10	10	3	10
Sago palm	10	10	5	10
Fruits	5	5	7	10
Vegetables	10	8	2	10
Spices	8	10	4	5

Source: Field study and focus group interviews (2001/02).

The above exploration shows the primary importance of non-timber forest products (NTFP) over timber. The NTFPs are available mostly from the secondary forest.

4. LIVELIHOOD UNCERTAINTY AND PEOPLE'S RESPONSES

4.1 Access to Land and Forest

The rights of indigenous peoples to land in Indonesia are only vaguely recognized by the state. This condition makes these rights fragile, particularly in areas with abundant and convertible resources for mining and forestry, because the people may be expelled from their land. Customary law does not provide a powerful basis for land ownership and access to land. This fragility is well illustrated by examples from each community.

In Muara Jawa, people's rights to the land are still based on customary law. The system of ownership is divided into two: individual/household land and open common land. The former applies to finder's rights, first clearance, and other traditional rights. The latter is also based on customary law (called *hukum adat*) and refers to the virgin forest in which there is no claim of ownership (except limited claim based on finder's rights). All individuals have access to the virgin forest. Even the clearing of forests for swidden agriculture was not prohibited under customary law.

But since logging operations began in 1971, a prohibition on the clearing of virgin forest has been imposed by the logging company, because the village is in its concession area. The prohibition is still in effect even though the forest no longer being logged. Only after the forest fire of 1982, in which a large section of forest was destroyed, did the company allow the indigenous ethnic people to clear the burnt, logged-over forest areas for swidden cultivation.

In Tanjung Jaan, the people's access to land has been traditionally based on customary law, as in Muara Jawa. In reality, since 1996 access to the land has become a problem because of the advent of an oil palm estate. The history of the community from 1997 to 1999 was characterized by intense land disputes between the majority of the community members and the oil palm estate. The estate confiscated about seven hectares of the village's land, but then its activities were stopped because of very strong community opposition. Had it not been for the opposition, more than half of the village land would have been taken by the company, which would have been a big loss for the people. If that had happened, the community of Tanjung Jaan would no longer have been able to practice swidden cultivation in a sustainable cycle and the pineapple cultivation would not have been able to support the people's economy as it does now.

The experiences of these two communities reveal some lessons. First, the customary-based right to land is always threatened when there are conflicting interests between the community and outside parties. The status of customary land is fragile in areas with rich natural resources (i.e., the primary forest in Muara Jawa and the land in Tanjung Jaan). Similar situations have occurred in other communities which have abundant resources (i.e., gold, coal, bird's nests). Second, power determines the actual access to land. The people in Muara Jawa were not able to counter the power of the logging concessionaires, and thus their access to primary forest has been restricted for a long time. In Tanjung Jaan, on the other hand, a powerful oil palm concessionaire faced strong resistance from the local people who had received legal advice from a non-governmental organization (NGO).

4.2 Uncertainty and Response: The Evolution of Land and Forest Utilization

The life of the communities of Muara Jawa and Tanjung Jaan is centered on the practice of rain-fed rice cultivation called *uma.* The practice, based on slashing and burning the forest, has been the main mode of land use for generations. But the village's history shows an evolutionary change in land use, because people have had to adapt to the changing environment. Generally, rain-fed rice cultivation can only supply nine to ten months of their annual need for rice (cf., Goener 2000), and the recent recurrent dry seasons and pests have caused harvest failures, forcing the people to make adjustments to their livelihood.

Table 5: Forest product collection, utilization, sales and importance as perceived by the men and women of Tanjung Jaan (1=lowest; 10=highest).

	Men			
Forest Products	**Collected**	**Used**	**Sold**	**Importance in household economy**
1. Rattan	6	8	2	8
2. Wood (construction material)	6	7	3	10
3. Fish (from forest marshlands)	10	2	8	10
4. Wild boar	7	5	5	5
5. Deer/antelope	5	2	8	5
6. Birds	3	4	6	3
7. Bamboo	10	4	6	10
8. Fruits	6	5	5	5
9. Turtle and monitor lizard	6	0	10	6
10. Snake (python and *besisi*)	6	1	9	6
11. Palm sugar	3	4	6	3
12. Medicinal plants	10	4	6	10
13. *Doyo* (*curculigo* spp.)	10	2	8	10
14. Firewood	10	10	0	10
15. *Kajang* (*pandanus tectorius)*	2	8	2	2
16. Wild taro (*Alocasia* sp.)	5	1	0	9

	Women			
Forest Products	**Collected**	**Used**	**Sold**	**Importance in household economy**
1. Bamboo shoots	6	10	0	3
2. Rattan shoots	10	10	0	6
3. Edible Ferns	10	10	0	8
4. *Doyo*	10	10	10	5
5. Firewood	10	10	0	10
6. *Betetee* (*Horn-stedtia scyphifera*-vegetable)	10	4	6	3
7. *Keniih* (*Garcinia cowa)*	10	3	7	7
8. Rattan fruit	10	4	6	2
9. Wild banana core	3	10	0	1
10. Jackfruit	2	10	0	3
11. *Kangkung* (*Ipomoea aquatica*-watery vegetable)	10	10	0	7
12. *Kajang* (*Pandanus tectorius*).	10	3	7	5
13. *Biru* (*Licuala valıda).*	10	10	0	2

Source: Field study and focus group interviews (2001/02).

4.2.1 Muara Jawa

In Muara Jawa, the uncertainty in the people's livelihood has been caused by recurrent droughts and two ecological disasters—the forest fires in 1982 and then again in 1997/98. Since the forest was largely destroyed by the fires, people's reliance on it has been reduced. Instead, they have tried to develop new strategies, such as growing rubber and opening up wetland cultivation by converting marshlands. There has been a reduction of primary forest cover, relating to the increase in the human population, the growth in areas opened up for rain-fed rice cultivation, the increase of critically degraded land, and rice production. Although the trend analysis does not use exact measurement of numbers, it indicates the people's views and understanding of what has happened around them.

Figure 4 illustrates how the size of primary forest has been sharply reduced over 40 years. It was assumed that in 1960 most of the area was covered by primary forest (20,000 hectares, or 89 percent). Within a decade the size was reduced to 67 percent in 1970 and to 44 percent by 1980. The deforestation within the range of 20 years (1960–1980) was mainly caused by exploitation of the primary forest for agriculture. A sharp drop began when a long drought and large forest fire flared in the area in 1982. In 1990, the primary forest cover was 27 percent of its size in 1960. The forest fire of 1997/98 exacerbated the situation. After the fire the size of remaining primary forest is estimated at only 4.4 percent of what it was in 1960.

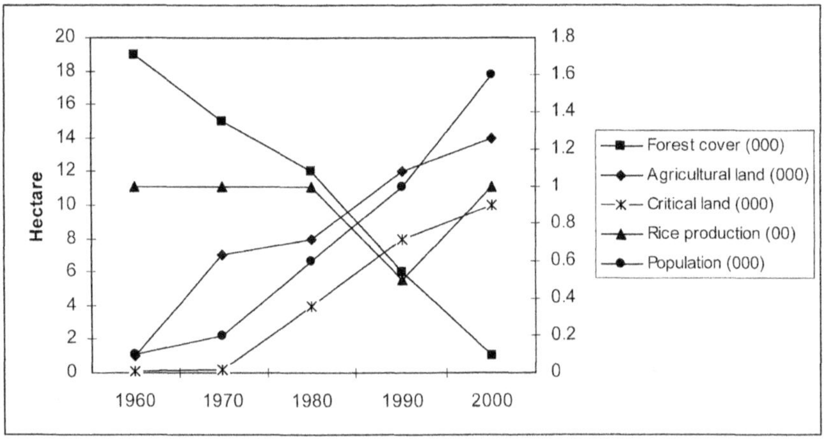

Figure 4: Trend line of forest cover, agricultural land, critical land, rice production, and population of Muara Jawa.
Source: Field study and focus group interviews (2001/02).

The human population has steadily increased. In 1960 the population was about 100 people. After the advent of logging companies in 1970, the population drastically increased. Many people from outside moved to the village as laborers for the companies. In most cases they decided to stay in the village. The significant increase of population to the present 1,600 people has contributed to the increase in demand for land for settlement and agriculture.

The area cleared for agriculture also increased significantly. Between 1960 and 1970 clearing of the forest for swidden cultivation was rampant. The reason is because there was no sense of crisis; the forest was abundant. Clearance of forest increased until the year 2000, but it was mostly secondary forest. Clearance of primary forest was only a small portion, particularly due to the restrictions imposed by the logging companies.

Another factor analyzed is the change in the size of critical land. By "critical" the people mean the land dominated by *Imperata* grass, where other plants can hardly grow and the land is difficult to convert to agricultural purposes. Even though there was a constant increase of critical land starting 40 years ago, a drastic increase can be noticed over the last 20 years, particularly after the 1982 and the 1997/98 fires. Agricultural productivity has decreased under these circumstances. Rice production per hectare was estimated at 100 cans (1,100 kilograms) during the first two decades (1960–1980), but saw an almost incredible drop to only 25 cans in 1982, 50 cans in 1990, 25 cans in 1998, and then returned to the normal 100 cans in 2000.

The situation, particularly the hardship brought about by the two fires and annual droughts, has forced the people to seek a buffer income. The lesson learned from the previous droughts and fires indicates that people can rely on rubber, rattan, and wet rice cultivation; so many households have developed small-scale rubber plantations. The vast area of marshland is also going to be converted for wet rice cultivation. This way the people respond to the changes in their physical environment. The response implicates a change in the utilization of land and forest, that is, from heavy reliance on rain-fed swidden rice cultivation and forest products to a more stable cultivation of perennial crops and wet rice.

4.2.2 Tanjung Jaan

The situation in Tanjung Jaan has both similarities and differences compared with Muara Jawa. The similarity lies in the fact that the people have adapted to the changes in the physical environment. As in Muara Jawa the mode of land use in Tanjung Jaan has been influenced by the changes in the physical environment brought about by recurrent droughts and particularly by the forest fires in 1982 and 1997/98, which devastated not only the area of Tanjung Jaan but its neighboring villages, too (cf., Inoue and Nanang 2003). It has also been influenced by the recent history of conflict over land with the oil palm estate. Intense conflict started when an oil palm company tried to confiscate the land from the community in 1996. This conflict with the company has also caused internal conflicts within the community itself, that is, between proponents and opponents of the company. A handful of people strongly supported the company for their own gain, but the majority of villagers strongly opposed it for trying to take their land. Although the government gave strong support to the company, by the time of our fieldwork, the village had successfully secured its land. But then the primary forest was completely devastated by the rampaging fire in 1997/98. Most villagers believed that the fire was intentionally set by the oil palm company to secure the takeover of their land; the fire occurred during the tensest conflict between the two parties.

More detailed analysis of the changes in the environment is presented in Figure 5 below. The figure shows the trend of several sectors in the history of the community over 40 years (1960–2000). The clearance of forest for agriculture has decreased, because most people have engaged in sedentary perennial crop cultivation (pineapple and rubber). The clearance of land for gardens is relatively constant, showing no significant increase or decrease. The size of land cleared for swidden rain-fed rice cultivation decreased, which shows that this mode of making a living and land use has become less important. On the other hand, the area of critical land has increased over a short time. The population has also increased, but this has not been followed by increases in the clearance of new land and forestland.

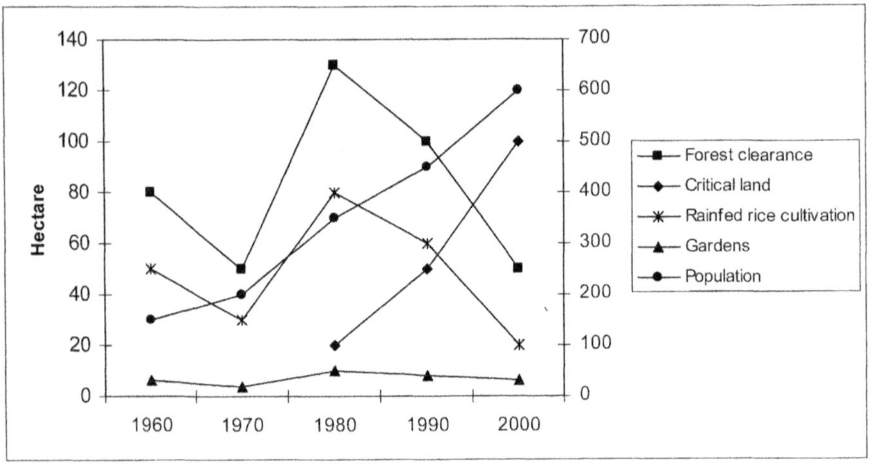

Figure 5: Trends in the size of forest clearance, critical land, rain-fed rice cultivation, and sedentary perennial crop gardens in Tanjung Jaan.
Source: Field study and focus group interviews (2001/02).

The above trends highlight two tendencies. First, once sedentary agricultural activities become profitable, the people tend to abandon traditional swidden rain-fed rice cultivation. In recent years, the people of Tanjung Jaan have gained more income from growing pineapples. In fact, pineapple has become a primary product of the community. Second, once sedentary agriculture is practiced by most of the population, the clearance of land for agriculture is likely to be reduced, even if the population doubles.

5. CONCLUSION AND POLICY IMPLICATIONS

The analysis of forest values reveals a number of facts. First, in the people's cognitive world the forest is highly-valued, but the values are not necessarily the same between communities. Basically, the values have their roots in the people's direct experience of interaction with the forest, although some may be derived from learned knowledge. It is clear that even between two communities with very similar culture and tradition, forest values may differ or have some variants, depending on their actual experience with their forest environments. Second, the collection of forest products indicates that the importance of the forest for the economy relates to the level of collection, but not necessarily with cash income, because they are used for domestic needs as well. Most of the products are available from secondary forest, indicating that for local people not only is pristine or primary forest important, but secondary forest as well.

Even though the forest is viewed as important by the people, in situations of livelihood uncertainty or crisis, in which swidden rain-fed cultivation fails and the forest offers little benefit, people tend to seek more sedentary agricultural activities by cultivating perennial crops. This new way of making a living implies reduced pressure on the forest. In comparison with swidden cultivation, it reduces the level of forest clearance, even if the human population increases. Thus, the crisis situation and subsequent adaptation may result in an unintentional protection of the remaining forestland, even though the people's actions are not guided by the perceived values mentioned above. The significance of such actions will be higher if the majority of the community members take the same action.

The unintentional protection of forest can be made intentional in order to have greater impact, in particular by supporting the cultivation of perennial crops. For that purpose access to land should be secured, because secure access to land ensures stable agricultural activities. The government, with its power and authority, needs to take the necessary measures to secure access to the land for the villagers.

REFERENCES

Gentner, Dedre and Albert L. Stevens. 1983 *Mental Models*. Hillsdale, N.J.: Lawrence Earlbaum Associates

Goener, Christian. 2000. *Resource management in a Dayak Benuaq village: Strategies, dynamics and prospects. A case study from East Kalimantan, Indonesia.* Eschborn, Germany: GTZ GmbH.

Inoue, Makoto and Martinus Nanang. 2003. Indonesia. In *The state of the environment in Asia 2002/2003,* ed. Japan Environmental Council. Tokyo: Springer-Verlag.

Kempton, Willet, James S. Boster and Jennifer A. Hartley. 1999. *Environmental values in American culture.* Cambridge: MIT Press.

Lenski, Gerhard and Jean Lenski. 1982. *Human Societies: An Introduction to Macrosociology.* New York: MacGraw-Hill Book Company.

Moore, Jerry D. 1997. *Visions of culture: An introduction to anthropological theories and theorists.* Walnut Creek-London-New Delhi: Altamira Press.

Nanang, Martinus. 1997. Community decision-making process: A study of the impact of power structure on community participation in an East Kalimantan village, Indonesia. MA Thesis. Quezon City: Ateneo de Manila University, Philippines.

Chapter 15

POLICY OF PROTECTED AREAS AND LOCAL USE OF FOREST RESOURCES IN INDONESIA: A CASE STUDY OF A NATIONAL PARK IN WEST JAVA

Kazuhiro Harada
Institute for Global Environmental Strategies

1. INTRODUCTION

Tropical countries have experienced the disappearance of forests on a global scale, and the world is rapidly losing valuable tropical forest ecosystems. Protecting biodiversity in tropical areas is one of the most important themes recently under discussion internationally. In response, there has been a rapid increase in the number of protected areas (PAs) established, in order to conserve biodiversity in tropical forests by legally protecting expansive, biologically rich areas. Currently, there are 12,754 PAs of sizes more than 1,000 hectares each throughout the world, including terrestrial and marine areas, covering over 1.3 billion hectares, or 8.81 percent of the Earth's land area (IUCN 1998). If PAs of less than 1,000 hectares each are included, the number of PAs reaches close to 30,350.

Protecting biological diversity has global benefits. When discussing the importance of biological diversity resources, biological values such as genetic biodiversity, species diversity, and ecological diversity are repeatedly pointed out. Apparently, the individual components of biodiversity, such as genes, species, and ecosystems, provide society with a wide array of goods and services (National Research Council 1999). But the widespread degradation of forest ecosystems has continued. In addition, the goals of conservationists have become inextricably embroiled with local communities who are often forced to accept conservation measures without compensation. To be successful in the long term, conservation must be incorporated into every aspect of the natural development process, including a new national culture, rather than just the responsibility of park authorities conserving specific wildlife (McNeely 1989). The

M. Inoue and H. Isozaki (eds.), *People and Forest—Policy and Local Reality in Southeast Asia, the Russian Far East and Japan, 231–247.*

global benefits of conservation are not meaningful to local residents faced with the demands of daily life, who insist on the immediate use of forest resources rather than their long-term conservation. There is often little incentive for local communities to conserve biodiversity and natural living resources without economic benefits (Tisdell 1995). Consequently, conflicts can arise between local people, who are forced to hunt, poach, and expand their land holdings just to meet their basic needs, and government authorities, who use coercion to protect large tracts of land and resources. Direct benefits for local inhabitants, such as new alternatives and possibilities, must first be identified in order to solve the problem of protected areas (Hough 1988, 1991; MacKinnon et al. 1986; Nepal and Weber 1995).

This chapter summarizes the forest policy and politics of biodiversity conservation at the national level. Furthermore, the local reality of communities and environments in Gunung Halimun National Park (GHNP), situated in West Java, Indonesia, are introduced to emphasize the historical problems of landholdings between the local people and the government, and the customary land use of the local people, including their use of forest resources. Finally, the role of non-governmental organizations (NGOs) as third parties is considered in encouraging the objectives of both community-based management and policymakers, and reconciling the latent conflict between them.

2. THE POLICY ENVIRONMENT

2.1 The Rights of Local People in Forest Areas

The Law of the Republic of Indonesia No. 41 of 1999 on Forestry, which was revised based on Act No. 5 of 1967 concerning the Basic Provision on Forestry, emphasizes "*adat*" (customary law) and "participation." The law stipulates that as long as customary laws exist and their existence is recognized, the customary law community shall have the rights to collect forest products for the daily needs of concerned communities. In addition, communities can be informed about plans of forest allocation, forest product utilization, and forestry information, and they can provide information, suggestions, and considerations for forest development, and undertake supervision regarding the implementation of forest development, either directly or indirectly. Furthermore, the Law of the Republic of Indonesia No. 23 of 1997 on Environmental Management encourages sustainable development and stipulates that the government determines national policies on environmental management and spatial management, while taking into account religious values, culture, traditions, and the living norms of the community.

The major premise of the laws, however, is that the government is responsible not only for the management of the relations between the people and the state in terms of forest ownership and utilization, but also for everything related to forest management and conservation. That is, all forests, irrespective of type, are managed and operated under the state's control. Consequently, customary rights have to be based on the state's purposes and cannot contradict national interests. In addition, the law declares that the concept of the existence of customary laws is not defined explicitly and can be interpreted expediently by the government. The areas in and adjacent to PAs are obviously under state ownership.

2.2 Policy of Protected Areas

The political issues of use and conservation of forest resources have been dominated by the Indonesian government authorities for a long period. The vast area of forests within PAs emphasizes the enthusiasm of the government to accomplish integrated biological conservation. With the United Nations Conference on the Human Environment, held in Stockholm in 1972, interest in environmental issues grew, and the number and area of PAs around the world rapidly increased accordingly. In Asia, the number of PAs began to increase in the 1970s, especially in China and Indonesia (Harada 2003). In Indonesia, both the number and area of PAs began to increase rapidly in the latter half of the 1970s (Figure 1). In particular, in recent years a significant number of protected areas have been established. In April 2001, the area of PAs encompassed 17.9 million hectares of forests—22.8 million hectares when marine PAs are included—for a total of 386 PAs (Departmen Kehutanan 2001).

The Indonesian government authorized the creation of PAs with Act No. 5 in 1990, the Conservation of Living Organisms and their Environment Act. The PAs were split into categories of strict nature reserve, wildlife sanctuary, national park, grand forest park, natural recreation park, hunting park, and protection forest.[1] Subsequently, the Law of the Republic of Indonesia No. 41 of 1999 on Forestry detailed the concept and the category of forests. In addition, many living resources were protected in extensive protection forests, limiting levels of utilization. These areas are destined to remain under forest cover for environmental protection purposes (Ministry of Forestry 1995). Table 1 lists the activities permitted and prohibited in the PAs. It can be summarized that almost all local human activities in the use of lands and forest resources for daily needs (A) are strictly forbidden. Large-scale human activities (B) are also forbidden. On the other hand, environmental management by human intervention (C) and use for

---

[1] This act defines two broad classes of reserves, which include these seven categories of PAs. The following are the definitions of these categories (Ministry of Forestry 1995).
Sanctuary reserves: A sanctuary reserve shall be a specific terrestrial or aquatic area having sanctuary as its main function, preserving biodiversity of plant and animals as well as an ecosystem, which also acts as a life-support system.
(1) Strict nature reserve: A strict nature reserve shall be a sanctuary reserve area having a characteristic set of plants, animals, and ecosystems, which must be protected and allowed to develop naturally.
(2) Wildlife sanctuary: A wildlife sanctuary shall be a sanctuary reserve area having a high value of species diversity and/or a unique animal species, in which habitat management may be conducted, in order to assure their continued existence.
Nature conservation area: A nature conservation area shall be a specific terrestrial or aquatic area whose main function is to preserve diversity of plant and animal species, as well as to provide a sustainable utilization of living resources and their ecosystems.

tourism and research (D), which are directly related with the development of biodiversity, are comparatively permitted.

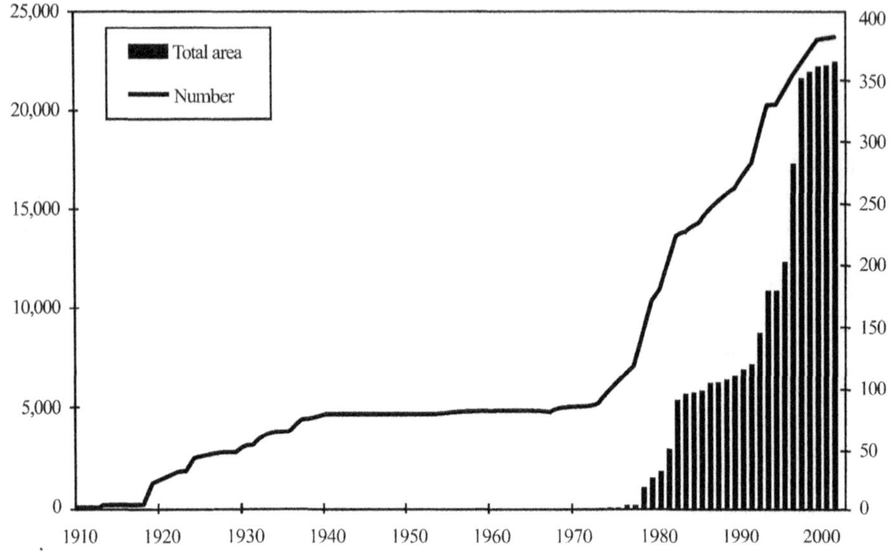

Figure 1: Changes of the total area (left axis, thousands of hectares) and number (right axis) of protected areas in Indonesia.
Source: Based on data from Departmen Kehutanan (2001).

Thus PAs contribute to conserve biological diversity by maintaining the integrity of the area at a high level without interruptions from outside human factors. Biodiversity includes the world's species from local habitats to whole continents or oceans (National Research Council 1999). In these areas, there is no space left for local people. Consequently, the trend of biologically focused conservation and the political pressure from vested-interest groups have contributed to the cultural and socio-economic marginalization of many people living within or close to PAs (Tacconi 2000). Global concerns have surpassed and coerced local necessities and, in the process, local residents have been eliminated from these areas.

On the other hand, there is the ideal paradigm, which has learned from much criticism of the disregard of local livelihood by government authorities and their negative attitudes towards residents' involvement in PAs management. The exclusive management models are expected to be shifted to the inclusive models that include local participation, recognizing the link between nature and culture, and employing collaborative resources management (Laird and Lisinge 2002). Governments and protected area managers are encouraged to incorporate customary and indigenous tenure and resource use by promoting full participation, so that protected areas can survive and biodiversity conservation can be enhanced. (Beltran 2000). Biodiversity conservation has strong links with cultural biodiversity, including a variety of languages and cultures (Brush 1996; Maffi 2001; Posey 2001).

But in reality, the significant problems faced by all PAs in Indonesia are certainly caused by disregarding the local rights.

Table 1: Activities Permitted (O) and Prohibited (X) in Different Categories of PAs.

Activity	Strict Nature Reserve	Wildlife Sanctuary	National Park	Natural Recreation Park	Hunting Park
A. Growing food crops	X	X	X	X	X
Growing tree crops	X	X	X	O	O
Collecting herbs and firewood	X	O	X	X	X
Hunting	X	X	X	X	O
Fishing	X	X	O	X	O
Human settlement	X	X	X	X	X
B. Collecting rattan & poles with permit	X	X	X	X	O
Commercial logging	X	X	X	X	X
Mineral exploration	X	O	O	X	O
C. Wildlife control	X	O	O	O	O
Active habitat management	X	O	O	O	O
Non-exotic introduction	X	O	O	O	O
Exotic introduction	X	X	X	O	X
D. Camping	X	O	O	O	O
Visitor use	X	O	O	O	O
Scientific collection with permit	X	O	O	O	O

Source: Ministry of Forestry (1995).

3. CASE STUDY IN GUNUNG HALIMUN NATIONAL PARK: LOCAL MANAGEMENT AND GLOBAL AUTHORITIES

In this chapter, GHNP is introduced as a case study of the complicated situation of the national park, where the use of lands and forest resources by the local people and biological conservation by government authorities exist together. The historical process of land occupation by the government, the customary land tenure, and the use of forest resources are discussed.

3.1 Study Site

GHNP, situated in West Java, Indonesia, encompasses three districts: Bogor, Sukabumi, and Lebak, covering 40,000 hectares (Figure 2). The park is located at an altitude between 500 to 2,000 meters above sea level, which includes three main vegetation zones: colline, montane, and submontane (Departmen Kehutanan 1998). GHNP possesses enhanced natural characteristics due to its high altitude and biodiversity. Up until now, 500 plant species, 53 mammal species, and 203 bird species have been identified, including rare animals such as the Javanese gibbon (*Hylobates moloch*), and Javanese hawk eagles (*Spizaetus bartelsi*) (ibid.). It is also estimated that in the three districts (13 sub-districts) that the park straddles, there are approximately 160,000 people living in 46 villages (BCP-JICA 1999).

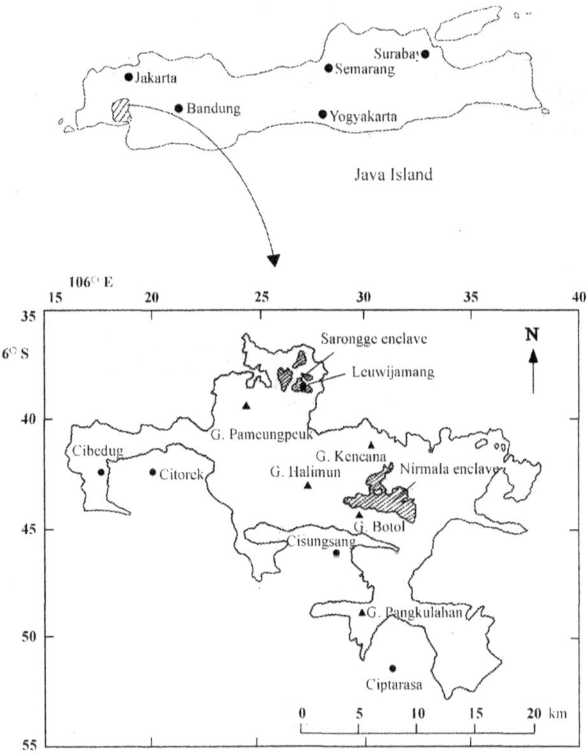

Figure 2: Map of study sites in and around Gunung Halimun National Park.

3.2 Methodology

This study was conducted in the GHNP from 1998 to 2000. Villages in and adjacent to the park are divided into three categories, based on the administration of park authorities: villages situated in areas adjacent to the park (adjacent), villages situated in areas that are inside the park geographically but outside of it administratively (enclaves), and villages illegally situated within the park (encroachment). Three villages were selected for the survey from each category, including the area adjacent to the park: Ciptarasa village (adjacent), Leuwijamang village (enclave), and Cibedug village (encroachment). Informal interviews were conducted with key informants with approximately six people, including customary village leaders, official village leaders, and elderly persons, to obtain information about the history of these areas and customary land-use systems. Formal interviews were also added to identify the private landholdings with 74 households in three villages. Furthermore, an ethnobotanical survey was conducted with two or three local villagers who have abundant knowledge about plants. Informal interviews were conducted to determine the local names and uses of plants that were collected in the field. Additionally, participant observations were conducted in each village for thirty days each to identify the inhabitants' dependency on plant species for the purposes of food and fuelwood, and to determine the places where the villagers routinely collected these resources. Seventy-six respondents were selected for this survey.

3.3 History of Forest Occupation

Until the late nineteenth century, forests in Java were abundant on both the hillsides and plains (Oki 1988). With increased human expansion, people cleared forests and scrub land, burned and slashed vegetation, and planted dry rice and other crops (Elson 1994). When these lands became exhausted after two or three years, people were forced to abandon their crude hamlets in search of other cultivation sites, where the whole process of settlement and reclamation was repeated. In the nineteenth century, the population of Java grew rapidly, resulting in increased land reclamation (ibid.). Rapid deforestation in Java, especially in the late nineteenth century, inevitably caused the decline of shifting cultivation in favor of more intensive agriculture on permanent farmland (Oki 1993).

Abundant forests still exist in the GHNP. In this area, small-scale swidden cultivation is still practiced by the Sundanese. The forests and lands around this area have been subject to several legislative changes (Figure 3). During the period of the kingdom of Pajajaran, the population of the Gunung Halimun area increased.[2] In the

[2] The Sundanese kingdom was established at the end of the seventh or eighth century, when Hinduism (Indian) was introduced to Indonesia (Ekadjati 1995). The fourteenth century saw the establishment of the Sundanese kingdom of Pajajaran. The capital, Pakwan, was situated near the present-day town of Bogor (Guillot 1990). Pajajaran was ruled by King Prabu Siliwangi (Wessing 1993). The Hindu kingdom of Pajajaran was flooded by the force of Islam in the sixteenth century and Muslim seized control in 1527 (Guillot 1990). According to the myth, Prabu Siliwangi left his court and entered the Sancang Forest on the south coast of Java because his court rejected the new religion (Wessing 1993). A more detailed history of the Banten area is described by Djajadiningrat (1983). People in Pajajaran reportedly settled in the southern part of the GHNP to escape from Islamic forces, and some locals living in and around the GHNP are believed to be descendants of the Pajajaran kingdom.

twentieth century, the Dutch colonial government invaded the area,[3] and divided it into protected forests, lands managed by local communities, plantations, private property, and local government lands.[4] Local people were forced to guard these forests. If other local people entered the protected forest or engaged in other destructive activities, they were punished in accordance with the Dutch colonial forest laws.[5] After the Dutch left in the 1940s and the Japanese arrived, the native population increased. Because the Japanese government did not adopt the Dutch forest management system, local people tried to reclaim their land. Consequently, protected forests, community forests, and plantations were all managed by locals.

After gaining independence, the Indonesian government devised a new forest management system, under which local community forests became protected forests (*hutan lindung*). In 1977, the government of West Java transferred the control of all protected forests in the province to the Directorate of Nature Conservation (PPA). In the same year, the PPA temporarily established a large system of nature reserves (over 40,000 hectares) in this area, and the designation of "protected forest" was changed to "strict nature reserve" (*cagar alam*). Until 1979, strict nature reserves were divided into three categories: strict nature reserve, plantation, and production forest (*hutan produksi*), all owned by the State Forestry Corporation (*Perum Perhutani*). On 26 February 1992, the area was designated as the Gunung Halimun National Park (*Taman Nasional Gunung Halimun*). At that time, some parts of the plantation and the production forest were included in the park.

In the historical process of encroachment, many community lands have been included in the park, plantation, and State Forestry Corporation areas (Table 2). Such unjust forest management by the state is the legacy of colonialism, and local people continue to suffer from state jurisdiction.[6]

---

[3] Colonial states often appropriated large tracts of land for forests and agricultural plantations, usurping prior systems of land rights and establishing new land-use and resource laws (Peluso 1992).

[4] In the nineteenth century, bureaucratic lands were delineated, and political forces were established by colonial law to restrict public access to trees for forest conservation (Peluso 1992). The 1927 laws define the state forestlands of Java and mediate an ideology of state legitimacy, justifying state control of one-quarter of the land area of Java; they are still in effect today (ibid.).

[5] Regulations and criminal action are detailed in Departmen Kehutanan (Departmen Kehutanan 1986).

[6] The early period of Indonesia's independence, from 1947 to the enactment of the Basic Agrarian Law of 1960, was characterized by a dual system of land laws (MacAndrews 1986). One law, which governed Indonesians, followed the customary laws (*adat*), concerned with unsurveyed, unregistered, and untitled lands. The other law, carried over from colonial times, served non-Indonesians and Western Europeans with holdings that were titled under Western law procedures. The legacy of colonialism meant that both customary and introduced laws governed land ownership and diversity, rather than homogeneity, characterized the tenure systems (Cleary and Eaton 1996). This duality of lands still remains at the heart of contemporary land legislation and administration, including the situation in and around the GHNP, even after the declaration of the Basic Agrarian Law of 1960.

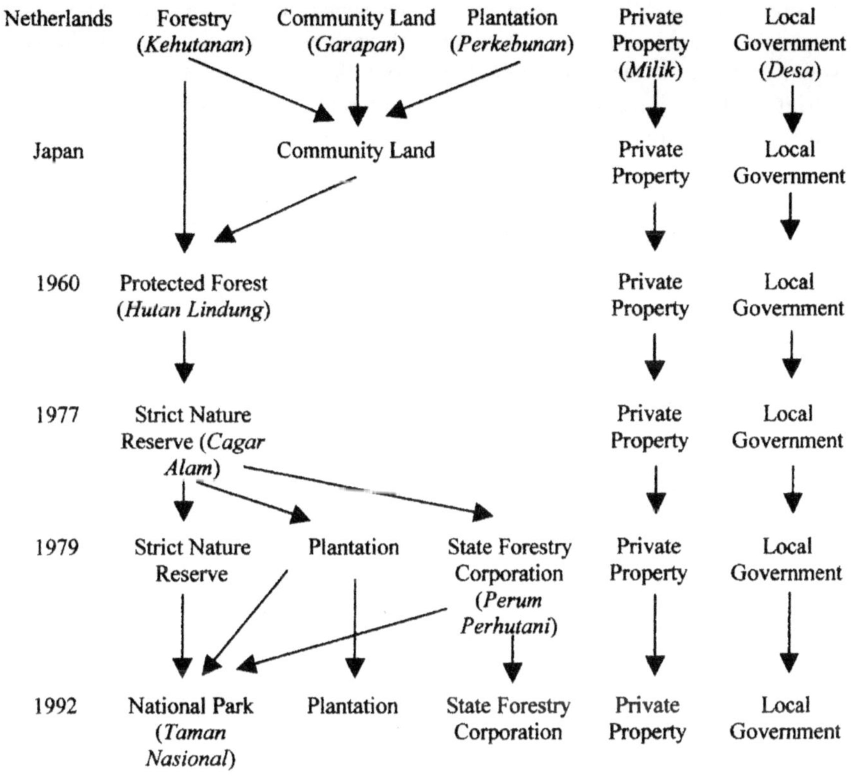

Figure 3: Historical changes of forest divisions in and around the GHNP.

3.4 Mixed Use of Resources

Many people still live their lives based on the subsistence activities involving several land-use patterns, such as gardens (*kebun*), mixed gardens (*kebun talun*), tree gardens (*talun*), paddy fields (*sawah*), and swidden cultivation (*huma*). Additionally, they divide the surrounding natural environments into primary forests (*leuweung kolot*), secondary forests (*reuma kolot*), and scrub land (*reuma ngora*). Primary forests are those with huge trees which have never been felled by people. Scrub land is indicated by the appearance of dense young bushes and grasses, which have been left alone as fallow lands for one or two years after swidden cultivation. Forests left untouched for more than five years after turning into scrub land become secondary forests. Additionally, old secondary forests have much larger trees than secondary forests and have almost the same scale as primary forests. Their main daily subsistence activity is practiced in paddy fields, which are characterized by sublime scenes of large tracts of rice terraces. Their yearly activity largely consists of labor in the paddy fields, from sowing to harvesting wet rice grains, and practicing harvest ceremonies in order to acquire their staple food. On the other hand, swidden cultivation plays only a small part in their activities, characterized as a minor subsistence activity that occupies a small area of land or is often situated in the corner of gardens. Dry rice grains are harvested for a slight increase of yields for each household. Annual crops, such as cassava and beans, fruit such as bananas and jackfruit, and other useful trees are planted and managed in gardens, mixed gardens, and tree gardens, based on the household labor system.

Table 2: Proportion of land acreage in each area per household (percent).

	Private Property			Plantation			State Forestry Corporation			National Park		
Village	**Pad.**	**Dry**	**Tot.**	**Pad.**	**Dry**	**Tot.**	**Pad.**	**Dry**	**Tot.**	**Pad.**	**Dry**	**Tot.**
Ciptarasa	17.9	3.4	13.0	0	0	0	81.5	96.6	86.6	0.6	0	0.4
Leuwijamang	54.2	25.9	42.2	15.9	7.6	12.3	0	0	0	29.9	66.5	45.5
Cibedug	1.8	3.9	2.4	0	0	0	7.7	0.7	5.7	90.5	95.4	91.9

Source: Author.
Note: Pad. = paddy field, dry = dryland, tot. = total.

The lands that include forest areas surrounding communities can be put into two categories: areas already owned privately, and areas which have no private ownership at all. The former includes paddy fields and many kinds of gardens, scrub land, and secondary forests. The latter embraces all natural habitats, such as scrub land, secondary forest, old secondary forest, and primary forest, located in the areas surrounding privately owned lands. Table 3 describes the human interrelationships within communities concerning the use of resources in each land category. In this table, plans are divided into two categories: "managed" and "non-managed" plants.[7] Managed plants are those which are managed deliberately by planting seeds or seedlings, while non-managed plants are those existing in the fields and left without any human intervention. Paddy fields are strictly protected against outside interruption by exclusive individual ownership rights. Picking rice in others' paddy fields is not allowed at any time. In gardens, mixed gardens, and tree gardens, even though people are basically prohibited from gathering vegetables in others' fields, small amounts of planted vegetables are allowed to be picked without permission by close friends or relatives. Grasses and trees, which are deliberately managed by owners, are recognized just like planted vegetables—these are forbidden to be cut down or picked from without permission. Dead trees which are managed can be used freely. Other plants under non-management conditions, both grasses and trees (including dead trees), are allowed to be used by others without an owner's permission on all privately owned land.[8] On the other hand, because the property and utilization rights of scrub land, secondary forests, old secondary forests, and primary forests are beyond customary local laws, anyone in the communities can freely practice subsistence activities access to communal lands are open to all.

[7] This division is based on the individual preference. Therefore, the same species can often be divided into both "managed" and "non-managed" plants.

[8] In one village, non-managed trees are only allowed to be used with the owner's permission.

What can be explicitly explained is that the land-use style of local people in and around the GHNP is characterized by a mixed use of resources based on the co-existence of members within communities. In almost all land categories, except paddy fields, the customary law emphasizes sharing of resources within local communities, based on well-defined individual land ownership rights. On the contrary, paddy fields are specified as sole property. Simultaneously, however, paddy fields can contribute to a variety of collaborative labor systems, harvest distributions, and lending and borrowing systems of private land in order to mitigate the land and property strain and form more sophisticated and organized communities.[9] The problem is that currently some agricultural fields and many forests have been included within the park, and as a result, people are suffering from a lack of resources and land to meet their needs. Consequently, they are inclined to expand their territories outside of their privately owned areas. These areas are considered as open access areas from the local point of view, but most of them are designated as "holy" natural habitat.

Table 3: Mixed property and use of resources.

		Paddy field	**Garden, mixed garden, tree garden**	**Scrub land, secondary forest***	**Scrub, secondary forest, old secondary forest, primary forest**
Managed	Rice, vegetables	Difficult to use	Possible to use with relative freedom	–	–
	Grass and trees	Possible to use with owner permission	Possible to use with owner permission	Possible to use with owner permission	–
	Dead trees*	Possible to use without owner permission	Possible to use without owner permission	Possible to use without owner permission	–
Non-managed	Grass	Possible to use without owner permission	Possible to use without owner permission	Possible to use without owner permission	Open access
	Trees	Possible to use without owner permission/ Possible to use with owner permission	Possible to use without owner permission / Possible to use with owner permission	Possible to use without owner permission / Possible to use with owner permission	Open access
	Dead trees	Possible to use without owner permission	Possible to use without owner permission	Possible to use without owner permission	Open access

Note 1: The degrees of freedom of use are ranked as follows: Open access > possible to use without owner permission > possible to use with relative freedom > possible with owner permission > difficult to use.

Note 2: "Dead trees" are mainly used for fuelwood.

Note 3: "Scrub land and secondary forest" are those areas that have clear ownership.

Source: Author.

[9] The detailed information is introduced in other articles.

3.5 Use of Forest Resources

Local people use many kinds of plants to meet their daily needs. It was revealed that the total number of species used by local people was up to 462 species, consisting of more than 100 plant families. Plants used for many purposes were morphologically classified as trees, shrubs, herbs, vines, mushrooms, and others (palms, bamboos, ferns, bananas, and so on). The Sundanese are characterized as a tribe who are accustomed to having many kinds of uncooked young leaves as sub-dishes. In addition, wild fruits, young rattan shoots or canes, mushrooms, and bamboo shoots are indispensable parts of the local diet. The materials for traditional houses are gathered from natural resources, as well. Many tree species can be used for house poles. And the walls of houses are woven from several types of bamboo, such as awi andong (*Gigantochloa pseudoarundinacea*), awi tali (*Gigantochloa apus*), awi terum (*Schizostachyum brachycladum*), and awi mayang (*Gigantochloa verticillata*). Lashing for construction materials come from eleven species of rattan: hoe cacing (*Calamus javensis*), hoe dawuh (*Calamus blumei*), hoe pelah (*Daemonorops rubra*), hoe sampang (*Korthalsia rigida*), and so forth. The roofs of houses are made from seven species, including tepus (*Alpinia* sp.), mareme (*Glochidion* sp.), patat (*Phrynium pubinerve*), pindung totat (*Hornstedtia paludosa*), kawung (*Arenga pinnata*), salak leuweung (*Salacca cf edulis*), and kiray (*Metroxylon* sp.). Many kinds of daily technological implements are made from natural plants: furniture (tables, chairs, stools, and wardrobes, etc.), agricultural tools (handles and cases for hoes, sickles, and small hand-knives; ox plows, hilts and cases for small swords, etc.), cooking implements and wares (for making sugar, ovens, rice streaming baskets, rice bowls, etc.), convenience goods (mats, hats, brooms, pipes, etc.), musical instruments (*gamelan*, *angklung*), and so forth. Although local people often depend on commercial medicines rather than traditional medicines, because of their convenience traditional medicines still play an important role in supplementing the commercial medicines and meeting urgent necessities, especially in the areas far from cities. Forty-three different disease symptoms are cured by traditional plant medicines;[10] most of these plants (27 species) were used as eye drops for eye disease, such as awi tali (*Gigantochloa apus*), ki beunteur (*Leucosyke capitellata*), seuseureuhan (*Piper aduncum*) etc., and as tonics for fatigue, like huru sintok (*Cinnamomum javanicum*), ki sampang (*Evodia latifolia*), and takokak (*Solanum torvum*) and so forth. In addition, 24 species were used to treat diarrhea, such as awi andong (*Gigantochloa pseudoarundinacea*), hamerang (*Ficus padana*), hantap (*Sterculia rubiginosa*) etc., and 21 species for medicines for eczema, scabies, leukoderma, boils, itching, and fungus, like babanjaran (*Eupatorium inulifolium*), kakalapaan (*Licuala pumila*), manggu leuweung (*Garcinia dulcis*) etc., 19 species for coughs, like bingbin (*Pinanga coronata*), eurih (*Imperata cylindrica*), kapipingkel (*Polygala venenosa*) etc., and 21 species to help with delivery and the postpartum period, like cente (*Lantana camara*), jonge (*Emilia sonchifolia*), rukem (*Flacourtia rukam*) etc., amongst other uses. The importance of plants as fuelwood is obvious, because gas stoves aren't prominent in this area. In addition, many species are used for ritual ceremonies, pest control (teureup

[10] The detailed information is introduced in Harada et al. (2002).

[*Artocarpus elasticus]*), preventing soil erosion by planting ki sabrang (*Gastonia serratifolia*), for pesticides (Suren [*Toona sureni*]) etc., toys for children (ki mokla [*Knema cinerea*]) etc., and so forth.

Table 4 shows the relative dependence of the villagers on wild, semi-domesticated, and domesticated plants as food.[11] The proportion of wild, semi-domesticated, and domesticated plants used for each purpose is based on the total number of days spent gathering each plant within a thirty-day period. Table 4 describes that the local people tend to use domesticated plants rather than semi-domesticated or wild plants as food. On the other hand, people in Ciptarasa and Cibedug largely depended on wild species for fuel wood (54.0 percent in Ciptarasa and 79.1 percent in Cibedug). Table 5 and 6 show the places where they gather food and fuel wood, respectively, over a thirty-day period. In every village, a large proportion of food was gathered in areas used for swidden cultivation, gardens, mixed gardens, and tree gardens. Paddy fields also played an important role as a source of foods such as genjer or hermit's water lily (*Limnocharis flava*). The total proportion of food gathered in cultivated fields was more than 80 percent. The dependence on forests was not very high in any of the villages. On the other hand, local people largely depended on forests (Table 6). Some people in Ciptarasa and Leuwijamang tried to plant and manage small amount of species in their gardens, gathering fuelwood from their gardens (65.1 percent in Ciptarasa and 50.3 percent in Leuwijamang). But the dependence on forests, including primary forests, old secondary forests, secondary forests, and scrub land, for fuelwood was still high (34.9 percent in Ciptarasa, 34.8 percent in Leuwijamang, and 75.0 percent in Cibedug). These results reveal that local people still depend on forest resources for daily fuelwood. Commonly, the habitat of many useful plants, ranging from scrub land to primary forests, can be found inside as well as outside the park. Consequently, local residents are forced out of desperation to intrude into the park to meet their needs. Collecting fuelwood especially can be a daily routine, and wood cannot be substituted with other materials. This activity carries a high risk of threatening and collapsing the integrity of the ecosystems of the park, considering that the people's requirements are extremely high and that tactical attempts to solve the lack of resources have not been put into practice until now.

4. CONCLUSION

Local people, living in areas that have traditionally contributed to their food and material supply, have been deprived in the past during the process of enclosure of natural resources by the state under the slogan of biological diversity. The government attempted to ensure the state authorized domination over natural habitats as conservation. On the other hand, local residents have concentrated on adapting their

[11] Wild species are those that only exist in natural habitats. Domesticated species are those that don't exist in natural habitats and cultivated in the cultivation areas. Semi-domesticated species are those that originally exist in natural habitats, but also exist in the cultivation fields, where local people deliberately manage these species.

lives to their surroundings and view surrounding ecosystems as human-centered nature. The dispossessed people dwelling near the ecosystems are forced to use "the forbidden nature" within the park to meet their daily needs. The basic needs of the locals cannot be easily changed, however, without future perspectives for planning alternatives for local communities geographically and putting political pressure on the park authorities. Through local knowledge, the reality of plant and human interrelationships—embedded in the dynamic ecosystems of nature and society—and the waning and insufficiencies of the current state-centered policy can be pointed out. Tropical forest peoples are historically acknowledged practitioners; they hold the key to understanding, utilizing, and protecting tropical plant diversity (Plotkin 1995). Therefore, strong emphasis has to be concentrated on the policymakers to make them understand and respect local living rights.

But these two actors continue to pursue their different and individual self-interests in Gunung Halimun National Park through the park's natural ecosystems. The consequent lack of compromise has failed to produce any win-win situations or conflict resolution. This fundamental local situation makes it difficult to practice co-management or collaborative management of resources in the park. Consequently, the historically state-centered resource management can cause inconsistencies in achieving biological diversity, and explains the social dilemma between local realities and global authorities. It is probable that the establishment of PAs without democratic agreement among related actors paradoxically accelerates the total destruction of forests. Effective protected area management must achieve an appropriate balance between the different interests of the various stakeholders (McNeely 2001). Serious considerations have to be strongly emphasized for the strategies of biodiversity conservation to work.

Table 4: Relative dependencies on wild, semi-domesticated, and domesticated plants for use as food and fuel wood (in percent).

Name of village	Food			Fuel wood		
	Wild	Semi-Domesticated	Domesticated	Wild	Semi-Domesticated	Domesticated
Ciptarasa	21.5	18.5	60.0	54.0	19.4	26.6
Leuwijamang	18.6	17.5	63.9	37.0	4.0	59.0
Cibedug	12.8	20.7	66.5	79.1	9.7	11.2

Source: Author.

Table 5: Places to gather food (in percent).

Name of village	Primary forest, old secondary forest	Secondary forest, scrub land	Swidden cultivation, garden, mixed garden, tree garden	Paddy field	Others
Ciptarasa	11.4	8.5	67.1	13.0	0
Leuwijamang	4.6	8.7	74.8	10.5	1.4
Cibedug	15.1	4.2	63.1	17.5	0.1

Source: Author.

Table 6: Places to gather fuel wood (in percent).

Name of village	Primary forest, old secondary forest	Secondary forest, scrub land	Swidden cultivation, garden, mixed garden, tree garden	Others
Ciptarasa	22.6	12.3	65.1	0
Leuwijamang	12.9	21.9	50.3	14.9
Cibedug	41.0	34.0	24.8	0.2

Source: Author.

The use of conflict management is encouraged to make the transition from competing or contentious situations to collaborative or problem solving situations. Problem solving is encouraged when actors have a strong interest in both their own and other's outcomes (Rubin et al. 1994). Unless the conflict between the two sides is confronted with solutions to break the deadlock, however, how can collaborative management be realized? In this situation, the intervention of a third party is indispensable. Third parties do not have direct stakes in the conflict and can contribute to the resolution of misunderstandings between actors. Third parties can play the important role of facilitator or catalyst, a role characterized as acting in the neutral position, as much as possible. In PAs, the positive influence of local NGOs can be expected, because they have deep understanding of both local reality and global issues, so they can contribute to balance this dualism. First, while the grassroots strength of NGOs is based on flexible adaptation to local contexts, the diversity of organizational forms has enabled them to retain a measure of autonomy from bureaucratic control (Eldridge 1995). NGOs are encouraged to play a role as service deliverers to supplement the lack of implementation by policymakers. In addition, local participation, which was apt to be disregarded in the management of PAs by the state, is an important issue for NGOs. As outsiders, NGOs tend to develop bonds with the people they serve, and they work with community groups as partners, emphasizing local self-help initiatives and local control of programs (Cerenea 1988). Second, the function of advocacy is also stressed in conflict management. It requires a radical change in the functions of NGOs, from welfare and service delivery to advocacy, lobbying, and even political action, in order to play a role in transforming the growth-oriented agenda toward a sustainable and more people-centered development (Heyzer 1995). NGOs are encouraged to represent local voices against government policy and empower the local capacity to participate equally. This collaborative two-way action characteristic of NGOs in working between the local people and the government is surely an essential key to the successful management of protected areas.

REFERENCES

Beltran, J, 2000. *Indigenous and traditional peoples and protected areas: Principles, guidelines and case studies.* IUCN, Gland, Switzerland and Cambridge, UK and WWF International, Gland, Switzerland.

Biodiversity Conservation Project (BCP-JICA). 1999. *Data Sosial Ekonomi dan Sarana Pendidikan di Desa Sekitar Taman Nasional Gunung Halimun.* Departmen Kehutanan dan Perkebunan.

Brush, S. B. 1996. Whose knowledge, whose genes, whose rights? In *Valuing local knowledge: Indigenous people and intellectual property rights,* ed. S. B. Brush and D. Stabinsky, 1–24. Washington, D.C.: Island Press.

Cerenea, M. M. 1988. *Nongovernmental organizations and local development.* World Bank discussion papers. Washington, D. C.: The World Bank.

Cleary, M. and P. Eaton. 1996. *Tradition and reform: Land tenure and rural development in South-East Asia.* Kuala Lumpur: Oxford University Press.

Departmen Kehutanan (Indonesia). 1986. *Sejarah kehutanan Indonesia II.* Jakarta: Departmen Kehutanan RI.

———. 1998. *Rencana pengelolaan Taman Nasional Gunung Halimun 1998-2003 Buku II rencana kegiatan pengelolaan*, proyek konservasi keanekaragaman hayati kerjasama antara Ditjen PHPA, LIPI dan JICA, Balai Taman Nasional Gunung Halimun, Direktorat Jenderal Perlindungan Hutan dan Pelestarian Alam, Departmen Kehutanan, Bogor.

———. 2001. *Data kawasan konservasi Indonesia S/D April 2001.* Bogor: Departmen Kehutanan, Directorat Jenderal Perlindungan Hutan dan Konservasi Alam.

Djajadiningrat, H. 1983. *Tujuan kritis tentang sajarah Banten: Sumbangan bagi pengenalan sifat-sifat penulisan sajarah Jawa* [translated from Joh. Enschede en Zonen, Haarlem. 1913. *Critische beschouwing van de sadjarah Banten: Bijdrage ter kenschetsing van de Javaansche geschiedschrijving*]. Jakarta: Djambatan.

Ekadjati, E. S. 1995. *Kebudayaan Sunda: suatu pendekatan sejarah.* Jakarta: Pustaka Jaya.

Eldridge, P. J. 1995. *Non-government organizations and democratic participation in Indonesia.* New York: Oxford University Press.

Elson, R. E. 1994. *Village Java under the cultivation system 1830–1870.* Sydney: Allen & Unwin.

Guillot, C. 1990. *The Sultanate of Banten.* Jakarta: Gramedia Book Publishing Division.

Harada, K. 2003. Protected areas. In *The state of the environment in Asia 2002/2003*, ed. Japan Environmental Council, 255–260, Tokyo: Springer-Verlag.

Harada, K., M. Rahayu and A. Muzakkir. 2002. *Medicinal plants of Gunung Halimun National Park, West Java, Indonesia.* Bandung: PALMedia.

Heyzer, N. 1995. Toward new government-NGO relations for sustainable and people-centred development. In *Government-NGO relations in Asia: Prospects and challenges for people-centred development*, ed. N. Heyzer, J. V. Riker and A. B. Quizon, 1–13. London: Macmillan Press Ltd.

Hough, J. L. 1988. Obstacles to effective management of conflicts between national parks and surrounding human communities in developing countries. *Environmental Conservation* 15(2): 129–136.

———. 1991. Social impact assessment: Its role in protected area planning and management. In *Resident peoples and national parks: Social dilemmas and strategies in international conservation*, ed. P. C. West and S. R. Brechin, 274–283. Tucson: The University of Arizona Press.

IUCN (The World Conservation Union). 1998. *1997 United Nations list of protected areas.* WCMC, WCPA & IUCN, Gland, Switzerland and Cambirdge, UK.

Laird, S. A. and E. E. Lisinge. Protected area research policies: Developing a basis for equity and accountability. In *Biodiversity and traditional knowledge: Equitable partnerships in practice*, ed. S. A. Laird, 127–176. London & Sterling, VA: Earthscan Publications Ltd.

MacAndrews, C. 1986. *Land policy in modern Indonesia: A study of land issues in the New Order period.* USA: Oelgeschlager, Gunn and Hain.

MacKinnon, J., K. MacKinnon, G. Child and J. Thorsell. 1986. *Managing protected areas in the tropics*. IUCN and UNEP, Switzerland, UK: Page Bos (Norwich).
Maffi, L. 2001. Introduction: On the interdependency of biological diversity. In *On biocultural diversity: Linking language, knowledge, and the environment,* ed. L. Maffi, 1–50. Washington, D.C. and London: Smithsonian Institution Press.
McNeely, J. A. 1989. Protected areas and human ecology: How national parks can contribute to sustaining societies for the twenty-first century. In *Conservation for the twenty-first century*, ed. D. Western and M. Pearl, 150–157. New York: Oxford University Press.
———. 2001. Roles for civil society in protected area management: A global perspective on current trends in collaborative management. In *Biological diversity: Balancing interests through adaptive collaborative management*, ed. L. E. Buck, C. C. Geisler, J. Schelhas and E. Wollenberg, 27–49. Florida: CRC Press.
Ministry of Forestry (Indonesia). 1995. *National conservation plan for Indonesia, Volume 2: National issues, goals, objectives and guidelines.* Jakarta: Ministry of Forestry, Directorate General of Forest Protection and Nature Conservation.
National Research Council. 1999. *Perspectives on biodiversity: Valuing its role in an everchanging world.* Washington, D.C.: National Academy Press.
Nepal, S. K. and K. E. Weber. 1995. Managing resources and resolving conflicts: National parks and local people. *International Journal of Sustainable Development and World Ecology* 2:11–25.
Oki, A. 1988. Deforestation in Java: Changes in the ecological environment and their negative effects upon the life of the population. *The Yachiyo Journal of International Studies* 1(1.2): 471–498.
———. 1993. Forms of forest exploitation: A comparative study of shifting cultivation in Java and Japan. *South Asian Studies* 30(4): 457–477.
Peluso, N. L. 1992. *Rich forests, poor people: Resource control and resistance in Java.* California: University of California Press.
Plotkin, M. J. 1995. The importance of ethnobotany for tropical forest conservation. In *Ethnobotany: Evolution of a discipline*, ed. R. E. Schultes and S. von Reis, 147–156. Oregon: Dioscorides Press.
Posey, D. A. 2001. Biological and cultural diversity: The inextricable, linked by language and politics. In *On biocultural diversity: Linking language, knowledge, and the environment,* ed L. Maffi, 379–396. Washington, D.C. and London: Smithsonian Institution Press.
Rubin, J. Z., D. G. Pruitt and S. H. Kim. 1994. *Social conflict: Escalation, stalemate, and settlement.* USA: McGraw Hill, Inc.
Tacconi, L. 2000. *Biodiversity and ecological economics: Participation, values and resource management.* London and Sterling, VA: Earthscan Publications Ltd.
Tisdell, C. A. 1995. Issues in biodiversity conservation including the role of local communities. *Environmental Conservation* 22(3): 216–222.
Wessing, R. 1993. A change in the forest: Myth and history in West Java. *Journal of Southeast Asian Studies* 24(1): 1–17.

Chapter 16

IMPACTS OF INDUSTRIAL TREE PLANTATIONS ON THE INHABITANTS OF SOUTH SUMATRA, INDONESIA

Yasuhiro Yokota
Forestry and Forest Products Research Institute

1. INTRODUCTION

The government of Indonesia has been promoting industrial tree plantations for over a decade, in order to meet the demands of a timber industry that has seen explosive growth. As only a few species of trees are planted in this kind of large-scale initiative, there have been concerns that the natural environment and the communities near the project sites will be markedly influenced (Photo 1). The objective of this chapter is to clarify how industrial tree plantations in Indonesia are operated and outline their impacts and influence on local communities. Some reports have been published on the impacts of industrial tree plantations on the inhabitants, especially the negative ones (Tasaka 1992; Carrere and Lohmann 1996; World Rainforest Movement 1999), which can still be observed in various places. As the democratization and decentralization of Indonesia progresses, forest policy is under review. Moreover, the carbon absorption properties of afforestation are receiving attention worldwide as one of the measures for controlling global warming. So as not to repeat the problems that industrial tree plantations have caused so far, they should be examined closely.

M. Inoue and H. Isozaki (eds.), *People and Forest—Policy and Local Reality in Southeast Asia, the Russian Far East and Japan*, 249–259.

2. INDUSTRIAL TREE PLANTATIONS

2.1 Industrial Tree Plantations in Indonesia

The size of Indonesian forests has generally decreased over the years because of commercial cutting, plantation exploitation, and conversion to other land types. Since the 1980s, the government of Indonesia has encouraged the timber-processing industry as an important export sector in order to reduce its dependence on the oil sector. Securing a timber supply, therefore, has become a matter of great importance. Industrial tree plantations (*hutan tanaman industri* in Indonesian) were started in the mid-1980s, and became an important national project after the issuance of Government Regulation No. 7 in 1990 (Peraturan Pemerintah No.7/1990). According to the Ministry of Forestry, about eight million hectares of tree plantations were originally planned (Sato 2000), and by the end of 2000 about 2.5 million hectares had been planted (Forestry Planning Agency 2001).

Photo 1: Burning after clearing a forest for land preparation (Photo: Y. Yokota).

Industrial tree plantation projects, or hutan tanaman industri (HTI), are implemented by companies that have obtained a HPHTI concession (*hak pengusahaan hutan tanaman industri*) from the government. The land targeted for plantations is "non-convertible forest" (*hutan produksi tetap*) that has low stumpage volume; grassland or bush constitute examples of this type of land.[1] Although the government defines non-convertible forest as forest for timber production, some of these areas have insufficient amounts of timber. One of the guiding principles behind HTI involves the idea of reducing pressure on natural forests by utilizing such areas. Fast-growing trees are planted, such as *Acacia mangium* and *Eucalyptus* spp., or trees with high commercial value, such as *Tectona garandis*, and used as raw materials to produce pulp, plywood, or furniture. The maximum area of a concession is 300,000 hectares and the minimum cutting period is seven years. Clear-cutting is allowed as a way to prepare sites. Although the deliberate setting of fires to clear land is prohibited by the government, many companies set fires in order to reduce their site-preparation costs and to promote the growth of planted trees, as well as to prevent tree diseases and insect pests. Setting fires was one cause of the uncontrolled wildfires of unprecedented extent and intensity that occurred in Indonesia in 1994 and 1997/98.

2.2 The Industrial Tree Plantation of Company M in South Sumatra

The HTI project site of Company M in the province of South Sumatra was the object of a case study of industrial tree plantation in the 1990s.[2]

Company M obtained a concession area equivalent to 296,400 hectares (Company M 2000). The net area of plantation activities turned out to be 193,500 hectares (ha), because land was excluded in areas of protected forest (80,374 ha), construction sites (9,150 ha), buffer zones or boundaries near rivers (6,076 ha), agricultural areas used by inhabitants (4,300 ha), and provincial government tree plantations (3,000 ha) (Company M 2000). Plantation activities began in 1990, and by March 1994 an area of 137,5951 hectares had been planted (Company M 2000), when this study was conducted (Photo 2). At the time, this plantation was one of the largest in Indonesia and amounted to approximately 8 percent of the total 1,659,738 hectares planted (Estate and Forestry Statistics Division 1997). Company M finished its initial planting by 1998 (Company M 2000). Another company (T), belonging to the same conglomerate (B), built two giant pulp mills. Each mill was designed to have a pulp production capacity of 450,000 tonnes per year, making them some of the largest mills in Southeast Asia (Company M 1994). Company T was to contribute to the Indonesian economy by exporting pulp. Company M was formed as a joint corporation by Conglomerate B with a government

---

1 All forests in Indonesia are state owned. Forests are classified into five categories based on function: protected forest (*hutan lindung*, 33.5 million ha, 27.85%), parks and forest reserves (*suaka alam dan hutan wisata*, 20.1 million ha, 17.03%), limited production forest (*hutan produksi terbatas*, 35.2 million ha, 29.25%), non-convertible forest (*hutan produksi tetap*, 23.1 million ha, 19.16%), and convertible production forest (*hutan produksi yang dapat dikonversi*, 8.1 million ha, 6.71%) (BPS - Statistics Indonesia 2001).

2 This field survey was carried out from 1993 to 1994, and was sponsored by the Carbon Sink Project implemented by the Japan International Forestry Promotion and Cooperation Center.

company in order to provide raw material for Company T, as well as to utilize grassland and bush and contribute to the local economy by providing jobs.

Photo 2: HTI project site of Company M. (Photo: Y. Yokota).

A program of transmigration (*transmigrasi HTI* or *trans-HTI*) was begun in order to secure a labor supply for plantation activities (Photo 3). The trans-HTI program was implemented by the Department of Transmigration (now the Department of Transmigration and Forest Squatter Resettlement).[3] Each resident of a transmigration village is provided, in principle, with a plantation job, a house with a well, a small home garden (0.25 ha) and rubber plantation (1 ha), food assistance for one year, vegetable seed, public facilities, village institutions, and education and health services, among other services. The residents are migrants from Java as well as local inhabitants. A total migration of 15,000 households was planned, and by 1993, 4,253 households had migrated to 11 transmigration villages (Company M 1994). The government took responsibility for paying much of the cost of transmigration, while Company M was responsible for providing job opportunities and small rubber plantations for each of the households.

---

[3] Transmigration is a policy that resettles people from Indonesia's inner islands, such as Java, to the less densely populated parts of the archipelago. It also promotes agriculture. This policy aims to balance population distribution and agricultural production, which has been centered on the islands around Java. By fiscal 1997, about 1,670,000 households had moved.

Photo 3: Trans-HTI village. (Photo: Y. Yokota).

3. THE LOCAL INHABITANTS' VIEWS OF THE HTI PROJECT

3.1 Significance of the Forest to Local Inhabitants

3.1.1 Landscape (Land-use Pattern)

Before the start of the HTI project, the land-use pattern at the project site was roughly classified as agricultural land mixed with bush (64.5 percent), forest (23.3 percent), grassland (8.1 percent), residential areas mixed with agricultural land (2.6 percent), and others (1.6 percent) (Kumazaki and Lilik 1994). The agricultural land mixed with bush consisted mainly of swidden agriculture land and fallow land (shifting cultivation land), traditional small rubber plantations on fallow land, and secondary forest that had grown on fallow land. Traditional rubber plantations appear similar to secondary forest. The land-use pattern is dynamic and follows a continuous cycle, from swidden land to fallow land and grassland, to bush, and then to secondary forest or traditional rubber plantation. Not much in the way of wasteland is produced by swidden agriculture (Kumazaki and Lilik 1994).

3.1.2 Forest Utilization by Local Inhabitants

In this area, about 80 percent of the inhabitants are engaged in agriculture. About 60 percent of those, estimated at roughly 10,000 households (Yokota and Inoue 1996), use swidden agriculture methods (JICA 1990). Usually, each household has its own land and cultivates an average of 1.5 to 2.0 hectares each year, utilizing one site for a few years at a time. After harvesting the dry field paddy, residents plant rubber, orange, or jackfruit trees as a source of cash income. Many live in permanent residences, and their swidden fields are located within about 15 kilometers of the village.[4] Each household also has several permanent vegetable fields and fruit gardens near the village. Many households have small, traditional rubber plantations, because rubber is a very important source of cash. Those swidden cultivators lacking a permanent residence live in temporary huts in their swidden fields. These farmers use more land than do the villagers who live in permanent residences, and their fallow period is longer than that of the villagers.[5] Their land tenure is not a juridical right, but a customary one. Ownership is asserted by placing a fence around the swidden field, or by specially planted trees, such as rubber trees. Without such markings, anyone, even Company M, could occupy and use the land.

The local inhabitants usually pasture cattle and goats near their homes or fields, or on grassland if there is a large population of livestock. In the latter case, they burn the grassland regularly to secure fresh feed for their animals.

For the inhabitants, forests serve as a source of agricultural land for swidden fields, small rubber plantations, and grazing land, as well as a place to gather forest products, such as fuel wood, timber, wild vegetables, animals, and traditional medicines. Fuelwood is indispensable for their daily life. Unoccupied land serves as a reserve for agriculture or for gathering forest products for use in emergencies. In other words, forests support the local inhabitants, both in the present and as a resource for the future.

3.2 Negative Impact of the HTI Project on the Local Inhabitants

The biggest problem caused by the HTI project has been the land-use conflict that emerged between the local inhabitants and Company M.

3.2.1 Problems with Land Condemnation

The principle used by Company M to secure land for plantation activities was to first negotiate with the local inhabitants who held customary land tenure. These inhabitants were offered the opportunity to migrate to a trans-HTI village, or they were

---

[4] Although an exact count is impossible, according to the JICA data more than half of the residents who perform shifting cultivation are thought to have a permanent residence (JICA 1990).

[5] Although there was marked variation, the authors found that the period was about ten years when the owners had a permanent home, and 25 years, in many cases, when they did not.

offered indemnification if they were to "donate" their land. If they were reluctant to give up their land, the company excluded it from the HTI project. But some NGOs reported that some inhabitants were forced to "donate" their land, or that the inhabitants were accused of opposing a national project if they did not surrender their land. It was also reported that the amount of indemnity was lower than usual, or that no compensation other than the opportunity to move to a trans-HTI village was offered (e.g., JATAN 1999).

In some cases, land was planted without negotiation when there were no clear signs of land tenure, or when the tenure holders were difficult to identify. Moreover, there were also reports that some land had been planted without negotiation, despite the fact that signs of ownership and the identity of the tenure holders were both obvious. There were also cases in which outside laborers accidentally clear-cut the local inhabitant's land because of inaccurate maps. In other cases, local residents reported that Company M had planted on their grazing land, despite being requested not to, and that their rubber tree bush forests has been cut during the night (Inoue and Yokota 1994).

3.2.2 Problems with Planted Forests

While the HTI project produces forests with a high stand volume, such forest is very different from the traditional forestland that has long been utilized by local inhabitants. Although residents are allowed to conduct *tumpang sari* (Indonesian agroforestry cultivation) in artificial forests, they are not given land tenure, and have to stop cultivating the land within a few years as the planted trees grow and shade their crops.[6] There is also no certainty that they will have the opportunity for tumpang sari on the same land after the trees have been harvested. They are not allowed to plant rubber trees after tumpang sari, and the rubber plantations that Company M prepares for migrants to Trans-HTI villages are just one hectare in size. Grazing is allowed in the plantation forests, but burning to secure fresh feed is prohibited. Although the residents can gather fuelwood, they cannot harvest timber for houses and fences, and there are few natural medicinal herbs to be found. While traditional forestland has areas of spare land that anyone can exploit, the concession that Company M holds means that it owns all the plantation forest. In other words, the amount of forest that is of use to local inhabitants is decreasing and is being replaced by "useless" plantation forest.

3.3 Positive Impacts of the HTI Project on the Local Inhabitants

One of the objectives of Company M in implementing the HTI project was to support local communities. Indeed, some positive effects in this regard can be identified, but the results have not always been satisfactory.

---

[6] *Tumpang sari* is a form of agro-forestry in Indonesia. Agricultural products, such as dry paddy rice, corn, cucumbers, and so on, are grown between planted trees after afforestation.

3.3.1 Increased Opportunities to Earn Wages

Few job opportunities were available before the HTI project, so, for this reason, the local inhabitants looked favorably on the HTI project. Nevertheless, a number of problems associated with this increase in job opportunities should be noted.

First, there was no job security. While labor opportunities with the HTI project did indeed increase, they did not meet demand. Consequently, the residents could not obtain continuous work. The supply and demand imbalance of labor opportunities was exacerbated by an increasing population of migrants from Java, who resettled in the trans-HTI villages, and by migrants from neighboring areas who sought jobs with the HTI project.

Second, the wages were low. When a group obtained a contract to provide labor for Company M, the amount of work and payment to the group were fixed, so daily earnings were higher if the job was finished quickly, and lower if it took longer. While the average daily pay for jobs throughout the HTI project site might be of the same grade as other jobs, the daily rate was often lower than the minimum wage of Rp. 2,300 (U.S.$1 to 2), as defined by Company M (*Kantor Wilayah Departemen Transmigrasi dan Pemukiman Perambah Hutan Propinsi Sumatera Selatan* 1993). In addition, wages were often paid months after the work had been finished, due to delays caused by office procedures. Moreover, jobs were often obtained through sub-contractors, who requested a high commission or who were reluctant to pay promptly.

3.3.2 Indonesian Agroforestry

Company M set out a support program for local residents to build good relations (Bagian Forest Development, Company M 1993). This consisted of providing opportunities for agroforestry (tumpang sari), supplying seeds for agriculture, supporting farmers' cooperatives, and so on. As of June 1994, however, only the opportunity to conduct tumpang sari had been provided; this enabled residents to legally farm government land. By fiscal 1993, tumpang sari had been practiced on a total of 5,127 hectares, equal to 3.8 percent of the total planted area, and more than 2,500 households had participated (Company M 1994).

The problem with tumpang sari, however, is that the cultivation period can be as short as a few years, and that land is not constantly available, making it difficult to secure farmland for the future.

Tumpang sari was used as a means of securing plantation sites. Company M was able to obtain local residents' traditional lands by drawing up tumpang sari contracts with the traditional landowners. As Company M only had to pay wages for clear-cutting or afforestation work to the contractors, tumpang sari could be used as a very cheap way to secure plantation sites.

3.3.3 Trans-HTI (HTI Transmigration)

The residents of trans-HTI villages were guaranteed employment by the HTI company, and were also given title to a dwelling, a home garden (0.25 ha), a rubber plantation (1 ha), and necessities, such as food for one year. Each village also had elementary schools, mosques, clinics, meeting halls, and so on. Furthermore, the Department of Transmigration supported each village by providing a variety of supplies and assistance for five years. Generally, the villagers' living standards improved within this period.

In practice, however, the guarantee of employment was not kept, and often the villagers were very dissatisfied. Some villagers were still unemployed six months after settlement. One reason for this was the supply and demand imbalance in labor opportunities, as noted previously. Furthermore, because trans-HTI started in 1992, two years after the HTI program started, a contract system where company M gave jobs to local inhabitants through a sub-contractor had already become established, and it was difficult for trans-HTI villagers to get jobs. The villagers were also very dissatisfied, because numerous clauses in their labor contracts were not being kept. Wages were often lower than the minimum wage specified by Company M, transportation between the village and the plantation site was not always provided, and compensation for industrial accidents was inadequate, among other complaints. A premise of the HTI project was that villagers could earn a living as laborers; however, these shortcomings meant that this project had a negative impact on many of their lives.

Moreover, with respect to their daily lives, infrastructure construction was often inadequate, so that the villagers were inconvenienced. For example, some wells dried up during the dry season.

Company M met with the villagers frequently and provided significant levels of support based on the results of these meetings. The villagers praised the company for financing cooperatives and providing generators. Nevertheless, the company still failed to guarantee labor or to improve working conditions, despite repeated discussions on this point.

At present, life in a trans-HTI village is far from being the rosy picture that was promised to people before settlement; their lives constitute a constant battle to make ends meet.

3.3.4 Compensation Benefits

Some local inhabitants praised the benefits that the HTI project had provided. Many of these people were those who had received benefits from the company, but who had been able to continue their traditional style of life because their land had not been seized. From their perspective, the HTI project had provided benefits by contributing positively to local society. Conversely, the inhabitants whose land had been expropriated and who had lost their livelihoods as a consequence of the project were very dissatisfied. As it had not provided sufficient jobs, and the wages were low, it proved difficult for a laborer to make a living. Although the opportunity to take part in tumpang sarı provided some short-term security, it did not guarantee the inhabitants a living, and there was no long-term guarantee of securing farmland. For the migrants, the opportunity to participate in trans-HTI provided short-term benefits, but labor opportunities in the HTI project were unstable, and life was generally hard. In general, the benefits provided by the HTI project were inadequate.

4. CONCLUSION: "REPRODUCTION OF A FOREST" VERSUS "DESTRUCTION OF A FOREST"—THE DIFFERENCE IN PERSPECTIVES BETWEEN BUSINESS AND LOCAL INHABITANTS

The government and the company saw the HTI project as a way of developing forests with high stumpage volume in areas covered with grassland or bush, that is, as an effective use of land or as the reproduction of a forest. From the perspective of the local inhabitants, however, the HTI project destroyed the forest that they had once utilized and valued as a resource, and forced them to change their lifestyles. These different viewpoints are the result, in part, of different sets of expectations in regard to the role of forests generally. The government and the company require raw materials for the timber processing industry to support national economic and forest resources policy. To them, the area used by the HTI project appeared to be land that was previously utilized ineffectively. The local inhabitants, however, require agricultural land and forest for shifting cultivation, growing rubber trees, grazing, and as a source of various forest products such as firewood, lumber, fruit, wild animals, and medicinal herbs. The project site consisted of gently sloping, hilly country with large areas of grassland and bush. There were some desolate areas that the local inhabitants could not use, but they considered most of the area useful. From their perspective, the useful forest was destroyed and replaced by a forest with low utility value.

In Indonesia, until 1997, the perspectives of local inhabitants were disregarded because of the policy of placing higher value on development. But things may change as democratization and decentralization progress in Indonesia following the retirement of President Suharto. It is clear that industrial tree plantations should respect and include local inhabitants in any future projects.

REFERENCES

Bagian Forest Development, Company M. 1993. Program Kerja HTI – Bina Desa Dan Bina Talang Company M. Palembang: Company M.

BPS Statistics Indonesia. 2001. *Statistical year book of Indonesia 2000*. Jakarta: BPS Statistics Indonesia.

Carrere, Ricardo and Larry Lohmann. 1996. *Pulping the South—Industrial tree plantations and the world paper economy*. London and New Jersey: Zed Books Ltd.

Company M. 1994. A Company M pamphlet, Palembang: Company M.

———. 2000. A Company M pamphlet, Palembang: Company M.

Estate and Forestry Statistics Division. 1997. *Statistics of timber estates 1994*. Jakarta: Central Bureau of Statistics.

Forestry Planning Agency. 2001. *Forestry statistics of Indonesia 2000*. Jakarta: Ministry of Forestry.

Inoue, M. and Y. Yokota. 1994. Relation between industrial tree plantation and local people—Case study in South Sumatra Province, Indonesia (in Japanese). *Study to promote the Carbon*

Sink Project—Annual report 1993. Tokyo: Japan International Forestry Promotion and Cooperation Center.

Japan International Cooperation Agency (JICA). 1990. Final report of the investigation into planned industrial tree plantations in Indonesia (in Japanese). Tokyo: JICA.

Japan Tropical Forest Action Network (JATAN). 1999. Progress with Sumatra pulp problems in Indonesia (in Japanese). *JATAN News* 38: 2–9.

Kantor Wilayah Departemen Transmigrasi dan Pemukiman Perambah Hutan Propinsi Sumatera Selatan. 1993. Naskah Kerja Sama Perjanjian Kerja Pengusaha / Pelaksana HTI Company M Dengan Tenaga Kerja / Pekerja HTI (Transmigran HTI). Palembang.

Kumazaki, M. and B. P. Lilik. 1994. Change of land use and carbon sink forestation in South Sumatra, Indonesia (in Japanese). *Study to promote the Carbon Sink Project—Annual report 1993,* 136–147. Tokyo: Japan International Forestry Promotion and Cooperation Center.

Sato, Y. 2000. *Problems with the wood industry of Indonesia, and policy trends* (in Japanese), 5–9. Mokuzai-Joho. Tokyo: Japan Wood-Products Information & Research Center.

Tasaka, T. 1992. *Eucalyptus business—Relations between Japan and deforestation in Thailand* (in Japanese). Tokyo: Shin-Nihon Syuppann.

World Rainforest Movement. 1999. *Tree plantations: Impacts and struggles.* Montevideo: World Rainforest Movement.

Yokota, Y. and M. Inoue. 1996. "Transmigrasi-HTI" in Indonesia; Focusing on a case study in Southern Sumatra (in Japanese with English summary). *Bulletin of the Tokyo University Forests* 95: 209–246.

Chapter 17

LOCAL FOREST MANAGEMENT OF NATIONAL BIODIVERSITY CONSERVATION AREAS IN LAOS

Kimihiko Hyakumura
Institute for Global Environmental Strategies

Thananh Khotphathoum
Faculty of Forestry, National University of Laos

1. INTRODUCTION

The creation of protected areas is being promoted all over the world as an effective measure to preserve biodiversity, rich ecosystems, and precious animals and plants. At the same time, the livelihoods of local people where protected areas have been established still depend on the rich ecosystems and various forest products from the forests and surrounding areas. After being established, entry into these areas and forest-use are often restricted—some local people may even be evicted. To them, the existence of a protected area is a source of inconvenience and obstacles, and they may consider it difficult to gain any direct benefits. It often appears that the protected area system puts the burden of restricted forest-use on the local people. There are many cases of harmful impacts suffered by local people after the establishment of a protected area—sometimes sowing the seeds of discord between them and government. New policies are needed in many tropical Asian countries that have urgent forest conservation problems, in order to relieve local people of such negative impacts and make implementation of protected areas smoother.

The introduction of a protected area management policy with less negative impacts on the livelihoods of local people is particularly necessary in Laos, with its rich forests and abundant biodiversity resources. As an effective measure to preserve forests and their biodiversity, the government has advanced the establishment of protected areas and various related legal systems, but there are many villages that overlap the protected

M. Inoue and H. Isozaki (eds.), *People and Forest—Policy and Local Reality in Southeast Asia, the Russian Far East and Japan, 261–274.*

areas or surrounding regions. This may cause conflict between local people and the government over land and forest use (Berkmuller et al. 1993).

How has the establishment of protected areas impacted the local people in Laos? And how do the government officials there conserve the protected areas? This report first describes the characteristics of protected area management in Laos and the issues of concern caused by these policies. Finally, it examines case studies of local realities and implementation of legal systems for protected areas.

2. SYSTEMIZATION OF PROTECTED AREA MANAGEMENT POLICY

Laos is a land-locked country located in peninsular Southeast Asia, with an area of 23.7 million hectares and a population of around five million. Population density is sparse, at about 200 people per square kilometer. It is an agriculture-based country, less developed economically because of its mountainous geography, and it has no big cities; many local people make their living in rice paddy fields or from swidden agriculture.

In recent years, the forests have been degraded quickly, worsened mainly by swidden agriculture and improper commercial logging. By 1989, forest cover had shrunk to about 47 percent from about 70 percent in 1940 (Manivong 1992). Although the rate of forest cover in the north is as low as 36 percent, because of the impacts of the civil war and swidden agriculture, in the south, which was somewhat spared, it is as high as 58 percent. The size of protected areas has reached about 3,300,000 hectares, equivalent to about 14 percent of the country's total area. More than 1,100 villages are included in a protected area and surrounding areas.

How was the protected area management policy advanced in this country? Laos, a socialist country since 1975, first adopted a new economic mechanism policy in 1986 called *Jintanakan Mai*, or "New Thinking." Three years later, the First National Conference on Forestry was held in Vientiane, Laos in May 1989 to discuss the rapid deforestation and environmental deterioration and to focus on forest resource management (Inoue and Hyakumura 2000). Afterwards, a legal system concerning land and forests was advanced, along with a protected area management policy. The first research report on biodiversity was published in 1986, and a report on candidates for protected area status came out in 1989. These reports lead to other publications related to protected areas by donors from foreign countries, such as the International Union for Conservation of Nature and Natural Resources (IUCN) and the Lao-Swedish Forestry Programme (LSFP).

Eighteen protected areas were set up in 1993 by the prime ministerial decree on "the National Forestry Reservation over the Country" as *Paa Sagwan Hensaat* (National Biodiversity Conservation Areas [NBCAs]). Another two areas were added in 1995 (Figure 1).

The standard of territory selection for establishing NBCAs, however, was left ambiguous. In some cases, disadvantages were imposed on the local people in swidden areas and areas where they had practiced traditional land uses, including the collection of non-timber forest products (NTFPs).

Figure 1: National biodiversity conservation areas in Laos.

When the first Forestry Law was enacted in 1996, the concept of zoning NBCAs came into existence. The NBCA territory is mainly separated into two categories—the core zone and the buffer zone—as demanded by purpose. The buffer zone is the adjoining area and the border area of a core zone, functioning to efficiently protect the core zone. Forest use in the buffer zone by local people was accepted in this decree, but restricted to sustainable uses.

In June 2001, the Ministry of Agriculture and Forestry enacted the Regulation on the Management of National Biodiversity Conservation Areas, Aquatic and Wild Animals. The item worthy of note in this decree is that the concept of a sub-zone in the buffer zone was clearly outlined, with the purpose of recognizing the rights of local people in forest and land use. Zones for villages, forest resource utilization, village agriculture, and forest production were included in the buffer zone. Designating this area for production activities suggested that the buffer zone area was not a "target of conservation," but an "area where productive activity would be carried out," as permitted on agricultural land and in production forests. Under this legal system, present land and forest use by the local people in NBCAs could be accepted.

2.1 The Land and Forest Allocation Program (LFA)

In Lao villages there is a long history of customary rights in land use; local people have long conducted rain-fed paddy field and swidden agriculture. The decree of the Ministry of Agriculture and Forestry, No. 822, in its "Instruction on Land and Forest Allocation for its Management and Use," and this land and forest allocation program go hand-in-hand. In it the customary right of land use by local people is permitted, and land may be allocated to households who do not have any agricultural land, allowing several rights of possession, use, and inheritance. In return, the local people must take on the obligation to manage forests surrounding the village (Namura and Inoue 1998). The land and forest allocation program contains very important elements that delegate the rights of land and forest use to the local people.

This decree should be adopted by all the villages in Laos, not just in villages concerned with an NBCA. One cannot argue for land use in NBCAs without this legal system in place. The government intended the implementation of land and forest allocation program to carry out the protected area management with this as a start. This would demand two seemingly contrary operations—in order to protect the NBCAs, the land and forest allocation program would allow the rights of land and forest use for local people inside the NBCA.

As of August 2001, this program had been carried out in 5,400 villages of the 12,000 villages in Laos since its start in 1996 (Ministry of Agriculture and Forestry 2001).

2.2 Problems with the Standard of Territory Selection for Establishing NBCAs

The standard of territory selection for establishing an NBCA was left ambiguous. Some cases occurred where local people in swidden areas and areas of traditional land use (including the collection of NTFPs) were disadvantaged.

To prevent the recurrence of this situation, the boundaries of NBCAs should be reconsidered. It is important to clarify the purpose for establishing a protected area, i.e., the object of conservation.

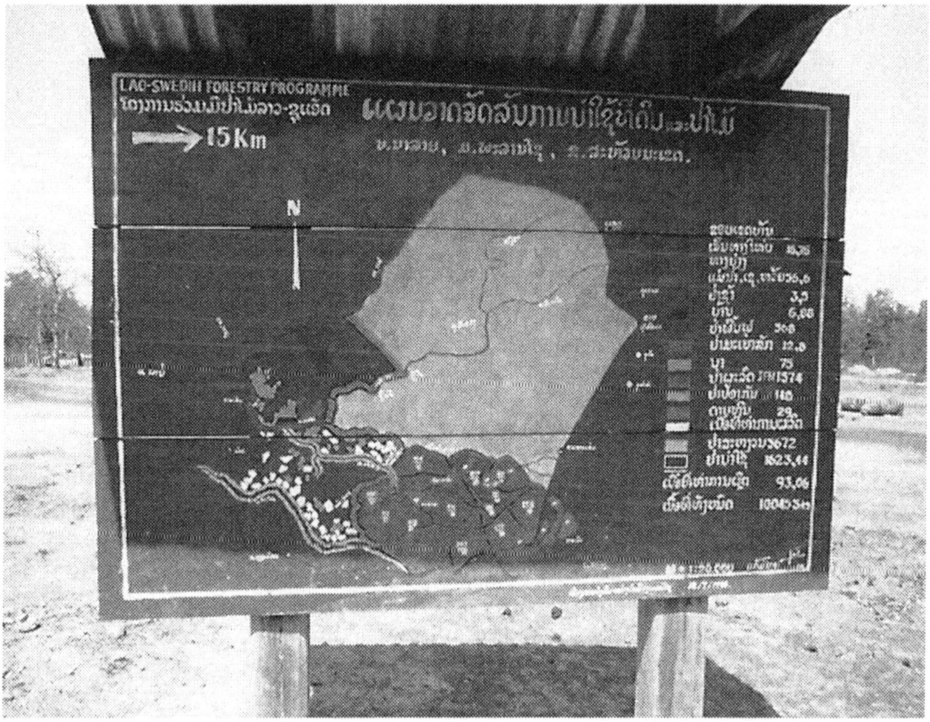

Photo 1: Land use map. After conducting the land and forest allocation program, DAFO places maps like this near or in the village territory. (Photo: K. Hyakumura).

There are other issues concerning the zoning of NBCAs. Most local government staff in forestry still consider that the core zone of an NBCA must be conserved totally, meaning that the concept of buffer zones has not yet fully permeated the local government. Government at the central level, however, did not regard the infiltration of the buffer zone at the local level as being important; it regarded the establishment of the NBCA as important. We cannot help but recognize that the government thinks little of the local people's rights of land use. For this reason, many villages within the territory of an NBCA are put at a disadvantage, with land and forest as an actual buffer zone regarded as a category of the core zone, with restricted use. When the land and forest allocation program is next reviewed, the matter of zoning processes should be re-examined.

3. THE PHOU XANG HE NATIONAL BIO-DIVERSITY CONSERVATION AREA AND RESEARCH SITE

This section provides an outline of the Phou Xang He National Bio-diversity Conservation Area and Research Site, which is located in the center of Savannakhet

Province in the south part of Laos. The Phou Xang He area is mountainous, with gentle slopes at altitudes of 700 meters or less, running east to west, and it's well forested. The Phou Xang He NBCA covers 109,900 hectares in total, with about 92 percent of the area covered by dry dipterocarp and mixed deciduous forests, providing habitat for about 30 protected mammals (Berkmuller et al. 1995).

Most of the local people who live in the Phou Xang He NBCA live in the hill land. These people belong to either the Bru ethnic group, who speak the Mon-Khmer language ethnic group belonging to the Upland Lao, or the Phu Tai ethnic group belonging to the Lowland Lao.[1] There are 117 villages in five districts located in the NBCA and surrounding area, including the Phalanxai district. The villages connected with the NBCA are far from the main road, so it is very difficult to access any administrative services. The villages concerned with the NBCA are in the least developed areas; in other words, abundant forest resources have remained in these areas precisely because they are so remote (Figure 2).

N is one of the villages of the Bru people, the so-called "people of the forest," who traditionally practiced swidden agriculture. Their religion is spirit worship (Photo 2). Presently, they carry out swidden agriculture and rain-fed rice agriculture, as well as collecting non-timber forest products. Raising livestock has recently become popular. The total population of the village is 714 persons living in 118 households.[2] It is 30 kilometers of unpaved roads from the local market and government office in the center of the Phalanxai district, meaning that it can be done in a one-day round trip on foot or by bicycle, but to get administrative service takes longer. It's rarely possible for local government staff to visit such remote areas and waste one day traveling.

The reason why N village was selected is that it is one of the few which have carried out the land and forest allocation program in connection with an NBCA, and the impacts on them can be examined.

4. FOREST USE BY LOCAL PEOPLE

How important are forest products to the livelihoods of local people, such as those in N village? To clarify this point, first, all the households were classified into three economic levels (rich, middle, and poor) using a wealth-ranking method (Inoue 1998). The main source of income of the richer level (11 households) is from paddy fields, from which they can provide rice for themselves throughout the year. On the other hand, the middle (42 households) and poor levels (65 households) run short of rice during the rainy season; it is very difficult to maintain their livelihood on just paddy fields and swidden agriculture. They supplement their livelihood by raising small livestock (pigs and chickens), wage-earning jobs, fisheries, and NTFP collection.

Next, an interview survey of their use of forest products, such as NTFPs and timber, was conducted. By measuring the quantities of forest products used, a clear dependence

[1] Ethnic groups in Laos can be classified into three categories: Lao Lum (Lowland Lao), 56 percent; Lao Theung (Upland Lao), 34 percent; and Lao Soung (Highland Lao), 9 percent.

[2] Phalanxai District Agriculture and Forestry Office.

on the NBCA by local people is revealed. The number of households interviewed was 25 percent of the total and were identified by key village informants, the most influential persons in a village, i.e., village leader, sub-leader, village elder group, or forestry volunteer.

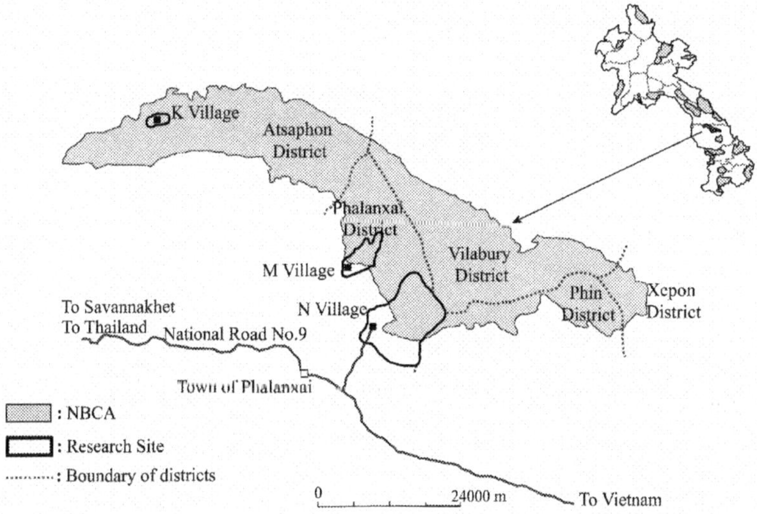

4.1 Income Generation From Non-timber Forest Products

At first, the degree of dependence on cash from forest products was calculated, based on the interviews. All forest products which could be selected by local people were non-timber forest products, i.e., cardamom (a raw material for medicine and spices) for transfer to China, yaang oil (a resin for torch and waterproof materials), small animals, and handicrafts for transfer to local markets.

The level of cash dependence ranged from 14 percent up to 33 percent (Table 1). Therefore, it is clear that local people of all levels are dependent on the forest as a source of income. On the other hand, the cash dependence from the NBCA was 10 to 18 percent. The degree of dependence on the NBCA at the poor level is less than the middle-income level, however, because there are fewer male laborers; in some households the head is a woman with many children. As a result, many poor households were limited in their ability to go into the forest to collect NTFPs.

4.2 Forest Use for House Construction

In N village, the people live in above-ground style houses (Photo 3). In this situation, we asked to what degree they depend on the forest for house construction materials. Each household interviewed was asked to calculate the amount of timber or trees used in all points of house construction—in the roof, the walls, the floor, and the pillars. In some cases, more than two kinds of materials were used in one point.

At each income level, the dependence on the forest for house construction materials was very high—over 90 percent, with half gotten from the NBCA (Table 2). Therefore, it is clear that a protected area management policy would have a significant effect on house construction for all households.

Photo 2: A spiritual forest located in the rich forests inside an NBCA. (Photo: K. Hyakumura).

5. IMPLEMENTATION OF THE POLICY

Communication from government to the village

It was not until 1998 that the local people in N village became aware of the new national biodiversity conservation area. They had not received any information since the NBCA was established in 1993, five years earlier, because of weak government implementation.

It was about 1996, when the Lao Swedish Forestry Programme was starting up, when the budget of the protected area was completely introduced into Savannakhet Province. Information on the protected area had not yet reached the district agricultural and forestry offices (DAFO) in affected districts. Besides this, the budget was still insufficient for implementing the DAFO activities. By the time the local people in N village were aware of the existence of the NBCA, it was time to implement the land and forest allocation program. Under the program, the land and forest use of the village were supposed to be discussed among local people and local government staff, so efforts were made for local notification of the NBCA.

Table 1: Cash income from the forest and the degree of dependence at each economic level.

Economic level	Cash income (kips/year)	Cash income from forest (kips/year)	Degree of dependence on cash from forest	Cash income from NBCA (kips/year)	Degree of dependence on cash from NBCA
Rich (n=3)	416,667	60,000	14.4%	40,000	9.6%
Middle (n=10)	285,600	74,500	26.1%	52,500	18.4%
Poor (n=17)	226,353	74,588	33.0%	29,882	13.2%

Source: Hyakumura (2001).
Note 1: "Forest" means forest including the NBCA, as well as scrub forest.
Note 2: At the time of this study, one U.S. dollar was equal to 7,700 kips.

5.1 The Process of Zoning in the Protected Area

The zoning of the NBCA in N village under the land and forest allocation program was carried out by the DAFO in Phalanxai, but the LSFP had also set up N village as a model of the program. It created a draft land-use plan, including a land-use map initially based on interviews and data collection from the local people, paying attention to the rights of traditional land use. During a three-week long workshop, the draft, the program, and its implementation were discussed.

The draft plan made by the LSFP, created by experts after conducting a survey in the village, used a top-down process. The plan was then discussed directly with the local people during the workshop, using the professional-guided participatory approach, and the plan to implement the program, with respect for traditional land uses, was finalized (Hyakumura 2001).

Originally, the boundary between N and the neighboring village was not supposed to be close or overlap, but at the suggestion of the LSFP, the boundary between those villages was joined together to expand the territory. The intent was that the LSFP was trying to prevent the trespass of local people into the core zone, managed by the government, and to expand the territory of the village, as well as to legally retain the land and forest by the local people. But all of the NBCA in the territory of N village is regarded as core zone, although the program can admit local people to use the forest in the area concerned. The LSFP drew the conclusion that the NBCA that overlapped with the territory of N village was essential to biodiversity conservation, because it including a drinking spot for elephants, inferring that the LSFP wavered between biodiversity conservation and forest use by the local people.

Photo 3: The above-ground style house in N village. Most house construction materials are made from forest products. (Photo: K. Hyakumura).

5.2 Improving the Land and Forest Allocation Program

In the land and forest allocation program, it was necessary to consider both biodiversity conservation and the forest use of the local village people where the rights of traditional forest use overlapped with the NBCA. At this point, the professional-guided participatory approach employed by the LSFP was very useful for both drafts of the plan, which were based on data analysis, including natural and socio-economic aspects and the possibility of adapting to people's opinions. On the other hand, this approach requires considerable data from the villages in advance, and it costs much more time and expense than the normal program. Therefore, increased budget allocation and capacity building within DAFO and with their staff should be considered. It is very difficult to implement this program at the district level, as well; it might be necessary to revise it to make it easier, with a less expensive system and simpler survey method. Creating a manual and providing training for local government staff is indispensable for capacity building.

Budget allocation should be considered a follow-up activity after the LFA is finished. Presently, the DAFO has no budget for ongoing activities after the program is completed. Therefore, DAFO staff cannot conduct regular follow-up activities, and can only visit the village when they have another obligation. It is a real disappointment that the activity to carry on with it could not be carried out, but the local peoples' awareness of the NBCA was high for a time. Particularly in the NBCA, where the rights of forest use and forest products have been changed under the LFA program, follow-up is very important.

Table 2: The number of points of house construction materials, number of points of forest product, and the degree of forest products dependence for house construction.

Economic level	All points, house construction materials	No. points, house construction materials from forest products	Degree of dependence for house construction	Number of points of forest products (from NBCA)	Degree of dependence for house construction (from NBCA)
Rich (n=3)	4.67	4.20	90.0%	2.33	50.0%
Middle (n=10)	4.40	4.20	95.5%	2.50	56.8%
Poor (n=17)	4.06	4.00	98.6%	1.94	47.8%

Source: Hyakumura (2001).

Incidentally, the LFA program should include reinforcement activities, such as workshops, where the government should clearly inform the local people of the existence of the NBCA and its use. In a word, the LFA programs should distribute consistent information to the local people and play a very important role in educating them on forest management policy, including the NBCA management policy.

Under the protected area management policy, the land and forest allocation program holds great significance. The LFA program is very highly evaluated as a participatory forest management system for use in Southeast Asian countries (Inoue 2000), so it would be expected that implementation takes local people into consideration.

6. THE IMPACT ON LOCAL PEOPLE'S LIVELIHOOD DUE TO THE INTRODUCTION OF THE NBCA POLICY

How did the introduction of the protected area management policy change the local people's use of agricultural land? The local people in N village have long practiced swidden agriculture in dense forest located in the present NBCA. There were swidden agriculture areas inside the dense forest visible in an aerial photo taken in 1992 in the present NBCA. Later, these decreased gradually after the heavy-duty decrease in swidden agriculture mandated by government policy, as well as awareness of the workload required by swidden agriculture by the local people themselves.

6.1 The Decrease in Swidden Agriculture Caused by the Land and Forest Allocation Program

According to our survey using the wealth ranking method, the LFA program caused the number of households that conducted swidden agriculture to decrease rapidly, dropping from 91 percent to 53 percent.

As for the number of households conducting swidden agriculture at each economic level, in the rich level it decreased from 91 percent to 36 percent. On the other hand, in the poor level, the decline was smaller, from 83 percent to 72 percent. The decrease of swidden land under the LFA program has been effective at only the rich level, but not for the poor who still must carry out swidden agriculture (Table 3).

Interviews with the local people revealed two reasons why swidden land area was decreasing. The first reason is the diversion and renouncement of swidden land. After the LFA program was implemented, land for cultivating rice as the staple food was to be acquired from swidden land or from new farmland. Most of the reasons for the decrease in swidden land are mentioned above. The second reason is the forfeiture of the rights of swidden land use. According to the government's instructions, swidden agriculture should not be allowed in the NBCA and production forest. Three households out of the thirty surveyed in N village had forfeited their right of land use in the swidden area because of the LFA program.

Table 3: Changes in agriculture area caused by the land and forest allocation program.

Economic level	Number of households practicing swidden agriculture				Number of households that acquired agricultural land	
	Before LFA program	(%)	After LFA program	(%)		(%)
Rich (n=11)	10	90.9	4	36.4	2	18.2
Middle (n=42)	40	95.2	12	28.6	18	42.9
Poor (n=65)	57	82.5	47	72.3	31	47.7
Total (n=118)	107	90.7	63	53.4	51	43.2

Source: Interviews in N village.

6.2 The Acquisition of Agricultural Land

Incidentally, in examining the differences at each economic level in the acquisition of additional agricultural land, 18 percent of households at the rich level acquired land, as did 43 percent of the middle level and 48 percent of the poor level (Table 3). The reason why so few of the households in the rich level got agricultural land is that they already had enough land for cultivation, so they did not need any more. On the other hand, there were many households from the middle and poor levels that got agricultural land, because of their lack of land for cultivating rice.

But more than half of the middle and poor households could not acquire any land. The land offered for cultivation are dry dipterocarp forest, with many trees still remaining. It is very difficult and requires much labor to reclaim land for paddy fields; it takes four laborers and one buffalo to cultivate 0.2 hectares per year. As a result, the households who do not have enough labor could not acquire the land to cultivate. In addition, taxes on such land are higher than for normal agricultural land without reclamation.

Among the poor households who could not acquire cultivated land, many households collected non-wood forest products for supplementary food. When a poor crop of rice occurred in the swidden lands, local people were forced to collect more NTFPs than usual, and this put the rich forest products in the NBCA at risk. The LFA program could not successfully produce the benefits needed by the poor households, and unfortunately, the impact on the forest resources in the NBCA still continues.

As mentioned above, it is clear that the benefits of the introduced policy did not reach the poor, who have the highest dependence on forest products. These programs have a lasting effect, particularly on local people who are most dependent on swidden agriculture.

7. CONCLUSION: PROTECTED AREA MANAGEMENT POLICY IN THE FUTURE

Local people have long used forest resources in areas now protected. With rich forests nearby, it is natural that local people would consider using these resources. If the government applies the old approach, which excludes the local people from the protected area, disputes may occur again. At the same time, incentives for the local people to conserve the protected area have not been strengthened.

Therefore, a system of protected area management that includes forest use delegated to the local people is needed. And local government staff in forestry would be engaged as protected area managers. It is also necessary to have facilitators who can share traditional forest-related knowledge (TFRK) with users of natural resources or local people, support the planning of protected area management by the local people themselves, with consideration for the local reality of natural, economic, and societal aspects, and transmit the agreement formation with the manager of natural resources or local government staff. The facilitator could be a resource person from some of the foreign donors and non-governmental organizations, but at the present time, in Laos, these organizations and individuals are not able to conduct these activities effectively enough. Therefore, volunteers who can act as facilitators should be attracted from local government staff and local people.

To effectively conduct the program in protected areas, the introduction of merits for the poor level of local people is a point to be considered. As there are many different stakeholders, even in a village, they must all be approached differently. The people who receive the first benefits from project activities are usually the rich in the village. There are two reasons why the poor might not receive benefits from the program initially. First, it is difficult to get program information to the poor level accurately. Second, it is difficult to participate in the program due to the lack of an adequate workforce and funds for conducting some of the activities. As mentioned above, an insufficient development program may increase the gap between the poor and rich among the local people.

The people in the poor level might need to continue to depend on the forest resources in the NBCA for their livelihood, although the programs are already underway. In this case, a system should be established so that local people can still make use of the forest resources in these areas.

REFERENCES

Berkmuller, K., B. Phanthavong and V. Vongphet. 1993. *Protected area system planning and management in Lao P.D.R: Status report to mid-1993.* Vientiane: Lao-Swedish Forestry Programme and IUCN.

Berkmuller, K., S. Southammakoth and V. Vongphet. 1995. *Protected area system planning and management in Lao P.D.R. Status report to mid-1995.* Vientiane: Lao-Swedish Forestry Programme and IUCN.

Hyakumura, Kimihiko. 2001. Issues of conservation forest management policy in Lao P.D.R. (in Japanese). *Forest Economy* 54(12): 22–33.

Inoue, Makoto. 1998. The measure of tropical forest conservation at local level (in Japanese). In *Global Environmental Studies Vol. 6*, ed. K. Takeuchi and M. Tanaka, 50–56. Iwanami Book Publish.

———. 2000. Institution and main actors of participatory forest management in Southeast Asian countries: An approach of forest sociology (in Japanese). *Journal of Forest Economics* 46(1): 19–26.

Inoue, M. and K. Hyakumura. 2000. Forest policy of Laos in terms of local participation. In *A step toward forest conservation strategy (1), IGES interim report 1998,* 300–308. Institute for Global Environmental Strategies.

Manivong, Kamphay. 1992. *Forest cover and land use in Lao P.D.R. Final report on the Nationwide Reconnaissance Survey.* Forest inventory report No. 5. Lao-Swedish Forestry Co-operation Programme.

Ministry of Agriculture and Forestry, Lao P.D.R. 2001. *Report on land and forest allocation and shifting cultivation stabilization in 1996–2000, and Plan on land and forest allocation and shifting cultivation stabilization in 2001–2005* (in Lao).

Namura, T. and M. Inoue. 1998. Land use classification policy in Laos: Strategy for the establishment of an effective legal system. *Journal of Forest Economics* 44(3).

Chapter 18

LOCAL FOREST MANAGEMENT IN THE RICE TERRACE AREA OF BANAUE, THE PHILIPPINES

Atsuko Hayama
Kyoto University, Graduate School of Asian and African Area Studies

1. INTRODUCTION

This chapter seeks to examine the effects of population growth, market forces, and local institutions (among many plausible factors) on the relations between forests and local people through a case study in northern Luzon.

The rapid and widespread deforestation that occurred in the upland Philippines after World War II was directly related to population growth and market forces (e.g., Repetto 1988; World Bank 1989). Population growth was manifested as an expansion of subsistence-oriented agriculture, including shifting cultivation and grazing, while market forces are visible in the market-oriented, large-scale logging operations, plantations, pastures, and cash crop cultivation.

At the same time, extensive forest cover still exists in some upland areas where human activities have long occurred. This simple observation poses the question: Why do some upland areas have wider forest cover, while many others have lost most of theirs? Another question concerns how population growth and market forces have influenced the relations between local people and forests in the upland areas that have wider forest cover.

A large number of empirical studies of common property arrangements all over the world show that local institutions created by self-organized resource users tend to control and mediate factors such as population increases and market forces, and thus prevent unsustainable resource use (in the case of forests, e.g., Brightman 1987; McKean 1992; Shigetomi 1996, 245–282; Agrawal and Yadama 1997). In this case, the term *institution*, as defined by Ostrom (1990, 51), refers to "the sets of working rules that are used to determine who is eligible to make decisions on some arena, [and] what actions are allowed or constrained." It is hypothesized in this study of the Philippine

M. Inoue and H. Isozaki (eds.), *People and Forest—Policy and Local Reality in Southeast Asia, the Russian Far East and Japan, 275–286.*

uplands where there is wider forest cover that the local institutions which have developed exert control over local people's forest use, thus preventing deforestation. If so, the question is how these institutions were able to control people's forest use in response to population growth and market forces under changing socio-economic conditions.

The answers here should provide some insights for examining local forest management, which the Philippine government, like many other tropical countries, has adopted as a national strategy over the last two decades in order to achieve the rehabilitation of deforested and degraded uplands, as well as to improve the well-being of forest-dependent people.

2. STUDY SITE

The study site is located in the province of Ifugao in the Cordillera, a series of high mountain ranges in northern Luzon. The omnipresent terraces in the higher elevations, flooded for rice production, are the most impressive feature of the landscape and have become a popular tourist attraction. Extensive forest cover in some rice terrace areas also makes a striking contrast with many other upland areas of the country.

The ratio of rice terraces to forested areas varies from place to place in Ifugao, and in the municipality of Banaue, a center of tourism, there is more forest than in other municipalities in the province. The study site is Barangay Poitan in Banaue.[1] (A *barangay* is the smallest administrative unit in the Philippines, and in Banaue it is almost synonymous with the traditional agricultural district sharing the same watershed [Conklin 1980, 6].) Banaue has an elevation ranging from 700 to 2,000 meters in which forest vegetation is classified as submontane and montane forest, represented by oaks (*Quercus* spp.) and the family Lauraceae (*Neolitsea, Cinnamomum,* and *Litsea*) (Science Education Center 1971, 32). Pine trees are dominant on the high ridges. The predominant grass is *Miscanthus* spp., which appears in swidden fallow (Conklin 1980, 9). Topographically, Barangay Poitan has gentle to very steep slopes, with altitudes ranging from 800 to 2,000 meters. Gentle slopes (15 to 25 percent grades) are located at 800 to 1,200 meters, where all the settlements and rice terraces are widely distributed.

3. THE DYNAMIC RELATIONSHIP BETWEEN FORESTS AND LOCAL PEOPLE

The relationship between the forests and the local people in the study area went through three phases: a period of increasing population pressure on the forests, a period

---

[1] Data for this study were collected in December 1998, January 1999, January 2000, July 2000, and March 2001. Interviews were done mostly with residents in Barangay Poitan. Municipal officers, government forest officers, and staff of the project, funded by the European Union and the Philippine government, were also interviewed.

of decreasing population pressure on forests with the area's integration into state and wider markets, and then a period of wide-scale tree planting with external assistance.

3.1 Period of Increasing Population Pressure on the Forests

Conklin, who conducted his research in the early 1960s, shows that forests occupied the largest part of about 40 percent of the total land area of Barangay Poitan, followed by rice terraces at 29 percent, and swidden fields and fallow lands at 27 percent (Conklin 1980, 8).[2]

Eder, who conducted his research in 1979, stated that the rice terraces suffered water scarcity during the dry season (March and April).[3] It is said that water shortages have increased since the 1950s and 1960s and that, as an immediate cause, deforestation expanded through land conversion from forests to swidden fields and rice terraces. Although selective, logging for firewood, woodcarving, and house construction also contributed to deforestation, with firewood extraction having a greater impact (Eder 1982, 107).

In the early 1960s, a family of five managed 0.5 hectares of rice terraces on average, producing about five months worth of rice (Conklin 1980, 10, 37). Food consumption for the remaining seven months, without rice, was supplemented by sweet potatoes grown in swidden fields (ibid., 10). Sweet potato was, therefore, an important staple food for the majority of people.

Elder residents in Barangay Poitan reported that new rice terraces have not been constructed since the late 1960s. This indicates that the growing population since that time has been mainly supported by swidden agriculture, resulting in the expansion of swidden fields and fallow lands.

Although population data for Banaue and Barangay Poitan in the pre-war and early post-war periods is not available, it is highly suggested from my field survey and the existing literature that population growth was a major driving force in deforestation; namely, that land was converted from forests to agricultural lands as long as local people depended primarily on the land and forest resources for their livelihood.

3.2 Period of Decreasing Population Pressure on Forests for Land Conversion

Available data on the population of Banaue in 1975, 1980, and 1990 shows that the population continued to increase,[4] but research conducted by Herzmann et al. reveals

[2] Conklin's research covered about 96 square kilometers, centering on the town of Banaue and including Barangay Poitan.

[3] Eder conducted his research mainly in Barangay Bocos, which constitutes the town of Banaue (1982, 102). Barangay Poitan neighbors Barangay Bocos.

[4] The population of Banaue in 1975, 1980, and 1990 was 14,005, 16,022, and 16,943, respectively, and the average annual growth rate from 1975 to 1980 and 1980 to 1990 was 2.72

that swidden fields and fallow lands were reduced from 6.7 percent and 34.5 percent, respectively, in 1963 (based on Conklin's research) to 1.1 percent and 13.5 percent in 1997, while forest cover and house compounds increased from 15.1 percent and 2.2 percent in 1963, respectively, to 34.3 percent and 7.6 percent in 1997 (Herzmann et al. 1998, 85–86).[5] This shows that population pressure on the forests turned out to be negligible.

Two reasons are given for the decline of swidden fields and sweet potato production. One is the flow of commercial lowland rice to the area, which replaced the sweet potato as a staple or supplementary food. The other is the increasing opportunities for income generating activities, particularly off-farm, which enabled local people to buy commercial rice. These two reasons are closely related to the construction of a major all-weather highway connecting Banaue to the lowlands in the 1970s and government efforts to promote the development of tourism in the area, and thus its integration into state and wider markets. Because only a small number of people are now engaged in swidden agriculture in the forestlands, this has lead to an increasing risk of crop invasion by rodents, wild pigs, and other wild animals, consequently requiring more labor to watch the fields. At present, sweet potato is grown on abandoned terraces and terrace embankments, but on a much smaller scale than before.

The main reason for their need to search for income generating activities lies in the comparatively low rice production yields (about 2.5 tonnes per hectare [Conklin 1980, 35; Barthelmes et al. 1988, 32]) and small land holdings (about 0.2 hectares on average as of 1998).

Two features characterize their off-farm income generating activities. One is the substantial role of wood-related craftwork. The other is the great number of seasonal migrant workers, especially those in their twenties. The most prevalent income generating activity is craftwork for tourists, namely, woodcarving by men and handloom weaving and wood/rattan handicraft making (such as baskets and vessels) by women.[6] About three-quarters of all the households in the barangay have members engaged in at least one of these craft activities, among which woodcarving is the major one. Procurement of the necessary wood is the responsibility of each craftsman.

---

percent and 0.56 percent, respectively (Provincial Planning and Development Office 1993: 23). The reason why the average annual growth rate form 1980 to 1990 is unusually low is not known.

5 Herzmann et al. conducted their research in part of Barangay Tam-an and a lower part of Barangay Poitan, covering 149 hectares (1998, 79). Barangay Tam-an constitutes the town of Banaue and neighbors Barangay Poitan.

6 In Ifugao, people had long practiced woodcarving, but doing it for market purposes started after WW II, when woodcarvers from Hapao (now one barangay in the municipality of Hungduan, neighboring Banaue) were employed by an American-owned souvenir shop in Baguio City. By the early 1950s, woodcraft markets in the U.S. had been established (Sajor 1999, 177). It is said that woodcarvers in Hapao brought their woodcarvings to Baguio City via Banaue, and people in Banaue, who noticed the economic opportunity, started woodcarving to sell in Baguio City or Metro Manila.

It is clear, therefore, that the pressure of population growth on the forests, particularly land conversion to swidden fields, was mitigated by the change of local livelihood activity—namely, the increase in off-farm income generating activities.

The effects of increasing off-farm income generating activities also appeared in the rice terraces. Herzmann et al. show that rice terraces declined from 34.5 percent in 1963 to 29.0 percent in 1997 (1998, 86). Most of the previous rice terraces were abandoned or left fallow for a long period of time, consequently turning into shrubland and young secondary forest. The main reason is lack of water in the terraces, which was caused by irrigation canals being cut off by landslides or over-grown grass. When the length of blocked canals is long, even water beneficiaries are reluctant to take on the burden of the immense cost and labor requirements without remuneration for repairs, opting for eventual abandonment of rice production.[7]

3.3 Period of Larger-scale Tree Planting

The fact that many woodcarvers work and procure wood outside Banaue indicates that the availability of suitable trees for woodcarving has declined in local forests. Therefore, while overall forest cover increased as a result of declining swidden agriculture, the local forest resources are insufficient to supply all the local people, especially for woodcarving.

Government support for reforestation first appeared in Barangay Poitan in 1999. Free distribution of seedlings by a project jointly funded by the European Union and the Philippine government encouraged people to plant tree seedlings in their private forests. In Barangay Poitan, about 30 percent of the total households showed interest in planting seedlings without remuneration. The seedlings distributed, about 200 each on average, were species of mahogany (*Swietenia macrophylla*), pine (*Pinus kesiya*), raintree (*Albizia saman*), narra (*Pterocarpus indicus*), and gmelina (*Gmelina arborea*), which were all requested by local people for woodcarving and house construction purposes.[8] They were planted for the enrichment of secondary forests or reforestation in former swidden fields.

The forest cover in Barangay Poitan at present is estimated to account for about 60 percent of the land area.

---

[7] Since 1995, when the rice terraces in Banaue were put on the UNESCO World Heritage List, repair work has been subsidized. It is said that subsidies were also allocated to Barangay Poitan, but were not sufficient to cover costs for repairs of all the damaged rice terraces.

[8] The limited number of seedlings available due to budget constraints did not meet the needs of all the people who showed interest in planting seedlings. Many believed that they had to pay for the seedlings and declined to participate.

4. ROLES OF LOCAL INSTITUTIONS IN PEOPLE'S FOREST USE

Local institutions (e.g., working rules) in forest use were observed in the study area. The effect of these institutions in local forest use in response to population growth and market forces is examined here.

The surrounding forests are customarily classified by local people into two categories: community forest (*inalahan*) and private forest (*pinugo* or *muyung*). "Community" here means a traditional agricultural district sharing the same watershed (Conklin 1980, 6), now called a barangay. Community forests are situated over the mountaintops, while private forests are scattered ubiquitously among the settlements and rice terraces.

4.1 Community Forest

The community forest in Barangay Poitan is located above 1,200 meters in altitude, extending to about 2,000 meters at the peak of the mountains. While the major role of the community forest is as a watershed, the lower portion is reserved for community swidden fields, and anyone in the barangay can cultivate crops. Recently, as already mentioned, no-one has been engaged in swidden farming in the community forest, and these abandoned fields are regenerating into shrubland and secondary forest.

One study in Ifugao found that anyone who tried to log in the community watershed was regarded as a community criminal (Guthrie 1964, cited in Klock 1995, 7). In Barangay Poitan's community forest there are no such regulations governing resource harvesting either by volume, species, or period of extraction; extraction of any resources is permitted to any barangay resident. Customary regulations governing resource use in the community forest are, nevertheless, observed. Cutting down trees for sale is forbidden, and members of other barangays are restricted in their extraction of resources, including the hunting of game. As Conklin (1980, 31) noted, in the past, "Regional trespass (by other community members), for hunting or woodcutting, can lead to bloodshed," while now, trespassers found hunting or logging are required to pay a fine to the barangay.

Although most of the large useful trees have been logged for house construction and woodcarving, the large-scale impact of deforestation in the community forest has been prevented. This is because of their practice of selective logging and the absence of roads in the community forest, which prevents the transport of large quantities of wood.

4.2 Private Forests

4.2.1 Formation and Evolution of Property Rights

Interspersed private forests protect rice terraces below from runoff and erosion and provide firewood to the majority of the local people. Conceivably, private forests are formed in two ways: encroachment of natural forests and succession from swidden fields.

Encroachment of natural forests took place when the number of residents was small and most of the forests were communally owned or not claimed by anyone. Respected elders divided up some forests for residents' families and kinsmen as private property (Hangdaan 1999, 8–9). Gradually, many parts of such private forests were converted into swidden fields and rice terraces. The remaining forests were left for the future generations of the families (Barton 1919, 40). This type of forest, which has never been converted to other land uses, is often found on hilltops.

Succession from swidden fields took place when a swidden cultivator invested his labor in fallow land to develop a private forest (Conklin 1980, 9; Hangdaan 1999, 11; Sajor 1999, 49). In the past, one could create swidden fields in unclaimed forests. Abandoned swidden fields were regarded as belonging to no-one or as communal property among community members. If someone, generally the swidden cultivator, invested labor in the land, by cutting grass to facilitate natural regeneration and/or planting tree seedlings, for instance, the land and subsequent established forest were recognized by other community members as the property of this enterprising individual. These efforts became the basis of the claim for one's private ownership. This type of private forest was planted with many kinds of trees for house construction and fruit production.

Compared with the firmly established inheritance system of rice terraces, i.e., non-fragmentation of a group of terraces and a weighted, bilateral form of primogeniture (Barton 1919, 40; Conklin 1980, 36), the situation for private forests is not as simple. This is because private forests are not regarded to be as economically valuable as rice terraces. This forms a striking contrast to a case in Kiangan, Ifugao (about 15 kilometers south of Banaue), where coffee trees are widely planted and the firmly established inheritance system of forestlands (coffee groves in this case), as used with rice terraces, is observed.

Three types of private forest can be identified based on the form of ownership: single-owned forest, sibling-owned forest, and more-than-two-generation-owned forest. Transferring private forest ownership varies; for instance, one's single-owned forest being transferred to one of the children (to become again a single-owned forest) or to all of them (to become a sibling-owned forest). It is notable, however, that the terms local people use do not necessarily indicate actual conditions. In the cases of a sibling-owned forest and a more-than-two-generation-owned forest, even if one claims his/her co-ownership of a forest, that person does not always hold rights equivalent to other co-owners in that forest. The most distinct right for private forest owners is to log trees.

Some features of private forests can be pointed out. One is that the local concept of privately owning a forest is akin to a use right. Claiming the actual ownership of a forest, particularly one owned by siblings or more than two generations of a family, is firmly related to how much labor one has invested in the forest. The other is, therefore, that many of the forests termed as being owned by siblings or several generations are actually single-owned forests in practice. The person who has invested the most labor to manage the forest would have the strongest sense of ownership compared to other co-owners, and these co-owners also recognize that they cannot claim the same use rights, particularly with regard to logging.

It is a general understanding among many of the people, nevertheless, that private forests, as well as rice terraces created by ancestors, are properties to be transferred to offspring.

4.2.2 Local Practices in the Use and Management of Private Forests

The right to log trees belongs exclusively to the owner who has invested labor in the forest, and nominal co-owners need permission from the actual owner, as noted above. However, non-owners are also allowed to use anyone's private forest as long as they do not log healthy trees. Non-owners who need firewood can extract branches from standing trees and from the forest floor without asking permission. Logging withered and crooked trees is also permitted. Therefore, even households who have no private forest, accounting for 6 percent of all the households in Barangay Poitan, can secure firewood from others' private forests in the vicinity.

A customary requirement of both owners and non-owners when logging trees is, as emphasized by the elders, to clean the forest floor by cutting surrounding grass and piling all the cut branches and leaves in one place. This is meant to protect and facilitate the re-growth of seedlings. At the same time, the piled branches are made available to the old women living alone who have to collect firewood by themselves.[9] In this sense, local institutions secure resource allocation for the socio-economically weak members of the community.

Frequent failure by a person to clean the forest floor after logging elicits a warning from owners and eventually a ban on lax users. Frequent visits to one's own private forests for firewood collection, usually once or twice a week, make it easy to monitor forest conditions. Upon the discovery that trees were recklessly logged, a sign of tree theft, the forest owner tries to track down the offender, even outside the barangay, and impose a fine.

Forest-related problems such as tree theft, disputed ownership of planted trees, or boundary conflicts (previously settled through consultation between the families concerned, accompanied by ritualists or "priests" [*mumbaki*] knowledgeable of the genealogy of community members) are now settled by a barangay captain and council. In the case of boundary conflicts or disputed ownership of planted trees, judgment depends on the recollections of the ritualists and the community members as to which party or which party's ancestors made efforts to foster the establishment of the forest or planted the trees in question.

4.2.3 Traditional Values in Establishing Local Institutions

Observing public order, namely, respecting local institutions, is partly attributable to the special topographical features of this area. Steep mountainous conditions, such as in

[9] Firewood collection is a task of men, since chopping down a tree, cutting it into firewood, and transporting it to the house requires considerable strength.

Banaue, impede the use of vehicles and draft animals on the narrow trails. Thus, manual labor is the only way to transport people and goods, and residents must depend on each other for assistance. The maintenance of public order to guarantee safety and security in daily life is the local people's shared norm. The fear of social reproach is usually an effective counter-force against the temptation to benefit from any violation of public regulations or to become a "free rider" (Hayami and Kikuchi 2000, 133).

Observing public order in this area is also attributable to the accumulated experience of rice production. Residents in the traditional agricultural district (now almost synonymous with a barangay territory) are organized on the basis of rice production, such as labor collaboration and water sharing. In this way, it is assumed that a long history of face-to-face interaction in daily life has nourished shared norms, leading to the establishment of local institutions.

4.3 The Effect of Local Institutions in Response to Changing Socio-Economic Circumstances

The conversion of forests to agricultural lands has not been controlled by local institutions, but they have mediated the effects of population increase (namely, the increasing demand for firewood) and market forces (the increasing demand of wood for woodcarving) by controlling the method of extracting trees by using rules of not impairing the natural regenerative capacity of the forests and not stealing trees.

The area's integration into state and wider markets, however, brought about differences in local forest dependency and changes of the traditional values in existing local institutions. The degree of forest dependency is generally related to household income.

The households in the barangay are divided into four groups, based on economic activities. The first group, those with the highest incomes, is comprised of those in which at least one household member is employed full-time, as a teacher or civil servant in Banaue or a factory worker in urban areas, for example. The second group includes those in which at least one household member is self-employed in an activity such as carpentry, merchandising in *sari-sari* stores (small-scale general goods), or working as a tricycle driver or handicraft middleman. Many household members in the first and the second groups are also engaged in making handicrafts. The third group includes those whose income generating activity is predominantly handicrafts. The fourth group consists of those whose income is generated largely from work as agricultural wage laborers. None of the household members in the fourth group are engaged in handicraft work. The ratio of each group in all the households in the barangay, from the first to the fourth, is 19.9 percent, 21.2 percent, 45.3 percent, and 13.6 percent, respectively.

Firewood is still the dominant energy source used for everyday cooking; 77.1 percent of all the households in the barangay depend solely on firewood. Although there has been an increase in the use of liquefied petroleum (LP) gas, most gas users belong to the higher income groups who have full-time employment or self-employment that is not handicraft-related. The majority of firewood users belong to households whose income sources are primarily from handicrafts and agricultural wage labor.

Using LP gas and less firewood and having no family members engaged in woodcarving leads to fewer visits to private forests, consequently decreasing the monitoring of forest conditions. When an owner rarely visits his forests, tree theft is often left undiscovered and unreported. In 1997 the barangay council passed an ordinance stipulating a fine of 18 pesos per board-foot of stolen trees, an amount that the councilors thought would discourage tree theft. They said that tree theft problems have not been reported since then. In some private forests that were seldom monitored, however, trees were stolen, cut, and branches and wood chips were scattered over the forest floor. This confirms that regulations per se (paying a fine in this case) without monitoring and sanctioning do not always deter people's action.

Meanwhile, a trend towards investing more labor in private forests, primarily by those who are engaged in woodcarving, has also been observed. The impetus was the distribution of a large number of free seedlings in 1999. The majority of seedlings were planted in single-owned forests or in sibling-owned and two- and three-generation-owned forests, where the planters are the actual owners and have the strongest claim over the resources. Planting a large number of seedlings and frequent visits to monitor the land further strengthen the consciousness of land and tree ownership.

It is, therefore, suggested that there are two trends appearing in local forest management at the study site: (1) leaving private forests less managed by households who do not depend on forests for their livelihood and (2) management by households who depend on them.

5. CONCLUSION: IMPLICATIONS FOR OTHER PHILIPPINE UPLANDS

Local forest management has been emphasized as a crucial forestry strategy in the Philippines. This case study shows local people's capacity to manage their surrounding forests. Four findings can be pointed out from the case study that suggest implications for other Philippine uplands.

1. Local people can take a substantial role in forest management, but this is not always totally efficient. In the study area, it was found that the local people could not afford all the costs of forest management (for procurement of a large number of tree seedlings or monitoring forest conditions, for instance) when they do not depend on forests for their livelihood. External support from either governmental or non-governmental bodies may be necessary for efficient local forest management. Such external support, however, must include an awareness of local conditions and the ability to identify the needs of local forest management in a specific arena.
2. Both institutional arrangements (such as regulations, monitoring, and sanctions) and local attitudes are crucial to local forest management. The case study indicates the community's capacity to adjust local institutions in response to changing socio-economic circumstances. This study suggests that face-to-face interaction and interdependence among community members,

along with shared norms, determine local forest management. Such community characteristics may be the social capital for maintaining local institutions. These characteristics, however, are formed over a long period of time and differ under varied conditions. For external intervention in forest management, therefore, it is important to pay more attention to local attitudes before identifying external means of support.

3. It is found that pressures of population increase and market forces are much more influential than the effect of local institutions on the relations between forests and local people. In the study site, if the people's dependence on forestlands continues without off-farm income generation activities, or if market forces encourage cash crop cultivation, local institutions will not be able to control land conversion from forests to agricultural lands. But the relationship between local people and the forests is dynamic, as shown by this case study, and it is necessary to identify an efficient combination of roles of local people (i.e., local institutions), external intervention, and markets under different socio-economic conditions.
4. Two types of transformation have occurred in the previous swidden fields in the Philippine uplands. The first is a transformation to commercial agricultural lands where market forces penetrate; such lands are de facto private space. The second is a transformation to abandoned lands that are neither used nor claimed by local people. Such lands are state land. This study is about the third case—where swidden fields turned into forests because of the changes in local livelihood activities and decreasing local dependence on forestlands. It shows that such lands are de facto private/community space. It is, again, vital to recognize where local people (or community), external intervention, and markets can play efficient roles in forest management regarding these three different cases.

REFERENCES

Agrawal, Arun and Gautum N. Yadama. 1997. How do local institutions mediate market and population pressures on resources? Forest *Panchayats* in Kumaon, India. *Development and Change* 28:435–465.

Barthelmes, R., C. Busemann, E. Heusinger, C. Ochs, B. Trosowski and K. Wellershaus. 1998. Banaue—Between yesterday and tomorrow (socioeconomic conditions). In *Rice terraces of Ifugao (Northern-Luzon, Philippines)—Conflicts of land use and environmental conservation,* ed. J. Settle, H. Plachter, J. Sauerborn and D. Vetterlein. UFZ-Bericht, Nr. 5.

Barton, R. F. 1919. Ifugao law. *University of California Publications in Archaeology and Ethnology* 15(1): 1– 186.

Brightman, Robert A. 1987. Conservation and resource depletion: The case of the boreal forest Algonquians. In *The question of the commons: The culture and ecology of communal resources,* ed. Bonnie J. McCay and James M. Acheson. Tucson: University of Arizona Press.

Conklin, Harold C. 1980. *Ethnographic atlas of Ifugao: A study of environment, culture, and society in Northern Luzon.* New Heaven and London: Yale University Press.

Eder, James F. 1982. No water in the terraces: Agricultural stagnation and social change at Banaue, Ifugao. *Philippine Quarterly of Culture & Society* 10(3): 101–116.

Guthrie, George M. 1964. *Impressions of Ifugao health and social activities.* University Park: Pennsylvania State University.

Hangdaan, Napoleon B. 1999. *The Muyung system of the Ifugaos.* Provincial Office of Ifugao.

Hayami, Yujiro and Masao Kikuchi. 2000. *A rice village saga: Three decades of green revolution in the Philippines.* Basingstoke: MacMillan Press.

Herzmann, S., A. Jost, T. Korbun, C. Willerding, J. Settele and H. Plachter. 1998. Ifugao rice terraces—Land-use changes in a traditional agroecosystem from 1963 to 1997. In *Rice terraces of Ifugao (Northern-Luzon, Philippines)—Conflicts of land use and environmental conservation,* ed. J. Settle, H. Plachter, J. Sauerborn and D. Vetterlein. UFZ-Bericht, Nr. 5.

Klock, John S. 1995. Agricultural and forest policies of the American colonial regime in Ifugao territory, Luzon, Philippines, 1901–1945. *Philippine Quarterly of Culture & Society* 23(1): 3–19.

McKean, Margaret A. 1992. Management of traditional common lands (Iriaichi) in Japan. In *Making the Commons work: Theory, practice, and policy,* ed. Daniel W. Bromley et al. San Francisco: Institute of Contemporary Studies.

Ostrom, Elinor. 1990. *Governing the commons: The evolution of institutions for collective action.* Cambridge: Cambridge University Press.

Provincial Planning and Development Office, Ifugao. 1993. *Socio-economic profile, Province of Ifugao.*

Repetto, Robert. 1988. *The forests for the trees? Government policies and the misuse of forest resources.* Washington, DC: World Resources Institute.

Sajor, Edsel E. 1999. *Upland livelihood transformations: State and market processes and social autonomy in the Northern Philippines.* Hague: Institute of Social Sciences.

Science Education Center. 1971. *Plants of the Philippines.* University of the Philippine Press.

Shigetomi, Shin'ichi. 1996. *Village organizations for rural development in Thailand* (in Japanese). Tokyo: Institute of Developing Economies.

World Bank. 1989. *Philippines: Environment and natural resource management study.* Washington, DC: World Bank.

Chapter 19

ADAPTABILITY OF COMMUNITY-BASED FOREST MANAGEMENT FOR EX-LOGGING WORKERS IN THE PHILIPPINES

Yoshiki Seki
Waseda University. Research Institute of Current Chinese Affairs

1. INTRODUCTION

Commercial logging operations have almost come to an end in the old-growth dipterocarp forests in the Philippines. In the region of Cagayan in Northern Luzon, the site of this study, there were 28 logging concessions covering 1.16 million hectares in 1987, but by 1995 these had shrunk to only five concessions covering 0.31 million hectares (DENR 1996). Some big problems and questions have arisen because of these changes, including who will manage the logged-over secondary forests and how they will be managed.

In 1995, the Philippine government launched its Community-based Forest Management (CBFM) program with Executive Order No. 263, which declared CBFM as "the national strategy to achieve sustainable forestry and social justice," and began strategically introducing it in logged-over concessions (DENR 1997). CBFM is a comprehensive forest management program run by local communities, consisting of agroforestry development, reforestation projects in open grassland, and selective logging in secondary forests.

While the concept of "community" seems straightforward, exactly what constitutes community in the Philippine uplands is seldom discussed. Upland communities that formed along with commercial logging operations are usually composed of various groups, including indigenous people, logging workers, and land-seeking migrants. In such a heterogeneous society, the methods of survival and utilization of forest resources are varied and fluid (Fujisaka and Wollenberg 1991; Walpole and ERDMO 1993; Nagano et al. 2000). It is very much debatable whether tightly structured communities, which manage their forest with common interests and views, can be formed in these former logging concession areas.

M. Inoue and H. Isozaki (eds.), *People and Forest—Policy and Local Reality in Southeast Asia, the Russian Far East and Japan, 287–298.*

Thus, in this chapter, a village where CBFM was implemented after the cancellation of a logging concession is researched, with a special focus on the logging workers who migrated from all over the Philippines during the period of commercial logging. They were not originally "local people," but many of them chose to remain after the cancellation of the logging concession, and thus became local people. A key question that must be asked is: Can ex-logging workers become participants in managing logged-over concessions in a sustainable way? In order to answer this, the following points must be clarified: first, the process of transforming the livelihood of logging workers after the cancellation of the concession; second, how ex-logging workers view the forest management of the former concession area; and third, whether there are contradictions between the government's CBFM strategy and the livelihood strategies of the ex-logging workers.

2. SETTING OF THE STUDY AREA

The Northern Sierra Madre mountain range spreads out along the eastern side of the Cagayan Valley. Commercial logging in the area was rampant in the early 1970s and widely operated until the early 1990s, after which most logging concessions were cancelled.

Field research was conducted in the *barangay* of Batong Labang, a municipality of Ilagan, in the province of Isabela (a barangay is the smallest administrative unit in the Philippines). Batong Labang is located at the edge of a former logging concession that once belonged to the ACME Plywood and Veneer Company, Inc. (ACME), which operated from 1970 to 1990. The logging concession was vast, covering an area from the western edge of Sierra Madre to the eastern coast of the island of Luzon (a total area of 89,275 hectares). In 1971, ACME's sawmill was established in Batong Labang, making this village a center of logging operations, which then saw a massive in-migration of logging workers.

The population of Batong Labang increased dramatically. According to the government's census, the population was 280 in 1970. During the 20-year logging operation (1970–1990), the population burgeoned to 2,357, with an annual growth rate of 11 percent. As of 1990, ACME's logging operation employed about 600 workers. Approximately, 85 to 90 percent of the total households (n=442 in 1990) in the village were connected to logging.

In those days, it was also considered the richest village in the municipality. The village economy was very commercialized, and bars and beer halls, frequented by logging workers, were everywhere.

3. EFFECTS OF THE LOGGING BAN

ACME's logging concession was cancelled in 1990 due to its illegal methods of logging. When the company moved out of the village, the prosperity of Batong Labang also came to an end. The end of the logging operation had a drastic impact on the village.

The cancellation of the concession meant the total lay-off of the logging workers. This prompted them to adopt any of four survival strategies: (1) leave and seek jobs elsewhere, (2) remain and convert to upland farming, (3) remain and continue logging, (4) or remain and engage in both logging and farming.

According to the national population census, Batong Labang lost 37 percent of its original population, decreasing from 2,357 in 1990 to 1,493 in 1995. These are the workers who went elsewhere to seek other jobs. But since most logging concessions in the country were already cancelled, there were few other concessions to work at, and about 60 percent of the logging workers were forced to remain in the area.

If they stayed in the village, the possible livelihoods were either upland farming or "illegal" logging. Out of the 38 ex-logging workers who remained in Batong Labang and were interviewed by the author in 1997, 15 had converted to upland farming, 10 were logging full-time, and 10 were engaged in both logging and upland farming. The other three were jobless or had started small businesses. Based on the author's research, the elder generation tended to become upland farmers, while the younger generation continued logging, reflecting the hard work and dangerous nature of logging, appropriate only for physically tough young men.

The logging methods available to villagers without a truck are either "water logging" or "carabao logging." Water logging utilizes rivers and creeks to transport harvested timbers. Carabao logging, on the other hand, uses water buffalo to transport timber down old logging roads. These activities are considered illegal by the government, but the loggers usually bribe the military and the government officials so that they are able to pass the check-points and bring the timber to market.

The loggers prefer to harvest narra (*Pterocarpus indicus*), which is in large demand by local furniture makers, and its price is higher than the ordinary dipterocarp species, such as red lauan (*Shorea negrosensis*). Narra extraction became rampant after the logging concessions were cancelled. Because the logging company had harvested mainly dipterocarp species for making veneer and plywood, the narra variety was still relatively abundant.

Loggers in Batong Labang now supply wood resources to the barangay of Marana, where many furniture makers are located. The number of local furniture makers doubled after the cancellation of the logging concession. There were about 50 furniture makers in Marana in 1997, while there were only about 25 in 1985. The timber supply for the local market even increased after cancellation of the concession and the imposition of the logging ban, activating the local timber industry.

Since the cancellation of the concession, the logged-over areas became a kind of open-access resource. While some ex-logging workers continued logging illegally, they also began to acquire farms in former concession sites as an alternative livelihood, because they felt logging was becoming difficult due to resource depletion and the intensification of government regulation.

Out of the 38 ex-logging workers interviewed in 1997, 65.8 percent had acquired farms, while 34.2 percent were still landless and had just continued logging. When the author again interviewed 33 ex-logging workers in 1999, 75.8 percent had acquired farms. Ex-logging workers gradually expanded their private farms inside former concession areas.

4. THE COMMUNITY-BASED FOREST MANAGEMENT STRATEGY AND EMERGING CONFLICTS

The CBFM was thus enacted in order to avoid the uncontrolled use of resources. Forestland tenure, which is called a community-based forest management agreement (CBFMA), is granted to a local people's organization. The CBFMA is a renewable 25-year land management title which can be utilized for farming and forestry objectives. It is a comprehensive forest management program composed of programs such as agroforestry development owned by individual farmers, contract reforestation in open grassland, and controlled timber extraction from secondary forests.

In the case of Batong Labang, the people's organization, the VIBANARA Multi Purpose Cooperative, Inc. (VMPCI), was formed in 1992 with technical assistance from a non-governmental organization (NGO) and the government with the participation of three other barangays. The VMPCI was given 4,845 hectares under a CBFMA to be managed under government supervision, with 273 hectares of already cultivated land designated for agroforestry development, 629 hectares of grassland and 1,425 hectares of shrubland allocated for reforestation projects, and 2,517 hectares of residual forest area for selective logging with assisted natural regeneration.

The Philippine government's Department of Environment Natural Resources (DENR) permitted the cooperative to conduct controlled logging operations, in recognition of the necessity to provide jobs for ex-logging workers. While water and carabao loggers exploit logs as an open-access resource, the cooperative's logging operation uses selective logging methods with assisted natural regeneration under government supervision.

In the case of the research site, however, although the government had permitted selective logging, the employment capacity of its operation was insufficient. The annual allowable cut (AAC) in 1997 was only 840 cubic meters, while in 1989 ACME's AAC was 65,420 cubic meters. With this logging capacity, the VMPCI was only able to employ about 60 workers (as of 1997), which is more or less 10 percent of the total employment in ACME's former logging operation. Thus, many ex-logging workers continued to log "illegally."

A conflict occurred between the illegal loggers and the VMPCI in terms of timber resources and timber prices. Table 1 shows the producer's price of a board-foot of red lauan. Red lauan harvested by water and carabao loggers fetches 12 pesos in the villages. The VMPCI's price, on the other hand, is 18 pesos, 50 percent higher than for illegally logged red lauan. As Table 1 shows, logging under CBFM requires higher operational costs, because of administrative costs and government forest charges (three pesos per board-foot). Thus, the members of VMPCI were troubled that they were losing to illegal timber in market competition. Because timber dealers much prefer the cheaper wood, legal timber resources are crowded out of the local market.

Table 1: Price formation of timber for red lauan (pesos per board foot).

(1) Water logging		**(2) CBFM logging**	
Chainsaw operation	3	Chainsaw operation	0.5
Water transport	5	Truck transport	3.5
Truck transport	2	Bulldozer operation	2
Middleman	2	Forest maintenance	3
		Sawmill operation	2
		Administration cost	4
		Forest charges	3
Total	12	Total	18

Note: 1 peso=U.S.$.0.026 (October 1997) and 1 board foot=0.00236 cubic meters. Hence 1 peso per board foot=U.S.$11.0 per cubic meter. (It must be noted that the peso has devaluated dramatically after the Asian currency crisis of the late-1990s.)

Conflicts have arisen over the government's land allocation strategy, as well. Another conflict occurred between the DENR-VMPCI, which wanted to allocate more open land for reforestation, and new land applicants wanting to expand their own farms. The government tries to limit individual landholding in upland areas to three hectares per head maximum and to allocate open grassland for reforestation purposes. The DENR targeted a total of 997 hectares for a large-scale reforestation project managed by the VMPCI, and it inevitably wants to minimize agricultural land use of individual households in order to maximize industrial tree plantation of fast-growing species. This land allocation strategy takes potential land away from the "landless" ex-logging workers, and the designation of land for reforestation prevents the future acquisition of land by landless households.

5. VIEWS OF LOGGING WORKERS ON FOREST MANAGEMENT

The goal of the DENR-VMPCI is to stop illegal logging and integrate loggers into CBFM projects, such as reforestation and legal logging, but conflicts have emerged between the VMPCI and the illegal loggers. In 1996, when the VMPCI tried to regulate illegal logging, a group of loggers attacked and burned a VMPCI truck.

The majority of water and carabao loggers are critical of the idea of regulation. In 1999, the author asked 24 water and carabao loggers, "Which do you support: to permit carabao and water logging or to prohibit it?" Twelve (50 percent) answered that water and carabao logging should be allowed. Another six (25 percent) answered that alternative jobs should be given before prohibition. The majority of loggers don't think they can stop their activities under present conditions. The following comments are taken from the author's previous study on people's views on CBFM (Seki 2000b).

The following two respondents had the same kind of background. Both of them were former chainsaw operators for the ACME company. While they were now engaged in illegal activities, they also had planted yemane trees (*Gmelina arborea*) on their farm lots, which were acquired from the former concession area.

The first one suggested that the government should legalize water and carabao logging and collect taxes from these activities, then use these funds for reforestation.

The second one felt it was a sin to be engaged in illegal logging, but according to him, he needs cash to develop a tree plantation. He expects that all logging activities will be banned in the future, so he's preparing a tree plantation as an alternative income source, but he doesn't have enough funds to last until harvest time comes.

> *In my personal opinion, why does not the government collect tax[es] from loggers? The government can monitor the people here, then make an agreement that people pay forest charges and the government permits the logging. I think it is possible. This will be a big income source for the government. Then the government can use this budget for reforestation projects. If the government does not like this kind of condition, people will just continue logging illegally. Why is the community still under the supervision of the DENR? I wish there was only the BIR (Bureau of Internal Revenue). We would just pay taxes to the BIR, then we can manage the forest. That's it.... Why is the DENR still collecting the forest charges? The DENR is not necessary. During the time of (commercial) logging, TLA holders did not follow the agreements with the government. The company bribed the evaluator from the DENR, then the evaluator reported to the office that "the company is very good."*
>
> (Male, 44, elementary school graduate)

> *When [commercial] logging stopped, people started illegal logging. But this is only minimum—unlike before. However, the DENR does not blame the logging company, but blames us as the destroyers of the forest. I agree with the opinion to stop illegal activity. If illegal activity is stopped, all people will concentrate on the CBFM project. Even if the income is small in legal activity, at least they can relax. [In the case of illegal logging], people can not relax even if income is bigger... I went back to illegal activity, although I do not like it. However, at least I have planted trees. I need cash to improve my farm lot. But some other water loggers do not do that. If the time comes that all logging stopped, they will be surprised. So we have to prepare now.*
>
> (Male, 39, elementary school graduate)

Some other loggers requested to be given alternative jobs before the government tries to regulate illegal logging. The following three loggers said that logging is getting harder every year due to the decrease of resources. The last interviewee was really concerned about the environment, since he had suffered from a flashflood in October

1999 that swept away 20 houses in Batong Labang. These three say that they are willing to stop if they have other job opportunities.

> *Before the government tries to stop water logging and carabao logging, they must observe our situation. We are impoverished while engaged in water logging. Nowadays they can not earn more than just enough for daily meals. If the government give us another job opportunity to the villages, there will be no water logging. Because logging is really getting harder. That is why their families are starved. They have to do it [water logging] even if they do not want to.*
>
> (Male, 53, elementary school graduate)

> *Okay, I agree to stop [illegal logging] if there is an alternative job. It is getting harder to engage in water logging. I want to have another job.*
>
> (Male, 46, third-year high school)

> *I support the policy [to stop logging]. Our watershed is heavily damaged now. During dry season, there is no water [in the river]. Then when rainy season comes, we have flashfloods. Mountains are already destroyed. Mountains are so pitiful. There are so many landslides. If there is a calamity, we are the one who are suffering from poverty. We are benefited from the forest. I wish logging could be stopped. If there is another job, it is possible.*
>
> (Male, 46, high school graduate)

The following respondent is a former carabao logger. He stopped because his two carabaos died from causes related to logging, illustrating the risks of carabao logging. According to him:

> *It is my personal opinion. It is better to stop (all logging operations). I was a carabao logger before. I stopped it because my two carabaos died from overwork. I was forced to work her [the water buffalo] when she was pregnant. Another male carabao was also dead by accident. Timber hit his leg in the mountain that made him a fracture. He was no longer able to work. So I killed him. They were so hurtful experiences for me. Then I stopped [logging activity]. Trees in mountains are not overexploited if there is no logging. Then people concentrate on the reforestation activity.*
>
> (Male, 33, elementary school graduate)

6. THE RISE OF INDIVIDUAL-BASED SPONTANEOUS REFORESTATION

Since 1996, individual-based reforestation in private lots has become popular. Ex-logging workers who acquire farm lots are spontaneously planting trees. During the author's first study, in October 1997, 26.3 percent of ex-logging workers (n=38) had already planted yemane or other tree varieties on their own lots. By August 1999 the percentage of tree-planters among ex-logging workers had risen to 75.8 percent (n=33), showing that illegal loggers had become tree-planters at the same time. In this discussion, a "tree-planter" is defined as one who had planted more that 50 trees on his or her own farm. Major crops in the area include maize, upland rice, and bananas. Now, yemane trees have emerged as a fourth "crop."

The reason for the yemane boom seems to be rooted in the changing economic and ecological conditions surrounding the ex-logging workers. In 1999, the author asked 25 tree-planters from among the ex-logging workers, including those who had converted to upland farming, why they had planted yemane. Of these, 64 percent answered that it was because yemane was profitable. (The price for yemane was almost the same as for red lauan.) Next to the economic reasons, 36 percent said that they did it to "alternate natural resources." It appears that the ex-logging workers consider man-made tree plantations as a necessary alternative to the diminishing numbers of natural trees. And because they are the ones that have been most heavily affected by resources depletion, they are more sensitive and concerned about the future sustainability of forest resources—and rightly so.

The timber resources in the natural forests have clearly been overexploited. Table 2 shows how the labor input for transporting timber from the mountains increased in the 1990s. According to the water and carabao loggers, narra trees were still bountiful when the logging concession was cancelled. They could haul timber down from the cutting areas within one day. But increasingly more loggers began cutting narra, resulting in the "tragedy" of the species. The rapid depletion of narra trees in the lower portion of Sierra Madre forced the loggers to cut increasingly farther away from the village. As the cutting advanced farther away, carabao logging became impossible, because it is only practical within the distance that the animals can go up and down the mountains in a day. Thus, water logging became the only method for transporting narra. As of 1998, the cutting areas had almost reached the mountain ridges, and water loggers would spend almost two weeks on just one operation. The labor input for transporting narra increased by more than ten times compared to eight years ago. The loggers camp for two weeks along the riverbanks, feeding themselves with crabs, small fish, and edible plants from the forest, such as rattan. Besides these hardships, they face the threats of malaria and physical injury. Although the price for narra wood has risen as transportation costs have increased, the activity has almost reached the limit of the loggers' physical abilities as of 1998. When the author revisited the site in August 1999, loggers were no longer able to obtain narra due to depletion of the resources.

The rapid exhaustion of narra trees forced the loggers to shift their sights onto the stock of dipterocarp species, which is in relatively better condition. Loggers are able to transport dipterocarp timbers from the cutting areas to the village by water within three days. But if overexploitation of water and carabao logging continues without any regulation, sustainable use of this resource will be difficult and the dipterocarp species may share the same fate as narra.

Table 2: Increases of transportation cost and timber price (pesos per board foot).

Species	Item	1990	1998
Narra (*Pterocarpus indicus*), native tree	Price	8 pesos	30 pesos
	Working days to transport	1 day	12 days
	Transportation	1.5 pesos	16 pesos
Red lauan (*Shorea negrosensis*), native tree	Price	8.5 pesos	12 pesos
	Working days to transport	1 day	3 day
	Transportation	1.5 pesos	4.5 pesos
Yemane (*Gmelina arborea*), plantation tree	Price	No	12 pesos
	Working days to transport	—	<1 day
	Transportation	—	<1.5 pesos
	Afforestation	—	0.6 pesos

Source: Seki (2000a).
Notes: The prices of narra and lauan are for illegally harvested timber. 1990 prices are based on the memory of local timber middlemen.

Thus, man-made tree plantations are being established by concerned loggers as alternatives to natural stocks of trees. The man-made yemane market has suddenly grown in recent years, while natural timber stocks have declined. Local furniture makers and timber dealers have begun to consider yemane as an alternative wood, because the resource shortage of natural species has become serious in recent years. This has encouraged the people to establish yemane plantations on their farms. With the market for industrial forest species finally established, farmers have responded sensibly. In 1997, it was widely observed that husbands, wives, and children were collecting yemane seeds, germinating them in their backyards, and then transplanting them on their farms and lots. Growing yemane does not require a large capital investment, just family labor and a parcel of land.

Yemane, called *gmelina* in the Philippines, is a fast-growing species and is used most frequently for reforestation. It takes only seven to eight years until it's ready to be harvested, and it's easy to collect seeds for germination. Compared to other fast-growing species, yemane is the most fire-resistant and is ideal for construction purposes.

As Table 2 shows, the transportation cost for red lauan is 4.5 pesos, while for yemane it's less than 1.5 pesos. The afforestation cost of yemane per board-foot is only 0.6 pesos. The cost of transportation plus the afforestation cost of yemane is 2.1 pesos, 2.4 pesos lower than the transportation cost of red lauan, producing 2.4 pesos of surplus profit for the tree-planter. Thus, regenerative forestry in man-made plantations has overtaken logging the natural forests.

While ex-logging workers continue to log illegally, they are preparing future income sources by planting yemane. When the time comes that they are able to harvest enough yemane timber from their farms, the illegal logging will stop.

7. CONCLUSION

Throughout this research, it was found that regenerative forestry based on man-made tree plantations is going to overtake exploitation forestry. If this trend continues, ex-logging workers will become the tenders of sustainably managed forests. If large areas of upland farm-lots were converted to yemane plantations, these would no longer be categorized as upland farms, but as upland agroforestry plantations.

Finally, it would be productive to compare these research findings with some previous studies from the economic, human ecological, and institutional points of view in order to generate topics for further study.

7.1 Economic Perspective

The process of transforming exploitation forestry into regenerative forestry is very interesting from the economic point of view. Ishiwata (1952) argued that as long as timber from natural forest supplies is steadily available, timber from man-made forests can't compete in price in the market because of the high cost of afforestation activities. Thus, he argued that man-made tree plantations can't be established without government subsidies.

Handa (1961), on the other hand, theorized on the process of how man-made forestry takes over from exploitation forestry under free market conditions. The depletion of natural forests drives cutting areas farther out to more marginal areas, resulting in a rise in transportation costs and increases in the market price for timber. When the transportation cost of harvested timber from natural forests exceeds the cost of afforestation near the village, surplus profit emerges from managing a man-made forest. Thus, regenerative forestry based on man-made plantations begins. This historical phenomenon is finally taking place at the research site. The process of transforming a forestry system found in this research seems to accurately conform to Handa's theory.

7.2 Human Ecological Perspective

The establishment of a man-made tree plantation is not only an economic phenomenon, where villagers seek to maximize their profits, it's also a human ecological one, where ex-logging workers are trying to adapt their livelihoods to the upland ecosystem. Planting yemane trees prevents soil erosion, flash floods, and is an alternate to the shrinking natural forests.

Fujisaka et al. already identified this kind of long-term evolutionary process through a series of studies in Laguna, a logged-over frontier community (Fujisaka 1986; Fujisaka and Capistrano 1986; Fujisaka and Wollenberg 1991). Fujisaka (1986) described this process as "co-evolution" between pioneer cultivators and a tropical forest ecosystem. Fujisaka and Wollenberg (1991) then proposed a dynamic model where farmers' practices evolve from an annual, crop-based, slash-and-burn (*kaingin*) system into a sustainable, perennial, crop based agroforestry system. According to the study, the people's small-scale timber extraction appeared as a transitional income-generating activity, which was used as a preparation for a perennial, crop-based, agroforestry system. This study has produced meaningful results, including a revision of the "destructive" image of upland communities.

The case of Batong Labang, Ilagan, Isabela also shows the dynamics of co-evolution, with concerned loggers adapting to the upland ecosystem by changing their livelihoods from logging to planting.

7.3 Institutional Perspective

Establishing small-scale tree plantations inside farm-lots is cost-efficient and requires only basic security and protection. Unfortunately, the CBFM program lacks the institutional arrangements to support the expansion of small-scale plantations. The government doesn't recognize the advantages of individual plantations to its full potential. It seems to be more interested in large-scale tree plantation projects managed by the cooperative, but this strategy has resulted in conflicts over land. Loggers want their own agroforestry farms, rather than being employed as plantation workers (Seki 2000b). While the government intends to regulate water logging and carabao logging, it has failed to create alternatives for loggers. While the role of "community" is much emphasized, the right of individual households to access the resources is somewhat slighted.

Inoue (1997) discussed two possibilities for communal forest management. One is "spatial" implementation. This is collective management of land and resources based on communal ownership of the land. The other is "conceptual" implementation. This type is basically individual management, but the use of resources is regulated and controlled by the community. In the research area of this study, it seems that these two methods of communal management are applied together.

In the case of the residual secondary forests, spatial implementation of management with a selective logging system must be applied. Present illegal logging is basically open-access exploitation without any regulation. If the cooperative organized all the illegal loggers, at least a selective logging system and controlled management would take over.

On the other hand, in the case of already denuded forest that turned to grassland, the author believes that the conceptual implementation of communal management must be applied. Large-scale reforestation projects are not worth being implemented considering the negative sentiments of villagers. Since villagers are now spontaneously planting trees on their own lots, grasslands can also be reforested, even if the land is divided up among households. This does not mean, however, that the cooperative has no role. Even if the grassland is divided up into individual parcels, the cooperative still has an important role in controlling and providing technical assistance for sustainable use. The harvesting and marketing of the timber is also important work for the cooperative.

REFERENCES

Department of Environment and Natural Resources (DENR). 1996. *1996 Philippine forestry statistics*. Quezon City: Department of Environment and Natural Resources.

———. 1997. *Moving CBFM into the 21st Century: DENR strategic action plan*. Quezon City: Department of Environment and Natural Resources.

Fujisaka, Sam. 1986. Pioneer shifting cultivation, farmer knowledge, and an upland ecosystem: Co-evolution and systems sustainability in Calminoe, Philippines. *Philippine Quarterly of Culture & Society* 14:137–164.

Fujisaka, Sam and A. Doris Capistrano. 1986. Upland development in Calminoe: The roles of resource use, social system, and national policy. In *Man, agriculture and the tropical forest: Change and development in the Philippine uplands*, ed. S. Fujisaka, P. Sajise and R. del Castillo (edited), 223–244. Bangkok: Winrock International Institute for Agricultural Development.

Fujisaka, Sam and Eva Wollenberg. 1991. From forest to agroforest and logger to agroforester: A case study. *Agroforestry Systems* 14:113–129.

Handa, Ryoichi. 1961. *Mokuzai-kakaku Ron* (On the price of timber). Tokyo: Chikyu-Shuppan.

Inoue, Makoto. 1997. Komonzu toshiteno Netairin: Karimantan deno Jirei wo motonisite (Examination of tropical forest as commons: Based on positive survey in Kalimantan, Indonesia). *Kankyo Shakaigaku Kenkyu* (Journal of Environmental Sociology) 3:15–30.

Ishiwata, Sadao. 1952. *Ringyo-Chidai Ron* (On forestry rent). Tokyo: Nourin Toukei Kyokai.

Nagano, Yoshiko, Atsuko Hayama and Yoshiki Seki. 2000. *Firipin no Kankyo to Komyunityi: Satou seisan to Bassai no Genba kara* (Environment and community in the Philippines: From the field of sugar production and logging). Tokyo: Akashi Shoten.

Seki, Yoshiki. 2000a. Firipin ni okeru saisyu-ringyo kara ikusei-ringyo heno tenkan katei (The change from exploitation forestry into regenerative forestry in the Philippines). *Ringyo-Keizai Kenkyu* (Journal of Forest Economics) 46(1): 51–56.

———. 2000b. Internal constraints on community-based forest management in a post-logging, upland community. *Philippine Quarterly of Culture and Society* 28:399–437.

Walpole, Peter S. J. and Environmental Research Division of Manila Observatory (ERDMO). 1993. Upland Philippine communities: Guardians of the final forest frontiers. Research Network report number 4. Berkeley: Center for South East Studies, University of California.

Chapter 20

FOREST UTILIZATION BY INDIGENOUS PEOPLE: A CASE STUDY OF THE HUNTING PRACTICES OF THE UDEGHE PEOPLE IN THE RUSSIAN FAR EAST

Shiro Sasaki
National Museum of Ethnology (Japan)

1. INTRODUCTION

Deforestation is one of the most complex problems facing the Russian Far East, especially in areas inhabited by indigenous people who are largely dependent on the forest. It has deprived them of the territory and resources required for their traditional practices of subsistence and production (i.e., hunting, fishing, and collecting), and has played a decisive role in the collapse of their culture and society, which have long contributed to their sustainable forest practices. At the same time, however, the indigenous people recognize the necessity of the timber industry, because it provides them with work and monetary income, and is important to the Russian economy. In the case of the Russian Far East, the discussion of forest conservation is very complex because of three kinds of intertwining problems: economic development, protection of natural resources and the environment, and indigenous rights in the territory, including the conservation of their culture.

The territories of indigenous peoples are generally located in very remote areas, where, more precisely, they were pushed out to and compelled to live; but the entire Russian Far East was originally their homeland. For example, the Tungus-speaking people (ancestors of the present Evenks, Nanais, Ul'chi, Orochi, Udeghe, and Negidals), the Nivkhi people, and a few Ainu lived along the Amur River and the Sikhote-Alin' mountains until the 1840s. The modern history of the Russian Far East, which began with Russian colonization in the 1850s, can be characterized by the deforestation and destruction of indigenous people's territories, as well as the history of development and progress in the region by the colonizers.

M. Inoue and H. Isozaki (eds.), *People and Forest—Policy and Local Reality in Southeast Asia, the Russian Far East and Japan, 299–319.*

In this chapter, I will describe the present state of forest utilization of the Udeghe people in the Primor'e region (Primorskii krai), one of the indigenous peoples in Far East Russia, focusing especially on the hunting activities of the villagers of Krasnyi Yar. Then I will explore the following questions: What kinds of animals do they hunt? What kinds of techniques do they use to hunt each type of game? What role does their hunting play in the village economy? How did they survive the most critical period after the collapse of the Soviet Union (1992–1998)? Is there any possibility for them to continue hunting under the difficult social, economic conditions in present-day Russia? The diversity of their products and the flexibility of their techniques and equipment will be important points in the discussion of these problems. All the materials analyzed in this chapter were collected during my field research in the village in 1995, 1996, 2001, and 2002.[1]

The village of Krasnyi Yar is located along the Bikin River, one of the tributaries of the Ussuri River, about six to seven hours by car from Khabarovsk and more than 12 hours from Vladivostok, the capital of the Primor'e region. The roads are not always in good condition. The population is presently at about 600. As to the ethnic component, people registered as Udeghe represent about 66 percent of the population (Table 1 and Figure 1). The remaining population is made up of the Nanai (one of the other indigenous ethnic groups of the region), the Evenks, Buryats, Russians, Ukrainians, Belorussians, and so on.

Table 1: Population of the village of Krasnyi Yar in 2001.

	Population	%
Northern minorities	523	82.1
Udeghe	418	65.6
Nanai	98	15.4
Orochi	4	0.6
Evenki	2	0.3
Chukchi	1	0.2
Other ethnic population	114	17.9
Total	637	100.0

Source: Author.

The economic conditions in the village are not good, similar to other indigenous villages in the Russian Far East, with few productive activities or jobs available. Many villagers are jobless, unemployed pensioners, or children. The standard income level is very low, while the price of commodities is comparatively high, because of the village's remoteness. The social infrastructure is not well equipped. Even the electricity supply is limited to only a few hours a day. The villagers recognize that logging gives them

[1] This research was financially supported by the Toyota Foundation in 1995 and 1996, by the Sumitomo Foundation in 2001, and conducted with the support of Russian collaborators. I am very grateful to Dr. M. Patrusheva, Dr. A. Startsev, and Dr. S. Bereznitsky from the Institute of History, Archeology and Ethnology of Russian Academy of Sciences in Vladivostok, and to Mr. A. L. Uza and Mr. V. A. Shilko, the former and the present presidents of the Ethnic Hunting Company OAO "BIKIN."

temporary but much-needed monetary income, yet they know that continued deforestation will destroy the very basis of their society and culture in the long run. The forest has long provided them with various types of game for hunting, edible and useful plants, shelter, and spiritual inspiration.

Hunting, the main topic of this chapter, is one of the traditional and most prestigious activities of the Udeghe people. Though the over-killing of animals, which sometimes happens, is often condemned as a typical cause of the destruction of the natural environment, hunting itself is beginning to be recognized as one of the styles of sustainable, non-timber forest use in the indigenous people's territories. In fact, indigenous hunters have demonstrated the sustainability of hunting for hundreds of years under various social and political conditions.

This chapter consists of five sections, including the introduction and conclusion. The next section describes the fundamental characteristics of the hunting practices of the Udeghe in the forests along the Bikin River to illustrate the diversity of their hunting techniques and products. The third section describes the process of the establishment of a hunting enterprise in Krasnyi Yar run by indigenous people, who are struggling with the present economic difficulties and making every effort to protect their forests. In this section, I outline the process of their survival in the economic crisis just after the collapse of the Soviet Union. The fourth section is an analysis of the data on this hunting enterprise and an evaluation of the possibility of sustainable forest use through hunting, fishing, and gathering using traditional and contemporary techniques. This section will show the future possibilities of the forest utilization activities of the Udeghe people in Krasnyi Yar.

2. HUNTING ACTIVITIES OF THE UDEGHE PEOPLE

2.1 The Most Prestigious Activity: Hunting

The forest is supremely important to the Udeghe people. It supplies them with materials for food, clothes, shelter, and spiritual and mental support. There are many dangers in their forest, not just physical but spiritual, as well. The Udeghe people believe that various kinds of spirits live there and that some are very dangerous. Before they go into the forest, they conduct rituals to negotiate with the spirits, escape from them, or overcome them with the help of shamans or supporting spirits. Nevertheless, they love their forest and are always eager to go there to track game, catch fish, and collect edible and useful plants. It revitalizes and brightens them, as if they enjoy having a type of "tension" in their relationship with the spirits. Even nowadays, though many Udeghe hunters pretend to be atheists, they still conduct the ritual in front of a sacred tree, which has a triangular mark (*piu* in the Udeghe language) carved on its trunk, standing at the entrance of the forest or their hunting territories. The tree represents the master of the forest. Hunters set out various foods and a cup of vodka on a small table to feed the master, asking for game and luck in hunting.

The Udeghe people generally recognize hunting as the most prestigious activity in their society, judging from various discourses with hunters. First, hunting supplies them with the most highly prized food—animal meat. Second, it provides the hunter's family with a large monetary income. Third, it demands keen perception and considerable skill, physical power, and precise knowledge about their own equipment and nature. Fourth, it provides them with frequent opportunities to access the spiritual world. Therefore, hunting implies various elements, not just material but social and spiritual, as well.

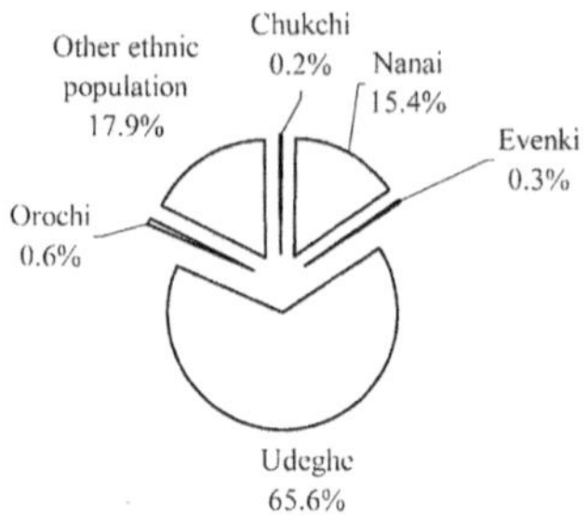

Figure 1: Population percentage of ethnic groups in Krasnyi Yar.
Source: Author.

2.2 Annual Calendar of the Udeghe Hunters on the Bikin River

Hunting resources are not equally distributed either in space or time. Natural conditions (geographic and climatic) and the characteristics of the animal (maturity, sex, and patterns of activity) limit hunting sites, seasons, and techniques. So hunters have their own calendar of activities, depending on the sort of animal and the purpose of the hunt. Many hunters say that the main hunting season begins in the middle of October, when the legal hunt for fur-bearing animals begins. Generally, autumn and the beginning of winter is the best time for hunters to capture many kinds of game. They have good sightlines in the forest after the leaves have fallen and, after the snow falls, animal tracks are easily spotted. The animals are very fat and healthy, having eaten extra to prepare for the severe winter, giving them considerable meat and excellent fur. Each type of game has its own hunting season, however, which varies. The hunting season for large mammals, for example, is generally longer than for smaller fur-bearing animals.

Red deer (*Cervus elaphus*) can be hunted in any season, but techniques vary, depending on the condition of the forest. For example, during winter, hunters can chase or track game by following an animal's footprints. But in spring and summer, when foliage hinders the hunters' sightlines, they will sneak up on game through a river or a brook on a dug-out canoe, or wait for them on a platform set in a tree near feeding or watering spots. In the latter case, hunters often attract deer with salt. During their mating season, from the end of August to the beginning of October, hunters use a special trumpet made of birch bark to lure the bucks. The hunting of red deer in summer and autumn plays a role in preparing food stocks for the intensive winter hunt.

During the Soviet regime, some Udeghe hunters hunted red deer for their soft, skin-covered horns from the middle of May to the middle of June. This was very difficult and processing the horns was complicated, but it provided them with large monetary income. The summer horns of the red deer are the source of one of the precious drugs used in traditional Asian medicine, so the state foraging farm (*gospromkhoz*) sold them to Chinese merchants for good prices. But since the collapse of the Soviet Union, no one has been hunting deer in summer for their horns, because the merchants switched to buying less expensive Chinese products of the same quality.

The hunting of bears sleeping in their dens begins in December and continues until spring, but the purpose is different between the beginning and the end of the season. In winter, hunters kill bears for meat, while in spring the main purpose is for bear gallbladders, the source of a very precious drug (more expensive than the red deer's soft horns) that is sold to Chinese merchants for very high prices. Nowadays, the Russian government prohibits the trade of bear gallbladders and the hunting of black bears, in compliance with an international treaty to protect wild animals (The Washington Treaty), while the regional government still licenses the hunting of brown bears for meat.

The end of the legal hunting season of fur-bearing animals is the middle of February. Towards the end of the season, hunters try to kill bears, red deer, and wild boars to provide enough meat for their families and relatives. When it ends, they gather their traps and game, pack their equipment and captured animals onto their snowmobiles, and return home. When they have too much equipment, it is left behind in their hunting huts and taken out later by motorboat when the river is ice-free. Once back at home, the hunters process the fur and meat and repair their equipment.

The river is liberated from ice in the beginning of April. Generally, spring is a rest season for the hunters, but it is not correct to think that they do nothing. This is the start of the main fishing season, a very important time for the Udeghe, because it supplies them with food for future use. They catch taimens, lenoks, pikes, carp, and so on, and dry or salt them to prepare them for storage.

Conditions in the forest during midsummer are not good for hunting. Leaves obstruct the hunters' sightlines, boughs and branches block their movements, hoards of mosquitoes swarm around them, and the animals go into the deepest parts of the forest. As H. Taguchi suggested, in general, summer is a difficult season in temperate and subarctic forests for foragers to acquire enough food (Taguchi 2000, 14). People survive by eating stored food or small amounts of fresh meat and fish, and wait for autumn, when the main foraging season begins again.

Nowadays, many hunters in Krasnyi Yar do not go out hunting during spring and summer. They are often employed and work in making roads or constructing buildings to make money. During holidays or on days off, they repair and clean their guns, traps, and other equipment, or tend their vegetable gardens, waiting for the next hunting season to begin.

2.3 The Different Techniques of Subsistence and Market-Oriented Hunting

The hunting activities of the Udeghe can be classified into two categories: subsistence hunting and market-oriented hunting. Though the targets and techniques sometimes partially overlap, their purposes are different. In subsistence hunting, hunters capture animals to consume all parts of their bodies for their own use. The meat is basically divided among all the participants of the hunt, and sometimes distributed among the villagers. The fur and head belong to the hunter, recognized as the owner of the game. Generally, he is the first to discover or shoot the animal. They sometimes sell the skin and some parts of the body, like the gallbladder, to merchants, but traditionally, the meat is rarely treated as a commodity. In contrast, the purpose of market-oriented hunting is to get precious commodities like fur, musk, soft summer horns, and gallbladders. Hunters are very careful to capture animals in their best condition, and have developed various techniques for this purpose.

The use of hunting territories is also different between subsistence and market-oriented hunting. In practice, territorial borders don't strictly restrict hunters for subsistence hunting, especially for tracking and chasing. Basically, a hunter can go into another hunter's territory to track or chase animals, provided some etiquette is followed, e.g., not touching an animal that other hunters have wounded, never stealing an animal from another hunter's traps, and so on. This may have something to do with the custom of distributing the meat among all the participants of the hunt or all the villagers. And at its basis, there may be a common recognition that the animal meat is a gift from the master spirits of the forest to all the hunters, and that no one can be its true owner.

In contrast, the territories for hunting fur-bearing animals are very strict. No one can set traps or track animals into another hunter's territory. If some hunters share a territory, they strictly divide it. This may be related to the fact that the income acquired from the trade belongs to a person or his family. Though they still recognize that the true owner of the fur-bearing animals is the master spirits of the forest, they believe that the game personally belongs to a hunter. These rules might have been established in the long history of the fur trade with Manchurian and Chinese merchants.

Generally, game targeted for subsistence hunting are large mammals which provide a large amount of meat, like deer,[2] elk, bear,[3] wild boar, and so on. Of these animals, the Manchurian red deer, wild boar, and bear are the most important game for the Udeghe hunters, because hunting these animals demands more power and skill, and provides

[2] There are several kinds of deer in the forests along the Bikin River: the Manchurian red deer (*Cervus elaphus*), the roe deer (*Capleorus capleorus*), and the musk deer (*Moschus moschiferus*).

[3] In the Primor'e region and in the southern part of the Khabarovsk region, there are two kinds of bears: the brown bear (*Ursuc arctos*) and the black (or Himalayan) bear (*Selenarctos thibetanus*). Though both kinds of bears were once game of the Udeghe hunters, state and regional laws concerning the protection of wild animals now prohibit the hunting of black bears, because they are in danger of extinction in these regions.

hunters with high prestige. Hunting bears and wild boars is especially dangerous (for details on techniques for hunting these kinds of animals, see Sasaki [2002, 90–91]).

Fundamental techniques for hunting red deer and wild boars are chasing, waiting, surrounding, stalking, and luring. Though bows, arrows, and spears are recognized as traditional hunting weapons, firearms like guns and rifles are widely distributed among the Udeghe hunters. For example, as mentioned above, when the Udeghe hunt the red deer during its mating season, hunters lure a buck from behind a bush or from another bank of the river with a birch-bark trumpet that sounds like a buck calling for a doe. When the buck hears it, he thinks that a rival buck is nearby. As soon as he appears near the hunters, they shoot him. In another situation, they hunt from the river in the early morning, when the deer comes to the riverbank to eat its favorite plants. A hunter will silently sneak up close to the deer in a small boat in the darkness. Once close enough, the hunter uses a light to spot it and shoots the moment it turns to the light.

The main techniques used in hunting wild boar are stalking and chasing. Though boars are tough and powerful, they are not good at running in deep snow, and hunters take advantage of this weak point. Usually, the Udeghe go boar hunting in a group of two or three hunters. When they find a group of boars, they sneak up from the leeward side and shoot at them the moment the boars are aware of them. When the animals run from them, the hunters give chase, and keep on running, switching off with each other, until the animals are exhausted. When the animals are too tired to run any further, the hunters shoot to kill. One old hunter, in an interview, said that in his childhood hunters used spears to hunt boars. They often drove the boars towards their hunting huts, making for easier capture and carrying out the meat.

As mentioned above, the bear-hunting season is in winter and early spring. Hunters seldom capture or shoot a bear out in the open, because it is very dangerous even when every hunter has a rifle. Usually, they go hunting in a group and try to find a bear sleeping in its den. When a hunter finds a den with a bear inside, he never tries to kill it by himself, but informs the other hunters. When all the hunters of the group have gathered, the bravest awakens the bear by poking it with a long stick. He repeats it several times until the angry bear jumps out of its den. The moment that it shows more than half its body, the hunters shoot it in the forehead or the heart. The bear belongs to the first discoverer of the den.

The Udeghe have always loved the bear hunt, regardless of the danger. It gives them the highest prestige, as well as much meat and a precious gallbladder. Nowadays, however, it is difficult to continue the bear hunt, because the number of bears is diminishing, especially the black (or Himalayan) bear, which is in danger of extinction in the Russian Far East because of deforestation and over-killing. The Russian government has officially prohibited its hunting.

In general, mid-size or small animals are targeted for their fur in the market-oriented hunt. Fur-bearing animals living along the Bikin River include sable, marten, stoat (ermine), river otter, fox, lynx, badger, raccoon dog, raccoon, muskrat, and mink. The last three animals migrated from North America during the Soviet regime. Fur production, being one of the most important export commodities of the Soviet Union to Europe and America, was protected in Siberia and the Far East by national policy, and foreign fur animals were even actively imported to boost productivity.

As mentioned in another article (Sasaki 2001, 15), before Siberia and the Far East modernized, the most important point of the hunt for fur-bearing animals was to capture the animals with the smallest amount of damage. In the pre-modern Asian world, consumers of the precious fur were royal families, aristocrats, and administrators of the highest class. They wouldn't accept any damage to the product and required hunters to capture and process fur of the highest quality. Hunters had to develop various techniques and equipment to satisfy this requirement (for more on traditional traps used for hunting fur-bearing animals, see Sasaki [2000, 2001]).

Modernization in these regions changed the situation. After the collapse of the Chinese and Russian empires, new consumers, rich people like the bourgeois, replaced the royal families and aristocrats. They wanted quantity, not quality, because the consumer population was much larger. Hunters had to increase their productivity to satisfy the demands of the newly rich. Following this change in consumer demand, hunting techniques and equipment changed too. Hunters began to give up using traditional traps, which inflict less damage but take more time to set, preferring to use jaw traps. Though some hunters from the older generation still use the traditional traps (deadfall traps like *Kafari* and *Langi* and snare traps like *Huka*), many younger hunters don't know how to set them anymore.[4]

Game captured with jaw traps or shot with a gun usually has some damage. Jaw traps bite an animal on its paw, often torturing it to death; the animal rages about and damages its fur. Animals shot with a gun have a small hole in the fur. Though sable hunters often train themselves to shoot an animal in the eye in order to protect the fur and fetch a higher price, they are sometimes unsuccessful and shoot the body. The traditional traps seldom produce such damage. For example, the deadfall type of trap (*Dui*, *Langi*, *Kafari*, and *Hadana*, see Sasaki [2000, 2001]) hits an animal on the head, neck, or the body, and kills it instantly. It does not torture the animal and inflicts none or very little damage to its fur, and does not allow the animal to escape. From the point of view of fur quality and the conservation of resources, traditional traps are much superior to contemporary jaw traps; but the latter is easier and faster to use, boosting productivity.

Hunters know that the price for damaged fur is lower. The Russian government has a ranking system for fur, based on quality and depending on the degree of damage. The less damaged the fur is, the higher rank the government gives it. Of course, hunting companies and fur traders pay better prices for the fur of higher ranks. The older hunters, using traditional traps, are proud of producing fur of the best quality. But the younger hunters compensate for the lower prices for damaged fur by trying to capture more animals. The tendency to value quantity and productivity over quality may be one of the characteristics of modernized societies; however, it might be one of the causes of the overexploitation of natural resources.

---

[4] Though younger hunters don't know how to set the traditional traps, some techniques, which were developed along with them, are still used and applied to setting jaw traps. See Sasaki (2002, 91–92).

3. FOUNDATIONS OF AN ETHNIC HUNTING COMPANY

3.1 Management of *Gospromkhoz "Pozharskii"* in the Soviet Regime

The main characteristics of the economic management of the Soviet Union were the planned economy and nationalized enterprises. The hunting and fishing territories, equipment, and organizations of the indigenous people in the Far East were the targets of collectivization and nationalization, as were the factories and farms in European Russia. The Soviet government began to organize indigenous hunters and fishermen into collective farms (*kolkhoz*) and national (or state) farms (*sovkhoz*) in the 1930s, and gave them production targets every year based on the annual state plan.

The organizations were integrated into several large state enterprises in the 1960s and 1970s, following the policy of expanding productive organizations practiced during the regimes of N. Khurshchev and L. Brezhnev. Under this process, the state foraging farm, *gospromkhoz* "Pozharskii," was established in Krasnyi Yar at the end of the 1960s, integrating all the small collective farms located in several villages along the Bikin River. At the same time, the people who lived in the various villages or hamlets along the river were gathered together into the village of Krasnyi Yar, which was established in 1957.

This state farm consisted of some productive sections, such as foraging, farming, processing (timber and other products), and services (support to researchers and tourists). The foraging section was the most important during the Soviet regime, producing fur, wild animal meat, fish, berries, and medicinal herbs. Fur sales represented 33.8 percent of annual farm sales in 1991, the final year of the Soviet regime (Table 2 and 3). In that year, 50 professional hunters were employed, who caught 2,206 squirrels, 607 sables, 198 weasels, 74 minks, 11 hares, a badger, a raccoon, and a marten, as shown in Table 4. The farm sold the fur of these animals and other animal products to the government (the state fur trading company) for 217,700 rubles, equal to U.S.$126,600 (according to the exchange rate at the end of that year, i.e., one dollar was equal to 1,719 rubles). As to the average unit price of the fur of a particular animal, sable was the best, at 213.18 rubles, or about $124. This price appears reasonable or even a little bit expensive, judging from the price for a sable boa of 1,000 yuan (about $120) in a department store in Beijing, China in 1993.

The proceeds from other hunting-gathering products, like meat, berries, and medicinal herbs, represented more than 40.7 percent of total sales. The farm sold a large amount of wild animal meat (3.2 tonnes), boars (215 heads), wild bracken (3.2 tonnes), ferns (5.7 tonnes), wild berries (31.2 tonnes), ginseng (403 kilograms), a large amount of *eleutherococcus* leaves and roots (3.3 tonnes and 8.2 tonnes, respectively), and other medicinal herbs (Table 4), with sales amounting to 261,800 rubles (sales of medicinal herbs were 84,000 rubles, while those of other hunting products were 177,800 rubles [see Table 2], equal to $152,300). In sum, the sales of the foraging section represented 74.5 percent of total sales.

Table 2: Sales of products from state farm (1991–1994, in 1000 roubles).

Item	1991	1992	1993	1994
Fur	217.7	1864.9	9130.0	29672.7
Medicinal materials	84.0	649.1	3620.8	5387.6
Other hunting products	177.8	401.4	1651.6	3705.8
Agricultural products	22.3	140.0	1123.5	0.0
Forestry products	70.0	2910.0	4304.9	18223.0
Processed bracken	15.4	101.7	229.3	268.0
Other products	42.6	113.1	243.4	152.0
Fee for service	13.6	924.6	21518.6	25789.3
Total	643.4	7104.8	41822.1	83198.4

Source: Author.

In contrast, timber sales were comparatively small, at only 10.9 percent of total sales, less than the percentage of sales of medicinal herbs, at 13.1 percent (Table 2 and 3). The output from timber production in 1991 was by no means smaller than in other years, but other products were produced in such huge amounts that it was not so obvious.

Table 3: Percentage of sales of products from state farm (1991–1994).

Item	1991	1992	1993	1994
Fur	33.8	26.2	21.8	35.7
Medicinal materials	13.1	9.1	8.7	6.5
Other hunting products	27.6	5.6	3.9	4.5
Agricultural products	3.5	2.0	2.7	0.0
Forestry products	10.9	41.0	10.3	21.9
Processed bracken	2.4	1.4	0.5	0.3
Other products	6.6	1.6	0.6	0.2
Fee for service	2.1	13.0	51.5	31.0
Total	100.0	100.0	100.0	100.0

Source: Author.

Products from the farming section consisted of vegetables and honey, with honey being one of the main products. (The state farm encouraged villagers to engage in apiculture.) The quantity of honey produced by the farm and other villagers amounted to 2.15 tonnes in 1991, with sales totaling 43,300 rubles (equal to $25,200), but it did not play an important role in total sales, because its unit price was restricted to low levels.

To summarize, the state farm depended largely on the foraging section. At the same time, however, the farm tried to diversify its product line and maintain a balance between them, which was very important for the farm to disperse the risk among various kinds of products.

Table 4: Outputs of products of the state farm and ethnic hunting company from 1991 to 2000.

Year	1991	1992	1993	1994	1995	1996	1997	1998	1999	2000
Fur										
Squirrel	2,206	1,340	4,597	5,884	1,393	1,221	535	468	333	328
Weasel	198	133	152	222	218	129	79	93	56	60
Mink	74	199	246	161	50	60	65	104	89	220
Sable	607	745	408	498	348	224	295	243	198	600
Racoon	1	1	6			2				
Badger	1	2	19	2						
Wolf		1								
River otter		1	1							
Hare	11	6	2							
Marten	1		2							
Muskrat			3	9						
Other animal products										
Bear gallblader			3							
Musk	3	2	3							
Wild animal meat (kg)	3,200	2,500	2,700	700						
Plant products										
Fern (kg)	3,200	3,000	3,500		72		321		210	193
Magnolia vine (kg)					3,456					
Leaf of *Eleutherococcus* (kg)	3,300	370	30				435			
Root of *Eleutherococcus* (kg)	8,200	18,400	520	6,320			1,425		4,000	10,000
Ginseng (kg)	0.403	1,151	1,345	980						
Celandine (kg)	60	20								
Leaf of cowberry (kg)	720	50							285	290
Aralia (kg)		4,300								
Bracken (kg)	5,700	3,100	130	40						
Berry (kg)	31,200	3,100								
Walnut (kg)		900	50	70						
Agricultural products										
Honey (kg)	2,150	700	1,650	820						
Vegetables (kg)		20								
Forestry products										
Round wood (cu m)		2,496	572	12	190	223	405	540	1,621	2,635
Fuel wood material (cu m)	1,602	2,654	972	395	895	351	1,382	807	3,778	3,920
Fuel wood (cu m)	1,964	2,047	1,028	522	442	292	884	1,638	2,461	4,132
Sawn wood (cu m)	85	413	287	113	137		316			
Lath (pieces)	5,500	2,500	21,000		9,000	5,000				

Source: Author.

3.2 Establishment and Management of the Ethnic Hunting Company OAO "BIKIN"

In the autumn of 1994 the state foraging farm, *gospromkhoz "Pozharskii,"* was privatized and reorganized into a joint-stock company named the Ethnic Hunting Company OAO "BIKIN."[5] The process of this reorganization was very complicated. The emergence of the campaign to restore indigenous rights in the Russian Far East and Siberia during the age of *perestroika* (1985–1991) played a preparatory role. The collapse of the Soviet Union at the end of 1991 and the following economic crisis woke up the people in Krasnyi Yar. The struggle against a deforestation project in the upper part of the Bikin River, planned by a Russian-Korean joint enterprise in 1991–92, strengthened the solidarity of the villagers. Although some dropped out after winning the struggle, the policy of privatizing state enterprises and support from some non-governmental organizations (NGOs) promoted the reorganization of the state foraging farm (Sasaki 2000, 498–499).

This reorganization could be seen as a successful example of the privatization of state farms in indigenous people's regions, because the company survived through a very critical period and still exists today, while many reorganized farms, co-operatives, and private companies collapsed or went bankrupt soon after being established. It was extremely difficult to manage a privatized hunting or reindeer-breeding enterprise, especially in Siberia and the Far East. These regions are so remote from state and regional centers that costs are very high for transportation and the construction of infrastructure. Moreover, there is poor demand for the products of foraging and reindeer breeding, such as the meat, skin, and fur of tame reindeer and wild animals. In general, farms and enterprises in the areas of the indigenous people in Siberia and the Far East have been fundamentally dependent on the support of the central and regional governments, but this is limited and restricted in the new market economy.

Krasnyi Yar is also far from the center of the region and village infrastructure is inadequate, as seen with the limited supply of electricity. Compared to other arctic or subarctic regions, however, Krasyni Yar's situation is a little better. It is accessible by train, bus, or car, while northern Yakutia or Chukotka can be only reached by airplane or helicopter. In fact, few reindeer-breeding or hunting enterprises still exist in these regions.

Why did the ethnic hunting company in Krasnyi Yar survive, even in the period of economic crisis after the collapse of the Soviet Union? Its location is not the only reason. Judging from the change in the main products of this company and the previous

---

[5] "OAO" means that the stocks of this company are opened. When the government decided to privatize former state enterprises and farms, it prepared some options. The establishment of joint-stock companies was one of them. In Russia there are two types of joint-stock companies: the open type and the closed type. In the former, the property of the former state enterprise or farm is once changed to stocks, which can be shared not only by the members of the company but also by other persons or organizations. In the case of OAO "BIKIN," the members of the company hold about 60 percent of the stocks, while it sold 40 percent to a securities company.

state foraging farm, it is obvious that the company tried to produce anything that it could sell to survive during the critical period. In 1992, soon after the collapse of the Soviet Union, the state farm strengthened its timber production. The share of timber sales compared to total sales rose from 10.9 percent in 1991 to 41.0 percent in 1992 (Table 3). In contrast, the percentage of sales of fur products diminished, even though the hunters captured many sables.

The sales of timber products decreased in 1993. For example, the percentage of sales of timber products diminished to 10.3 percent in 1993 and then went up a little to 21.9 percent in 1994 (Table 3), perhaps a reaction against the quick increase in 1992. Instead of timber production, the service section expanded its percentage of total sales. It was 51.5 percent in 1993, but diminished to 31.0 percent in 1994. Compared to the sales of these two kinds of products, the sales of fur products were relatively stable, ranging from 33.8 percent in 1991, 26.2 percent in 1992, 21.8 percent in 1993, and 35.7 percent in 1994 (Table 3). This indicates that the hunting section was the most reliable source of income for the farm, even during the most critical period.

In contrast to these three kinds of products or activities, other products quickly lost their significance in the management of the farm. Decreases of other hunting-gathering products, like wild animal meat and wild berries, were especially obvious in 1992, when the share of their sales dropped from 27.6 percent in 1991 to 5.6 percent in 1992, and have not yet recovered (Table 3).

As a result, one can draw the following conclusions upon analysis of the management of the state farm in Krasnyi Yar during the most critical period just after the collapse of the Soviet Union (1992–94). First, the balance between each productive section collapsed in 1992. Second, the main sources of the farm's income changed every year during this period, while the fur section maintained its position as the most reliable income generator. Third, the farm lost the diversity of its products. In other words, it survived the most dangerous period at the expense of its productive diversity. The percentages of sales of each product in 1994 looked as if the farm, which was reorganized into a private company in this year, tried to reestablish its productive system on the basis of three main sections (fur, timber, and services), but various conditions hindered it from doing so.

The percentage of fur products went down from 35.7 percent in 1994 to 7.9 percent in 1995, and since then has not recovered (Table 5). It may be inappropriate to compare the percentages in 1994 and 1995, because they do not share the same basis of statistical data. For example, total sales jumped from 83,198,400 rubles in 1994 to 244,106,800 rubles in 1995 (Table 5), although hyperinflation was diminishing in Russia in those years. The sort of statistical data gathered and their analysis might have also changed after the reorganization of the state farm into a joint-stock company. But the output and sales from the hunting of fur-bearing animals actually diminished. For example, the farm sold 498 pieces of sable fur and 5,884 squirrels in 1994, while it sold only 348 and 1,393, respectively, in 1995 (Table 5 and 6). The productive output of squirrel fur peaked in 1994 and gradually went down to 468 pieces in 1998. The sales of fur dropped from 29,672,700 rubles in 1994 to 19,219,900 rubles in 1995, regardless of the rise of the unit price of fur. For example, the price for sable fur rose from 20,490 rubles in 1994 to 40,340 rubles in 1995, as shown in Table 6-3.

The decrease of the output and sales of fur products in 1995 had much to do with the decline of fur production in Russia. One can point out several reasons. First, the economic disorder after the collapse of the socialist system damaged fur-producing enterprises (private companies, co-operatives, and collective farms). Second, the drop in demand for fur in the world market, influenced by anti-fur campaigns and the development of synthetic materials that replaced fur, pushed down prices to even lower levels. Third, the rivalry between domestic and foreign producers in the fur market in Russia also restricted fur prices. Fourth, hyperinflation rapidly pushed up the costs of fuel, hunting equipment, and transportation, and put a huge burden of expenses or debts on the backs of the enterprises. Under these difficult economic conditions, the more fur they produced, the more expenses and debts they incurred, so many enterprises, farms, and co-operatives in indigenous areas gave up on producing it.

Table 5-1: Sales of fur and timber products (1991-1998, in thousand roubles before 1998, roubles in 1998).

Year	1991	1992	1993	1994	1995	1996	1997	1998
Squirrel	8.5	34.0	1,719.6	7,461.2	2,611.6	3,528.8	2,374.5	2,918.0
Weasel	3.1	20.8	147.9	605.3	825.0	768.0	569.8	857.0
Mink	4.0	207.3	1,331.8	1,488.0	1,745.3	2,688.8	2,880.5	4,680.0
Sable	129.4	977.1	2,802.4	10,203.6	14,038.0	15,094.0	21,194.0	27,880.0
Other fur	72.7	625.7	3,128.3	9,914.6		85.5		
Sales of fur products	217.7	1,864.9	9,130.0	29,672.7	19,219.9	22,079.6	27,018.8	36,335.0
Sales of timber products	70.0	2,910.0	4,304.9	18,223.0	62,383.5	44,650.0		
Total sales all products	643.4	7,104.8	41,822.1	83,198.4	244,106.8	688,677.0	567,807.3	369,000.0
Fur share of total	33.8%	26.2%	21.8%	35.7%	7.9%	3.2%	4.8%	9.8%
Timber share of total	10.9%	41.0%	10.3%	21.9%	25.6%	6.5%		

Table 5-2: Proportion of fur product sales (percent).

Squirrel	3.9%	1.8%	18.8%	25.1%	13.6%	16.0%	8.8%	8.0%
Weasel	1.4%	1.1%	1.6%	2.0%	4.3%	3.5%	2.1%	2.4%
Mink	1.8%	11.1%	14.6%	5.0%	9.1%	12.2%	10.7%	12.9%
Sable	59.4%	52.4%	30.7%	34.4%	73.0%	68.4%	78.4%	76.7%
Other	33.4%	33.6%	34.3%	33.4%	0.0%	0.4%	0.0%	0.0%

Source: Author.

For example, in northern Yakutia (in the Eveno-Bytantai region in the Republic of Sakha), fur was one of the main products of the state farm, "Sovkhoz named Lenin." It had a hunting section and employed 15 to 29 professional hunters to carry out the state fur production plan. Moreover, it constructed and managed a fox-breeding farm. But the negative conditions surrounding fur production made it impossible for the farm to keep the hunting and fox-breeding sections. They were closed in 1996 and the hunters lost their jobs (Sasaki 1997, 182–188; 2002, 99).

The natural conditions of Krasnyi Yar are much better than those in the Eveno-Bytantai region, because the faunal diversity is much larger. It plays an especially decisive role in whether sable populations survive or not. Sable fur has not lost its value in Russia, even today. The government controls its production to conserve this precious natural resource. The Eveno-Bytantai region is located in the area outside the sable's habitat, which is the weakest point of the hunters there.

Table 6: Sales and unit prices of fur (1991-2001).

Table 6-1: Fur output (number of furs).

Year	1991	1992	1993	1994	1995	1996	1997	1998	1999	2000	2001
Badger	1	2	19	2							
Squirrel	2206	1340	4597	5884	1393	1221	535	468	333	328	320
Wolf		1									
River otter		1	1								
Racoon	1	1	6			2					
Hare	11	6	2								
Weasel	198	133	152	222	218	129	79	93	56	60	62
Marten	1		2								
Mink	74	199	246	161	50	60	65	104	89	220	15
Musk rat			3	9							
Sable	607	745	408	498	348	224	295	243	198	600	158

Table 6-2: Fur sales (in thousand roubles before 1998, roubles 1998–2001).

Year	1991	1992	1993	1994	1995	1996	1997	1998	1999	2000	2001
Badger		1.5	34.1	13.5							
Squirrel	8.5	34.0	1719.6	7461.2	2611.6	3528.3	2374.5	2918.0	1809.0	13520.0	25600.0
Wolf		0.1									
River otter		2.2	7.5								
Racoon	0.1	0.3	34.5			85.5					
Hare	0.0	0.0	0.1								
Weasel	3.1	20.8	147.9	605.3	825.0	768.0	569.8	857.0	1055.0	4,800.0	4,960.0
Marten			4								
Mink	4.0	207.3	1,331.8	1488	1,745.3	2,688.8	2,880.5	4,680.0	4,970.0	26,400.0	2225.0
Musk-rat			4.8	10.2							
Sable	129.4	977.1	2,802.4	10,203.6	14,038.0	15,094.0	2,1194.0	27,880.0	40,400.0	480,000.0	134,300.0

Regardless of possessing the sable resource, the ethnic hunting company in Krasnyi Yar has been shifting away from the fur trade to other products and activities since 1995, as if it had realized that hunting fur-bearing animals would not pay in the future. What will be its new main product? The role of timber products will be limited, though the output of the timber section in 1999 and 2000 was restored to the 1992 level (Table 4). The company will not adopt a management strategy largely depending on timber, because it would mean logging the forests of the hunting territories to increase productivity.

Judging from the output data of each section of the company, one can guess that it is trying to restore the diversity of its products to overcome the present difficulties and avoid future crises by conserving their forest resources. For example, it has increased the production of edible wild plants and medicinal herbs. According to recent data (Table 4), the enterprise produced dried ferns (210 kilograms [kg] in 1999; 193 kg in 2000), cowberry leaves (285 kg in 1999; 290 kg in 2000), and *Eleutherococcus* roots (4,000 kg in 1999; 10,000 kg in 2000). Though data on their sales was not available, these products may provide the company with a substantial income. Collecting useful plants from the wild can be identified as one of the third ways of forest usage. But one must always pay attention to the fact that wild plant collecting implies a danger of overexploitation of resources, as with hunting and timber cutting.

Table 6-3: Unit prices of fur (in thousand roubles before 1998, roubles 1998–2001).

Year	1991	1992	1993	1994	1995	1996	1997	1998	1999	2000	2001
Badger		750	1,795	6,750							
Squirrel	3.85	25	374	1,268	1,875	2,890	4,438	6.24	5.43	41.22	80.00
Wolf		100									
River otter		2,200	7,500								
Racoon	100.00	300	5,750			42,750					
Hare			50								
Weasel		156	973	2,727	3,784	5,953	7,213	9.22	18.84	80.00	80.00
Marten			2,000								
Mink	54.05	1,042	5,414	9,242	34,906	44,813	44,315	45.00	55.84	120.00	148.33
Musk rat			1,600	1,133							
Sable	213.18	1,312	6,869	20,489	40,339	67,384	71,844	114.73	204.04	800.00	850.00

Source: Author.

Ecological tourism can also be included in the third way of forest usage. The enterprise in Krasnyi Yar conducted tours to the upper part of the Bikin River in 2000 and 2001 with support from NGOs. The hunters took Japanese tourists to their hunting territories and introduced them to the nature of their forest. Of course, there are many problems with such tours, and project organizers should establish tour manuals to educate tourists about not disturbing or exploiting the forest.

4. PRESENT PROBLEMS CONFRONTING UDEGHE HUNTERS

As mentioned above, hunting by using the traditional techniques of the indigenous people can be one of the styles of sustainable forest usage. In fact, the indigenous hunters in the Russian Far East, including the ancestors of the Udeghe, have long demonstrated this. Though they caught a large amount of fur animals for the payment of tribute to Imperial China and for trade with merchants, the resources were never completely exhausted through overexploitation, as if the hunters understood the concept of "sustainable forest usage."

The situation has rapidly changed since the middle of the nineteenth century, when Imperial Russia modernized this region. As mentioned above, the policy of development of farms and industry accelerated the deforestation, and the quantity-oriented hunting boosted the over-exploitation of animal resources. Some species, like sable, musk deer, and bear, have often fallen into danger of extinction because of over-hunting, deforestation, and other ecological problems.

The socialist government, established by the October Revolution in Russia, could have adopted a policy of sustainable forest usage, because all productive activities were controlled by the state plan. If the government had established a hunting plan within the restorative powers of the forest, the resources would not have been overexploited or wasted. In fact, some fortunate forests and communities, like Krasnyi Yar, which were defined and preserved as areas of traditional foraging, survived, while the Soviet government destroyed many other forests and indigenous communities that were defined as targets of socialist development.

The collapse of the Soviet Union brought back primitive capitalism, much like that of Imperial Russia, even into the conserved forests and communities. Hunters had to capture as many animals as they could in order to survive during this critical period. The changes in the amount of fur-bearing animals sold clearly demonstrate the struggle of the hunters and the state farm against the economic crisis. As shown in Table 4, 5, and 6, the farm and company sold many squirrels up until 1996, peaking in 1994 at 5,884 pieces, while the amount of sable fur slowly declined after its peak in 1992 of 745. The sales of squirrel fur (7,461,200 rubles) approached the sales of sable (10,203,600 rubles) in 1994. From these data, one can make a conjecture on the following two things. First, the hunting section of the farm fell into crisis because of a shortage of precious sable fur, so it tried to make it up with the large amount of squirrel fur. Second, the results of hunting were actually poor after 1992, because of either disappearing resources or a lack of motivation on the part of the hunters.

After the reorganization in 1994, the output and sales of squirrel fur dropped rapidly, while those of sable and mink were beginning to return to 1991 and 1992 levels. The drop in output of squirrel furs was especially obvious in 1997, although the decrease in sales was comparatively slow because the unit price rose over the five years. Generally, the unit price for squirrel fur is so low that it produces little profit in either the capitalist or market economy. When there were no other alternatives, such as during the economic crisis between 1992 and 1994, the farm and the hunters had to put up with capturing and selling the less expensive fur in large quantities. But it is natural for them to capture and sell more expensive and profitable furs as far as ecological and economic conditions allow. The data of the changes of the output and sales of fur in these ten years show that the motivation of hunters for supplying the company with sable fur was restored several years after the reorganization of the state farm.

According to a report of the ethnic hunting company, the population of the sable is large enough to be hunted for fur. The company estimated that the population of sables in its territory stood at 10,620 in autumn 2001, and, as it limits the number of sable that can be hunted to 33 percent of the population, hunters could have captured up to 3,504 last season. But the actual number was much less than this. Even now, however, the sable is still in danger of extinction as long as hunters continue with quantity-oriented hunting. One of the most important tasks for the company is to create a system of economically and ecologically sustainable hunting that makes a profit.

The hunting section of the company now has another serious problem: the absence of successors of the older hunters. As shown in Table 7 and Figure 2, the number of hunters has decreased over the last ten years. For example, while 50 professional hunters worked in 1991, only 24 hunters, less than 50 percent, remained in 2001. It is said that there are many amateur hunters in the village and that they sometimes overexploit or deplete the animal resources in the hunting territories. Just after the collapse of the Soviet Union, especially when food shortages in the village were very serious, many people went into the forest to capture large mammals for meat, which promoted the overexploitation of the animal resources. But it is urgent now that the company employs well-trained hunters, who have good techniques and considerable knowledge about animals and the forest, if the company wants to strengthen its medicinal herbs and ecological tourism sectors. Only good hunters can safely guide the tours, ensuring positive experiences, and know the best places and seasons for collecting herbs.

Table 7: The population and percentage of hunters in the company.

Year	1991	1992	1993	1994	1995	1996	1997	1998	1999	2000	2001
Administration	10	7	7	9	11	11	10	6	13	13	13
Hunters (professional)	50	50	47	44	43	45	34	32	34	28	24
Hunters (non-professional)									29	29	29
Engineers, workers, and others	56	40	23	20	23	28	28	36	77	73	59
Total	116	97	77	73	77	84	72	74	153	143	125

Source: Author.

The absence of young, well-trained hunters is very serious. In an interview with a young professional hunter, he said that of the youth of his generation, he was the only one who became a hunter and that the others had left the village to find jobs in cities like Luchegorsk or Vladivostok.

In order to be a good hunter, it is necessary to have been taught since childhood by the nearest male relatives, such as a father, uncles, and elder brothers, by demonstrating and explaining their techniques, and sometimes letting them experience the hardships of the forest. Through these educational processes, young hunters learned various traditional techniques and knowledge that has been passed on through many generations. Establishing a good education system is also an urgent task for the company in the near future in order to develop ecological tourism, as well as to continue with foraging production.

Finally, one more problem should be raised here. What role can the traditional techniques and equipment of the Udeghe people play in establishing sustainable usage of their forests?

Though it is difficult to define what is "traditional," it is clear that we should reevaluate it, not from the point of view of cultural authenticity, but from that of the sustainability of forest use. One can identify various devices and knowledge for continuous fur production in the "traditional" hunting techniques and equipment. For example, an old hunter showed us that he could capture a fixed sort and size of animals with his "traditional" deadfall traps by controlling the size and the strength of the release mechanism.

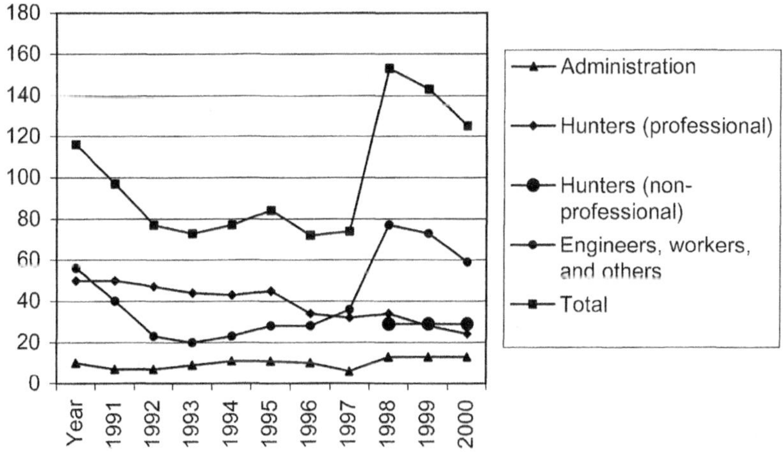

Figure 2: Population of the employees of the company.

Source: Author.

But it is problematic to say whether the hunters were conscious of sustainability or not. They might have tried to capture a fixed sort and size of animals only because the larger animals had fur of high quality and provided them with more income. When they had captured too many animals in a season, they might have had a serious crisis in the next year. Even if the selection of game through the control of the trap had prevented them from depleting the resource, and if they had been conscious of the importance of resource conservation, they may not have been able to scientifically explain the relationship between their techniques and sustainability. Therefore, the scientific analysis of their techniques and equipment, both traditional and present, is one of the more urgent tasks for researchers.

There are two directions for the analysis. One is the purely objective analysis on the basis of ecology, biology, forestry, technology, and economy. Quantitative analysis is inevitable in this case. The other is the analysis of the discourse of the Udeghe people. Researchers should clarify their recognition of nature, the forest, resources, techniques, equipment, and so on, through interviews and observation. It is a task for anthropologists and ethnologists, and it will be more effective with the support of ethno-zoology and ethno-botany. An interdisciplinary study will support the development of the study of hunting (or foraging as a whole) from the point of view of sustainability.

5. CONCLUSION

Nowadays, hunting does not have much significance in the economy of the villagers in Krasnyi Yar. Even the ethnic hunting company has given up managing itself on the basis of the hunting of fur bearing animals. But hunting still plays some important roles in food production, cultural activity, and mental and spiritual support of the villagers. It is, nonetheless, prestigious, because it supplies villagers with food, implies skills and considerable knowledge about nature and equipment, and provides opportunities to access the spiritual world.

It is a fact that hunting sometimes caused overexploitation of animal resources, especially in the modern age. It is often condemned as a primitive and savage activity that kills precious wild animals. As far as the case in the Russian Far East, however, the overexploitation of wild animals began only after modernization that demanded hunters to practice quantity-oriented hunting. On the contrary, indigenous hunters with the traditional techniques and equipment have long demonstrated sustainable use of animal resources.

In present days, hunting can be utilized, either directly or indirectly, for the conservation and sustainable use of the forest. Strictly controlled hunting of big game and fur-bearing animals provides hunters and the company with stable incomes and sales without damaging the forest resources. The large amount of knowledge of experienced and well-trained hunters about nature and animals can be utilized for collecting medicinal herbs and conducting ecological tours.

Sustainability, implied in the traditional knowledge and techniques, can only be empirically observed, but no one can scientifically explain it. Therefore, it is necessary to clarify the mechanism of sustainability through scientific research and analysis of traditional hunting techniques and equipment, and the role of knowledge for the conservation of forest resources. Results of such research and analysis can be utilized for the education of children in the village and the training of young hunters and forest rangers, as well as for the development of forest science and the enlightenment of people all over the world.

REFERENCES

Sasaki, S. 1997. Crisis of the fur animal hunting of the indigenous peoples in Siberia and Russian Far East. *People and Culture* 5(9), 173–191. Ansan: The Institute for Ethnological Studies, Hanyang University.

———. 2000 Fur animal hunting of the indigenous people in the Russian Far East: History, technology, and the economic effects. In *A step toward forest conservation strategy (1), IGES Forest Conservation Project interim report 1998*, ed. M. Inoue and H. Isozaki, 495–513. Hayama: Institute for Global Environmental Strategies (IGES).

———. 2001. Comparative study of hunting techniques of the native people of the Lower Amur basin and the Primor'e region: With the focus on traps for fur game animals. Proceedings of the 14th International Abashiri Symposium, ed. Hokkaido Museum of Northern Peoples, 11–21. Abashiri: Association for the Promotion of Northern Cultures.

———. 2002 Hunting activity and forest management of the Udeghe people in Krasnyi Yar in the Russian Far East. In *Russia country report 2001*, ed. H. Kakizawa and M. Inoue, 88–103. Hayama: IGES Forest Conservation Project, Institute for Global Environmental Strategies (IGES).

Taguchi, H. 1998. Life and hunting activity of the Udeghe people in the Primor'e region in Russia: Focusing on the influence of the fur trade on hunting techniques, especially on trap hunting (in Japanese). In *Hunting culture in Russia,* ed. H. Sato. Tokyo: Keiyusha.

———. 2000 The hunting and fishing activity of the indigenous peoples in the Amur basin (in Japanese). In *The report of joint research on the natural settlements in the Far East*, ed. T. Fujimoto. Niigata: Faculty of Humanities, Niigata University.

Chapter 21

FOREST MANAGEMENT AND WATER RESOURCE CONSERVATION: COST-SHARING SYSTEMS BETWEEN UPSTREAM AND DOWNSTREAM COMMUNITIES IN JAPAN

Masanobu Yamane
Kanagawa Prefectural Natural Environment Conservation Center

1. INTRODUCTION

As emphasized in other chapters, the proper valuation and recognition of the multiple functions of forests are essential elements of forest conservation in the Asia-Pacific region. Similarly, the participation of local people who benefit from forests is also important, since the beneficial functions of forests commonly extend far beyond the forests themselves. Who are the actual stakeholders and beneficiaries of forest functions, and how can we identify the actors for forest conservation?

Typical examples of such functions in the context of global warming are the climate control and carbon dioxide absorption functions of forests. An external effect, known as the "water resource conservation function" is another familiar example. This function is important, particularly for big cities, which have high demand for stable water supplies from their water reservoir areas. When we consider those functions of forests that extend far beyond the local forest area, it is easy to imagine that a participatory approach to conserve them would be different from one to conserve a forest mainly for local stakeholders.

The aim of this chapter is to study participatory approaches to conserving those external functions of forests that extend substantially beyond the local forest area. It examines the conditions required of a cost-sharing system for water resource conservation (WRC) between upstream and downstream communities, based on experience in Japan.

The framework of the study is as follows. The chapter starts by outlining the theoretical basis of the "benefit principle" of the external forest economy, and then

M. Inoue and H. Isozaki (eds.), *People and Forest—Policy and Local Reality in Southeast Asia, the Russian Far East and Japan, 321–336.*

moves on to analyze existing cost-sharing systems for water resource conservation in Japan, with reference to previous studies. Next, three case studies are introduced to illustrate different systems, focusing on the participation of local people in cost-sharing systems. The three systems selected have different origins but were all established by municipalities that had a large demand for a stable water supply from their headwaters. In addition, this chapter discusses the preconditions for reaching agreement on cost-sharing for water resource conservation between upstream and downstream municipalities.

In this study the following definitions of terms are applied, based on Kumazaki (1982) and Izumi (1996).

Reforestation for water resource conservation (RWRC): Tree planting on deforested or degraded land in secondary forests for the purpose of water resource conservation in water reservoir areas.

Forest management for water resource conservation (FMWRC): Tending forests in water reservoir areas by operations such as thinning, pruning, felling for regeneration, and reforestation for the purpose of water resource conservation or timber production.

Forest improvement for water resource conservation (FIWRC): This concept covers both RWRC and FMWRC. Thus, the protection of natural forests for the purpose of water resource conservation is also included here.

Cost-sharing: An arrangement by which downstream citizens or municipalities pay all or part of the costs of FIWRC to forest owners or municipalities in headwater or water reservoir areas for reasons of security of such functions as soil and water resource conservation.

2. OVERVIEW OF COST-SHARING SYSTEMS FOR WATER RESOURCE CONSERVATION IN JAPAN

2.1 Theoretical Bases

Forests can be seen as resources that have both economic functions, such as those that arise through timber production, and also a variety of environmental functions. From the perspective of utility to society, some functions of forests are compatible with each other and others are trade-offs with economic functions. It has generally been believed that the water resource conservation function is compatible with timber production (Clawson and Knetsch 1966). In this context, systems of FMWRC in Japan are generally based on a principle of "assumed harmony" (*yotei chowaron*) between timber production and water resource conservation (Izumi 1996), i.e., the assumption that forest management has to accomplish both goals.

FMWRC is based on the precondition that profits from timber production will be enough to compensate for past as well as future investments in reforestation. Generally speaking, "public benefit functions" such as water resource conservation do not have a feedback loop in which benefits are exchanged for monetary profit, nor are there proper devices for information exchange and control of supply and demand of the benefits (i.e., water)(Figure 1). This situation thus creates the need for policy instruments to internalize certain external economic factors and to close the feedback loop. When profits from timber production are sufficient, however, the need for such policy instruments is usually less of a concern.

When profits from timber production are not adequate, as in the case of a decline in timber prices or a rise in costs for forest management, harmonization between timber production and public functions becomes more difficult due to the shortage of funds to invest in forest management. Consequently, a cost-sharing system aimed at maintaining the forest's environmental functions—which are economic externalities for forest management—comes to the fore of policy matters.

Two different viewpoints exist with respect to payment for forest management costs to maintain such non-timber functions (Kumazaki 1977). One view is that forest owners should pay all management costs, including costs for environmental conservation; this approach makes it necessary to build in policy instruments such as regulations and penalties. Another view is that forest ownership extends to the external economy, and thus, beneficiaries must also share a certain portion of costs for forest management. In the case of water resource conservation in headwaters, downstream municipalities and citizens can be clearly identified as beneficiaries, and thus can be expected to share the costs of forest improvement for water resource conservation.

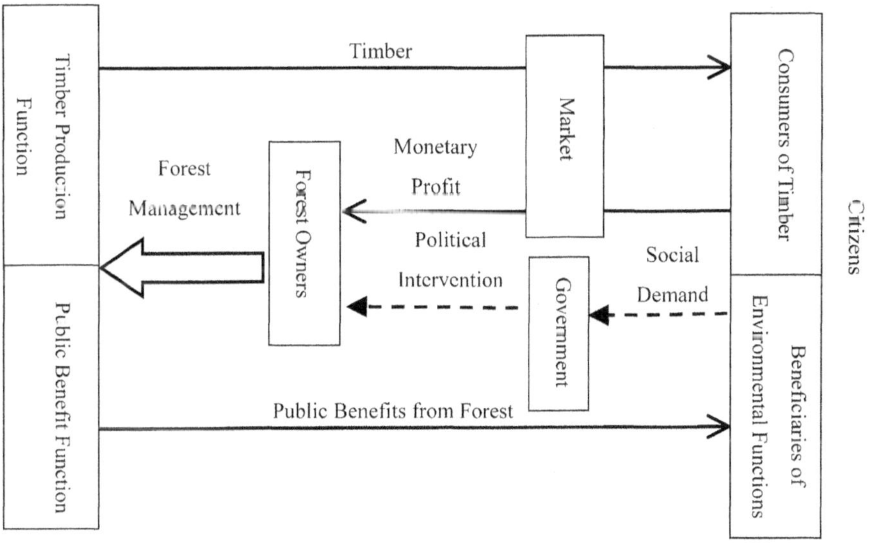

Figure 1: Outline of key factors in cost-sharing regarding forests for timber versus water resource conservation.
Source: Kumazaki (1977).

2.2 Classification of the System

In Japan, cases in which downstream municipalities cooperated in reforestation far upstream to conserve their headwaters can be found as early as the seventeenth century, but the use of such systems became more common in the late nineteenth century. Many types of cost-sharing systems for forest improvement for water resource conservation emerged in this period; examples include reforestation projects in which both upstream and downstream towns and villages cooperated, water-utilization cooperative associations, and town water supply agencies that diversified into direct management of headwater forests. Cost-sharing systems for water resource conservation between upstream and downstream municipalities appeared during this period (Kumazaki 1982). In the early 1980s, forest improvement programs for water resource conservation were re-activated, mainly in urban areas where the population density had grown rapidly. These programs promoted revenue enhancement campaigns for water resource conservation and emergency improvements of forests and rivers in water reservoir areas from 1985 to 1986, but in the end, no new taxes were realized. Since the beginning of the 1990s forest improvement for water resource conservation has been emphasized again. A variety of new projects and cost-sharing systems have since been launched and implemented by local authorities, after declining drinking water quality and serious droughts became major concerns in downstream municipalities.

Table 1 displays an analysis of existing cost-sharing systems between upstream and downstream communities for WRC in Japan, based on key measures, resources, and main actors (Kumazaki 1982; Izumi 1996).

3. CASE STUDIES

For case studies this paper uses three representative systems, established at different times in Japan. The first is the city of Yokohama's Water Resource Conservation Forest (YWRCF), which is one of the oldest official cost-sharing systems in Japan, and could be categorized as 4-a-ii (measure–resource–main actor) using the criteria indicated in Table 1. The second case study is the Fukuoka Water Resource Conservation Fund (FWRCF), which was established in the 1980s during a time of active debate about the principles and benefits of forest conservation in water reservoir areas in Japan—a debate that arose due to frequent water shortages in large cities. This system can be categorized as 2-c-iii. The third case study is the Kanagawa Water Resource Forest Conservation Project (KWRFC), which was launched in 1997. This system employed various measures for resource conservation in water reservoir areas.

In the following section, the main features are outlined, with a focus on the background and relationship of each system with the local community. Several changes in water supply and city dwellers' attitudes towards WRC are also indicated in order to identify whether these systems work as good water conservation mediums.

3.1 The Yokohama Water Resource Conservation Forest (YWRCF)

The city of Yokohama in Kanagawa Prefecture purchased headwater forests surrounding the village of Doshi, Yamanashi Prefecture in 1916. Since that time, the Yokohama City Water Supply Agency has managed these forests using revenues from the city. The YWRCF covers 2,873 hectares, equivalent to 36 percent of the total area of the village. The city also established a water resource conservation fund in cooperation with the village in 1997, with 1.01 billion Japanese yen (about U.S. $8.4 million) as a trust for natural environmental conservation and maintenance of the living environment of the village. Ninety-nine percent of the money was contributed by the city of Yokohama.

Table 1: Measures, resources, and main actors for upstream and downstream cost-sharing for water resource conservation in Japan.

	Key Measures		Resources		Main Actors
1.	Compensation for losses due to logging restrictions	a.	Subsidies from general accounts	i	Organizations owning water-use rights for agriculture
		b.	Taxes		
2.	Subsidies to forest owners for reforestation, forest management, extending cutting period	c.	Funds	ii	Water supply authorities in urban areas
		d.	Investments		
		e.	Providing manpower	iii	Municipalities in urban areas
				iv	Downstream prefectures
				v	Public corporations established by upstream and downstream municipalities
3.	Profit-sharing from reforestation, forest management				
4.	Acquisition of stands and forest land				

Source: Compiled by author from Kumazaki (1982), Izumi (1996).

3.1.1 Background

While in its early phases of development into a major port, Yokohama introduced several policy measures to address the problem of serious forest devastation in the watershed of the Doshi River. These included subsidies for tree-planting, designating conservation forests in water reservoir areas, and other measures. Results were disappointing, however, and conditions worsened. Eventually, in 1915 the city decided to seek to acquire forests in the Doshi village area from Yamanashi Prefecture, and thus embarked on the path of direct forest management. During negotiations to acquire the forests, the city did not make any particular effort to build consensus with the village, despite the fact that the forestland eventually sold to the city included around 40 percent of the village's land area. Thus, after the acquisition serious conflicts occurred repeatedly between the city and the village, as the city objected to and tried to prevent such development projects as waterpower generation and mining for mineral resources.

The city later made a number of commitments that showed better consideration of the local economy. For example, it assigned to the village 25 percent of the price paid for sales of natural forest stands, it provided preferential and fair price conditions to the village when forest products were sold, it employed local people preferentially in such projects as tree-planting and construction work, and gave more consideration to ensuring that some of the benefits of forest management went to the local people. In addition to these commitments, the city established a woodcraft factory, with the aim of expanding opportunities for cash income for the local people, and launched projects for their advancement and training. The ownership of the project was transferred from the city to the village after several years. In addition, the city frequently paid out large contributions from its waterworks budget to introduce mushroom cultivation and for the construction of public facilities.

3.1.2 Forest Management

After the forest acquisition, the management office for the YWRCF moved ahead with reforestation for water resource conservation, using Japanese cypress (also known as hinoki cypress, *Chamaecyparis obtusa*) and Japanese cedar (also known as sugi, *Cryptomeria japonica*), under a forest management policy that sought to strike a balance between timber production and water resource conservation. As a result, in total 1,239 hectares of manmade forests were established. The remaining 1,544 hectares, covered with natural broadleaf and mixed conifer forests consisting of Japanese fir *(Abies firma)*, Japanese hemlock (*Tsuga sieboldii),* and beech (*Fagus crenata*), were left unmanaged.

The reason that reforestation using *sugi* and *hinoki* was actively promoted was the "assumed harmony" between timber production and water resource conservation as goals for forest management (Izumi 1996). After 1990 this management principle was drastically changed to attain one single goal: emphasizing water resource conservation. The ninth forest management plan of the YWRCF, established in 1996, set as management goals the preservation of natural forest as well as the conversion of the uniform manmade forests into coniferous and broad-leaved mixed forests and multi-storied stands (Yokohama Water Supply Agency 1996). As a result, operations for timber production in these forests virtually ended during the latter half of the 1990s.

3.1.3 Local Involvement and Benefits

The YWRCF has created both direct and indirect employment, brought cash income to the local community, and attracted continuing and extensive economic benefits to the village.

To begin with, the YWRCF has resulted in the permanent hiring of staff at the management office as well as casual employment for forest improvement operations. The management office has nine permanent staff that belong to Yokohama City; all but the chief are from the village. Workers for forest improvement for water resource conservation activities, such as nursery work, reforestation, pruning, and thinning, were hired from the village, amounting to over 4,500 persons in total from 1985 to 1989. In addition, 2,000 workers were employed casually from 1996 to 2001. Until the mid-1980s, timber sales amounted to between 1,000 and 2,400 cubic meters per year, providing a certain level of income to the local community. Due to the changes in management principles described above, however, the benefits to local people from timber production have almost disappeared since the end of the 1980s.

During the 1990s, the city of Yokohama constructed a series of facilities in the village aimed at tourism development, such as sports grounds and outdoor recreation facilities. As a result, more than one hundred thousand people visit the village each year, providing a source of tourist income. The combined effects of the favorable impression of the water resource conservation forest and the recent outdoor recreation boom have brought a sharp increase in tourists and tourist income. The village's location, only two hours by car from Yokohama and the Tokyo metropolitan area, is another positive factor. The city of Yokohama organizes events periodically to enhance the village's drawing power, such as a tree-planting event in 2001 in which 1,200 people participated.

In addition to these benefits from the forest, various public projects financed by the Doshi Water Resource Conservation Fund also provided local hiring and economic benefits. In addition, the fund has created work for local construction companies, through projects to improve the local infrastructure (Kobayashi and Kono 1991).

3.2 Fukuoka Water Resource Conservation Fund (FWRCF)

Fukuoka Prefecture established a cost-sharing system in 1979, in cooperation with the cities of Fukuoka and Kitakyushu, and downstream water consumers, such as municipalities and major corporations in Fukuoka Prefecture, aiming to create new funds and provide finances for forest improvement for water resource conservation. The fund's initial endowment was 1.22 billion yen (about $5.5 million), to which the prefecture contributed 99.2 percent.

3.2.1 Background

The impetus to establish the system was the city dwellers' bitter experience of water shortages, when water-use restrictions in the city of Fukuoka lasted 287 days, from May of 1977 to March of 1978. Seven cities, eight towns, and one village near Fukuoka also experienced serious effects of drought in the same period. As a result, city dwellers began to recognize the importance of water resource conservation and to understand the necessity of cost-sharing for that purpose. At the same time, some stakeholders upstream expressed apprehension about the possible deterioration of water resource conservation functions, knowing that manmade forests in the water reservoir area, which had been planted during the late 1940s and 1950s, would soon be ready for final cutting. They started to call for an extension of the final cutting age and ask for restitution for forest owners to compensate for the reduced timber income that this measure would cause.

Against this background, the Fukuoka prefectural government established a new fund for forest improvement for water resource conservation, focusing on forest management, with funds raised by collecting contributions from municipalities in the prefecture.

3.2.2 Implementation

The prefecture designated forest stands around major dams, as well as existing and planned protection forests, as WRC forests and provided subsidies for forest improvements based on a given standard. In addition, it provided a subsidy on interest for loans required in relation to longer cutting periods of *sugi* and *hinoki* manmade forests, which were extended by 10 to 20 years. In sum, the system is a kind of public cost-sharing for forest management on the principle of "assumed harmony," i.e., that forest management focusing on timber production will lead to water resource conservation. The projects funded include the purchase of forests, studies, research, information dissemination, and employment projects.

The area assigned as water reservoir area in the initial phase was 86,426 hectares, or 43 percent of the area of private forest in the prefecture. In the first phase, from 1979 to 1983, 1.8 billion yen were spent to implement forest improvement operations on 64,485 hectares, or 75 percent of the designated water reservoir area. Thinning operations reached 30,726 hectares in total during the first phase, at 7,681 hectares per year, equivalent to about six times the annual average of 1,300 hectares before the establishment of the fund. In the second phase, the project implemented forest improvements on 60,124 hectares, spending 1.43 billion yen. In the succeeding two phases, the results of the project were down by half, and the expenditures also declined to around 0.5 to 0.6 billion yen in each phase.

3.2.3 Local Involvement and Benefits

During the initial phase after the fund's establishment, parties involved in forestry upstream spoke highly of the fund. Records show that forest owners welcomed the subsidies for forestry improvement for water resource conservation, because they helped cover the cost for forest management at times when the timber market was in a downturn. They also expressed satisfaction that downstream dwellers had started to recognize the importance of the forests and forest owners in the headwater areas as key actors in forestry and water resource conservation. In addition, because the projects were implemented through local forest owners' cooperatives, they enhanced employment opportunities and income by stimulating forest management and timber production. The main expenditures of the fund were subsidies for thinning operations in private forests and interest subsidies relating to the extended cutting periods.

3.2.4 Current Status and Problems

As in other areas in Japan, the condition of forest management in the water reservoir area are worse today than when the fund was established. Such factors as persistently low timber prices and rising management costs due to rising wages have made it more difficult for private owners to realize adequate profits through proper forest management, harvesting, and reforestation. The aging of forest workers and the shortage of successors are also results of the recent declines in performance of forest management. Amidst this background the prefecture launched new projects to enhance forest workers' skills and encourage new entries in 1994 and 1998, but forest improvement for water resource conservation has not recovered its earlier vitality.

3.3 Kanagawa Water Resource Forest Conservation Project (KWRFC)

Aiming to strengthen such environmental functions as water and soil conservation and promote proper forest management, Kanagawa Prefecture designated the headwaters of major dams in the prefecture as water reservoir areas and launched a major project to promote public management and support for private forests in the water reservoir areas, starting in 1997 and lasting until 2020.

3.3.1 Background

During the mid-1990s, the prefecture designed a forest management program named the Kanagawa Forest Plan and created three types of forest zones: recreation and amenities, water resource conservation, and forest ecosystem reserves. It also declared a policy to realize sustainable forest management by strengthening public support for private forest management in the water resource conservation zone and promoting public management in the zone for forest ecosystem reserves.

In 1996, the prefecture carried out a cursory survey on the actual state of forest degradation in the water resource conservation zone. It revealed that nearly half of the manmade forests within the water reservoir area were in bad condition and required proper attention. This situation originated in such complex causes as prolonged low timber prices, escalating management costs due to rising wages, an increase of absentee landowners, and the aging of forest owners. The prefecture concluded that private forest owners were virtually incapable of properly managing their forests on their own. Based on this analysis it started to consider public management and more active support for private forests in the water reservoir area, and introduced these concepts into a new project in 1997. The speed with which decision was made to launch the project was probably a result of the added impetus of two water shortages that struck Yokohama in April and June 1996.

3.3.2 Implementation

In total, 60,779 hectares in thirteen municipalities located in the water reservoir areas of major dams in Kanagawa Prefecture were designated as target areas for the project. A sum of 8.4 billion yen was earmarked for the project from 1997 to 2001, consisting of around 5.3 billion yen from the prefecture's general budget and 2.5 billion yen from its waterworks budget.

The project employs four key means to achieve forest improvements for water resource conservation: grants based on written agreements to cover a part of the cost of forest improvements, public management of young stands based on profit-sharing contracts, public leasing and management of stands based on agreements, and purchasing of forest stands for the purposes of water resource conservation. The condition of manmade forests in the water reservoir areas has improved thanks to operations such as pruning and thinning. The work has also induced the creation of three types of forest stand believed to be preferable for water resource conservation: multi-storied stands, long rotation stands, and mixed forests of coniferous and broadleaved species. Secondary broad-leaved woods are also being leased based on agreements and managed through operations such as the cutting of climbers (vines), selective thinning, and planting.

By the end of 2001, about 4,300 hectares of forest stands, amounting to around 10 percent of the private forests in the target area, had been signed up with the prefecture under this project, and a total of 2,700 hectares had undergone forest improvement operations. More recently, the pace of signing agreements has slowed somewhat in comparison, but forest improvement operations have increased significantly in area to 766 hectares in 2000 and 790 hectares in 2001, compared with 300 hectares per year in 1997 and 1998 (Table 2).

In addition to these forest improvement operations, the project has promoted various types awareness-raising activities. It has sponsored such year-round programs as tree planting and silvicultural work for citizens, and so far about 16,000 citizens have participated in these events.

Aiming to cement relations with groups who benefit from the water reservoir areas, the prefecture introduced a water resource conservation partnership program, in which beneficiaries such as industrial firms and community groups join in conservation activities by making a fixed donation for more than five years to support forest improvements in water reservoir areas. Participating groups are allowed to display their name in a designated forest stand and to use that forest as a base for conservation activities. As of December 2001, twenty groups, including Kanagawa Toyota Motor Sales and Kirin Brewery, had participated in the program.

3.3.3 Local Involvement and Benefits

Public management and support are key aspects of the projects and instruments such as profit-sharing, purchasing, and leasing of forests have been applied, with the aim of decoupling ownership and management of private forests in the water reservoir areas. There is little room to take into account individual owners' wishes regarding forest management, since it is more important that the operations be in line with the goals of enhancing water resource conservation.

Table 2: Actual achievements of the Kanagawa Prefecture Forest Conservation Project for water reservoir areas.

	1997	1998	1999	2000	2001	Total
Area contracted (ha)	872	1,261	779	692	684	4,288
- by purchasing	565	817	506	435	464	2,787
- by leasing	300	188	255	245	217	1,205
- by donation	7	255	18	12	2	294
Area improved (ha)	283	332	506	766	790	2,677
Road construction (m)	2,181	2,071	3,875	3,717	3,425	15,269

Source: Developed by author based on Web site of Kanagawa Prefecture.

In compensation for these constraints placed on them, forest owners who join in the projects are rewarded with a handsome economic allowance. For example, in the agreement for broadleaf forest improvement they can receive 27,000 yen per hectare every year during the term of the contract. In addition, the prefecture pays all costs for forest improvement. In the case of profit-sharing, 1.2 million yen per hectare are paid to a forest owner as the payoff for reforestation costs, and the return to the forest owner from final cutting is around 40 to 70 percent of the wholesale price. In the case of agreements to convert forests to multi-storied stands through purchasing, leasing, or profit sharing, the trees growing in the understorey are provided to landowners without any compensation in return when they are cut. In terms of the cooperation agreement for forest improvement for water resource conservation, the prefecture tacks a new subsidy onto any exiting subsidies, and thus contractees pay almost nothing for forest management operations.

Forest improvement for water resource conservation in the project also contains contracted forest management work, including site surveys, forest improvement operations, and monitoring of project sites. These tasks provide stable local hiring conditions, mainly for the local forest owners' cooperatives.

3.4 Performance of Water Conservation

3.4.1 Water Supply

Generally speaking, there are not enough examples to judge whether upstream water conservation resulted in stable and good quality water supply in downstream. The performances of water conservation in three case studies are also not clear. Kanagawa, including Yokohama, has experienced drought two times from 1996 to 2000 and is quite a contrast to the situation in Tokyo. Such differences might have occurred partly due to the upstream water conservation in Kanagawa. In contrast to Kanagawa, Fukuoka has experienced frequent droughts and the recurrence interval is very short, as with Tokyo (Table 3).

Many surveys indicated a growing recognition that the quality of drinking water has worsened, and recent significant increase of mountain spring water is obvious proof. Although Japan's consumption of mountain spring water in liters per capita is still small compared with Europe and the United States (Japan, 8.91; United States, 64.3; United Kingdom, 16; France, 16; and Italy, 138), it has increased more than four times in the last decade, from 2.26 liters in 1991 to 9.81 liters in 2001.

The quality of drinking water in the three case study areas do not have serious problems, but many residents have recognized the deterioration of it. A questionnaire survey by Kanagawa Prefecture in 2002 revealed that more than 40 percent of people use water purifiers for their drinking water and around 70 percent were unsatisfied with drinking water quality.[1] A questionnaire survey conducted by the Yokohama City Water Supply Agency in 2000 showed that around 33 percent of citizens use a water purifier for their drinking water and only 26 percent drink water without any conditioning such as boiling, purifying, etc.[2] There is no similar survey for Fukuoka, but the deterioration of drinking water due to frequent droughts has often led to serious questions.

3.4.2 Commitment of Downstream Dwellers

The understanding and commitment of people in Kanagawa Prefecture has grown significantly these days. According to the survey mentioned above, more than 80 percent of citizens recognize the forests upstream as water resource forests. Most people expressed willingness to accept a cost burden for water conservation. The survey by the city of Yokohama in 2000 also indicated that 13 percent to 42 percent of respondents requested improvements such as water quality conservation, safe and stable water supply, and water resource area conservation.

As for Fukuoka Prefecture, there is not a similar survey, but a questionnaire survey carried out in 1997 indicated growing expectations of water conservation as a key policy measure in the area, with 33.1 percent support from respondents.[3]

4. ANALYSIS AND DISCUSSION

4.1. Background of Establishing Cost-Sharing Systems

It is important to point out that existing cost-sharing systems for forest improvement for water resource conservation in Japan were not established on a robust and dependable assessment of the functions of water resource conservation and the contribution of forest management. The systems that exist today are the outcomes of

---

[1] The prefectural government surveyed 3,000 adult residents and the ratio of available respondents was 68.9%.

[2] Of 4,000 adult residents in Yokohama surveyed, the ratio of available respondents was 59.5 %.

[3] Of 5,720 adult residents in 22 municipalities in Fukuoka surveyed, the ratio of available respondents was 88.6%.

interplay between issues and expectations—held by people upstream and downstream—relating to forests that are in water reservoir areas. A common thread in all three case studies was the emergence of the issue of security about water resources, felt downstream due to such problems as declining water quality and severe water shortages. At the same time, declining forest conditions in the managed forests of all three water reservoir areas were also serious concerns. The systems in Fukuoka and Kanagawa were established without a hitch, because forest owners upstream were motivated to cooperate, as they were finding that timber production alone was no longer playing an adequate role as a driving force in the sustainability of the local economy. This change led to the smooth establishment of a cost-sharing system between upstream and downstream stakeholders.

Table 3: Frequency and interval of droughts from 1996 to 2000 in study areas and metropolitan Tokyo.

Municipality	**Frequency of droughts (1966–2000)**			**River basin**	**Year system introduced**
	Total	**After system introduced**	**Mean recurrence interval (years)**		
Yokohama	2	2	29	Sagami	1916
Kanagawa				Sagami and Sakawa	1997
Fukuoka	8	5	4	Chikugo	1979
Tokyo	1	1	-	Tama	1911
	13	-	2.3	Tone	2002

Source: Developed by author based on the Web site of the Land, Infrastructure and Transportation Ministry, Japan.

The actual design of the systems depended on the extent to which the stakeholders were motivated to avoid uncertainty in water resource conservation, as well as the financial capacities of the downstream municipalities (Izumi 1996). In circumstances where affluent downstream municipalities are confronted by serious water problems and have control of only a small water reservoir area, it is natural that they would seek to purchase forests in headwater areas and embark upon direct management of the forests, aiming to reduce the risks to their water supply. In such cases, there is probably less resistance to change the principle of "assumed harmony" between forestry for timber production and water resource conservation, as one can see with the YWRCF. On the other hand, if the water reservoir area is large and the financial capability of downstream municipalities is limited, measures such as subsidies and grants based on a fund could be employed, as in the case of Fukuoka, and it is probably more difficult to depart from a forest management goal based on "assumed harmony" between the two activities.

In summary, the existing cost-sharing systems for forest improvement for water resource conservation in Japan have been shaped by key elements, which include the water demand and financial capabilities of downstream municipalities, and the size of forest tracts in water reservoir areas, in addition to the characteristics of the stakeholders involved.

4.2. Local Benefits of Cost-Sharing System

The social and economic impacts of cost-sharing systems on people and communities within water reservoir areas have both positive and negative sides.

Positive effects include social and economic benefits that flow directly or indirectly from forest improvements for water resource conservation, as shown in the case studies. Among them, the most important economic benefits are the creation of stable employment and the improvement of forest quality as an economic asset. As shown in the YWRCF case study, when the social infrastructure is developed in a suitable way, one can expect a significant increase in tourist income. In terms of social benefits, it is notable that city dwellers have started to recognize the importance of water resource conservation and the role of forest owners in maintaining the functions of the water reservoir areas.

Negative effects of these cost-sharing systems include certain constraints on timber production. Even greater constraints would further reduce the timber production function and consequently lead to a greater shift in forest management goals that had originally been based on "assumed harmony," as seen in the case of the YWRCF. It is possible that such changes would further exacerbate the decline of local timber production activities. In addition, when the system employs purchasing and leasing of forest land as major instruments, the alienation of local people from their forests may emerge as an issue.

4.3. Key Areas for Improvement

Addressing the issues of whether, and if so, how, and to what degree there is truth that forests are a good conservation medium for water is a point for consensus building in downstream communities. The efforts of water resource conservation in Fukuoka do not show as clear in terms of effectiveness in drought prevention as in Kanagawa and Yokohama. Requests for a stable and safe water supply have grown steadily in the three areas, in spite of water resource conservation programs upstream. It seems that water quality has deteriorated. Peoples' concern and their willingness to support water conservation upstream have grown gradually. So far, policy decisions in this field appear to depend on unquantified myths, but soon rational accounts and perspectives will be requested. Thus, long-term monitoring and information on the benefits to water should be built into the system in order to justify the price of conservation.

When cost-sharing systems are applied to water resource conservation in the ways described above, important economic and social benefits arise for the local people and community, as well, through the internalization of external economic factors in forest management.

Thanks to these benefits to both downstream and upstream municipalities, such cost-sharing systems have great potential to be institutionalized in various regions. For example, Kochi Prefecture in Sikoku region has established a unique local taxation system for water conservation. Kanagawa Prefecture is also attempting to introduce a similar local tax in addition to the KWRFC. Gunma Prefecture, which is a major water reservoir area for metropolitan Tokyo, has established new subsidies in 2002 aimed at activating forest management with an emphasis on environmental functions.

When cost-sharing systems are being designed, a number of points regarding policy measures should be taken into consideration.

First, there should be flexibility in the design of supporting measures for the cost-sharing system, in order to respond to actual conditions both upstream and downstream.

Second, policies should build in peripheral measures such as labor security, the collectivization of forest operations, cost reductions in forest management and harvesting operations, and logging plans that ensure stable profits from timber production at the same time as sustainable forest management. These measures are often necessary, because the options of purchasing and leasing all the forests in a water reservoir area are not feasible in most regions.

Third, in the case that the goal of forest management is shifted from the principle of "assumed harmony" between timber production to non-timber production emphasizing water resource conservation, it is important to consider ways of breathing new life into the local economy. As mentioned, policy instruments such as purchasing and leasing of forests in water reservoir areas may lead to the alienation of local people from their forests, thus the development of new relationships between the local people and the forests, based on the forests' environmental functions, are essential. The efforts of the YWRCF provide a useful reference in this regard. More broadly, it is also desirable to ensure that funds are available for infrastructure development and the promotion of unique tourism features of the area; these will bring benefits to the local people. It is important to conserve the local charm and unique natural environment as the bedrock of policies and measures.

5. CONCLUSION

The rapid growth of water demand and frequent water shortages, due to the ever-increasing concentration of populations in large cities in Asia, have become serious issues. At the same time, forest loss progresses at a rapid pace, often leading to serious flooding in some regions. Water and soil conservation are thus sure to be key issues of forest policy in countries that have experienced a serious loss of forests in recent decades.

In the context described above, the analysis of such conditions could serve as a tool for rational consensus building, and the involvement of local people could play an important role in discussions about cost-allocation issues in forest management for water resource conservation in the Asia-Pacific region.

ACKNOWLEDGEMENTS

The Doshi Water Resource Conservation Forest Management Office of the City of Yokohama and the department for forest improvement for water resource conservation promotion of Kanagawa Prefecture provided documents and information concerning their activities for the case studies. The author expresses deep gratitude for their kind cooperation.

REFERENCES

Clawson, M. and J. L. Knetsch. 1966. *Economics of outdoor recreation.* Maryland: John Hopkins Press.

Izumi, K. 1996. Cost allocation issues in the management of forests in headwater areas (in Japanese). *Transaction of Japan Forestry Society* 107:73–76.

Kumazaki, M. 1977. *Forest use and environmental conservation: Basic philosophy of forest policy* (in Japanese). Tokyo: Nihon Ringyo Gijutsu Kyokai.

Kumazaki, M. 1981–1982. History of downstream cooperation on reforestation in water reservoir area: An approach to an issue on cost-sharing (in Japanese). *Suiri Kagaku* (I)(III):140–142.

Kobayashi, H. and A. Kono. 1991. Forest management in cooperation with upstream and downstream municipalities: Case study on interchanges between Doshi Village in Yamanashi Prefecture and Yokohama City in Kanagawa Prefecture (in Japanese). *Kaiho* 338:15–18.

Yokohama Water Supply Agency. 1996. *Printed explanation of ninth management plan for Yokohama City's Doshi Water Reservoir Forests* (in Japanese). Yokohama Water Supply Agency.

CONCLUSION: SUSTAINABLE FOREST MANAGEMENT THROUGH LOCAL PARTICIPATION— PROCEDURES AND PRIORITY PERSPECTIVES

Makoto Inoue
The University of Tokyo, Graduate School of Agricultural and Life Sciences
Institute for Global Environmental Strategies

1. INTRODUCTION

In its first phase (1998–2001) the IGES Forest Conservation Project (FC Project) aimed to identify principles or elements of sustainable forest management that would account for an important portion of an effective forest conservation strategy, based on experiences in the Asia-Pacific region. The research was carried out by four interrelated sub-teams which dealt with (1) structural analysis of forest destruction (ST sub-team), to provide basic information to other sub-teams; (2) participatory forest management policy (PM sub-team), to make recommendations covering local and national levels; (3) timber trade policy (TT sub-team), to make recommendations covering national and international levels; and (4) legal and administrative measures for forest conservation (LA sub-team). Their work was to elaborate principles and elements of sustainable forest management as a final outcome of the project. Target countries were Indonesia, Thailand, the Philippines, Laos, Vietnam, China, and Russia (Far East region).

As shown in Part 1 of this book, the ST sub-team reconfirmed that the leading root causes of forest destruction include an insufficient base of local participation and community rights, the impacts of market forces, a forest development paradigm with an industrial emphasis, and economic and political challenges. The PM sub-team analyzed and compared existing participatory forest management systems in Southeast Asian countries, aiming to clarify their characteristics, and placed them into categories based on their main actors, the legal status of forest land, and activities. The sub-team then made policy recommendations through the examination of internal and external constraints on participation. The TT sub-team mainly conducted time-series economic analysis (TEA) of the timber trade of both exporting and importing countries in the Asian region, as well as data collection for space-equivalent analysis (SEA) of the timber trade. The LA sub-team focused on international legal measures relating to forest conservation, international processes of policy dialogue on forest issues, and domestic legal and administrative measures relating to participatory forest management. The sub-team elaborated the principles and elements for sustainable forest management in cooperation with other sub-teams.

M. Inoue and H. Isozaki (eds.), *People and Forest—Policy and Local Reality in Southeast Asia, the Russian Far East and Japan, 337–356.*

The project successfully constructed a valuable network with researchers, non-governmental organizations (NGOs), local people, and government officials in the Asia-Pacific region. These interpersonal relationships with project members evolved into beneficial inter-organizational relations in the second phase of IGES research (2001–2004).

Because the major output of the project was to be a set of principles or elements of sustainable forest management, the main target group in the first phase was to be governmental authorities. Although we invited governmental officials to a series of regional workshops held in Jakarta, Vientiane, and Khabarovsk to discuss and examine our draft strategies, including policy recommendations, these efforts did not seem to be enough for the project to have an influence on the national forest policies in each country.

2. ON-GOING ACTIVITIES IN THE SECOND PHASE OF THE FOREST CONSERVATION PROJECT

2.1 Goal and Objectives

The goal of the Forest Conservation Project is to develop strategies for forest conservation and sustainable forest management. There are admittedly many possible approaches that should be taken into consideration to achieve this goal. We decided to focus on ways to promote the participation of local people in forest management—in ways that could help to alleviate poverty and achieve sustainable forest management at the same time. Thus, the development of the following guidelines and recommendations became the objective of our work:

1. **Village action guidelines** for villagers and other stakeholders at the village level.
2. **Local policy guidelines** for local (provincial or district) governments and other stakeholders at the local level.
3. **National policy recommendations** to ensure the effective application of international treaties on local participation in forest management at the national level. Local governments as well as stakeholders at local and national levels can also make full use of these recommendations.)

2.2 Two Approaches and their Integration

The research is being carried out by using two interrelated approaches. One is the "local approach" to elaborate *village action guidelines*, in which the focus of attention moves from the village level to the local level, and then to the national government level. We apply a methodology we call *participatory action research*—a process of inquiry through which people work together to bring about an improvement on issues they consider relevant. The other methodology is the "international approach" for elaborating *national policy recommendations*, in which the focus of attention moves from the international level to the national and local government levels. These two approaches are combined and synthesized in discussions at the local government level, especially in the process of elaborating *local policy guidelines*. Figures 1, 2, 3 and 4 illustrate the relationships between these elements.

2.3 Target Countries

In the Forest Conservation Project's second phase, the study targets three countries: Indonesia, Laos, and Russia (Far East). The significance of the strategic policy studies for these three countries lies in the political characteristics of each country and the actual state of their forests and forest management. There appears to be a relatively high possibility of our research outputs being applied in these three countries.

Indonesia is a country in transition to democracy. Specific features concerning the forestry sector are (1) the reforms of forest policy, in accordance with the concepts of participation and decentralization; (2) chaos, as a result of drastic decentralization; and (3) rapid deforestation and degradation of its forests, due to various factors such as forest fires, large-scale plantations, logging, and slash-and-burn agriculture.

Laos is a country in transition to the market economy. Specific features concerning the forestry sector include (1) insufficient measures and mechanisms to implement new forest laws, and (2) the recent reaction to and trend of efforts to transfer the responsibility of forest management to the local people.

Russia has undergone some reforms in national policy, but this country is also in transition to a market economy. Moreover, its boreal forests offer the use of a unique approach in the mode of participation compared with the tropical forests of Indonesia, Laos, and other tropical countries. Specific features concerning the forestry sector include (1) the dynamic reform of forest policy, (2) the strong impact of Asian countries on the management of local forests, and (3) changes occurring in local forest management, along with changes in the national economic system.

2.4 Expected Policy Impact

2.4.1 Impact of the Guidelines and Recommendations

The results of the project in the second phase of research of the IGES Forest Conservation Project, consisting of *village action guidelines*, *local policy guidelines*, and *national policy recommendations*, are expected to have impacts at both the local and national levels, as described below.

2.4.1.1 Influence at the Local Level

The activities of local forest management and local policymaking processes will be improved by applying the village action guidelines and the local policy guidelines. There is every possibility for the local people and local governments to apply these guidelines, because the local people, local NGOs, and local governments, as well as local researchers, would be considerably involved in the research process from the beginning, and a series of local workshops will be organized.

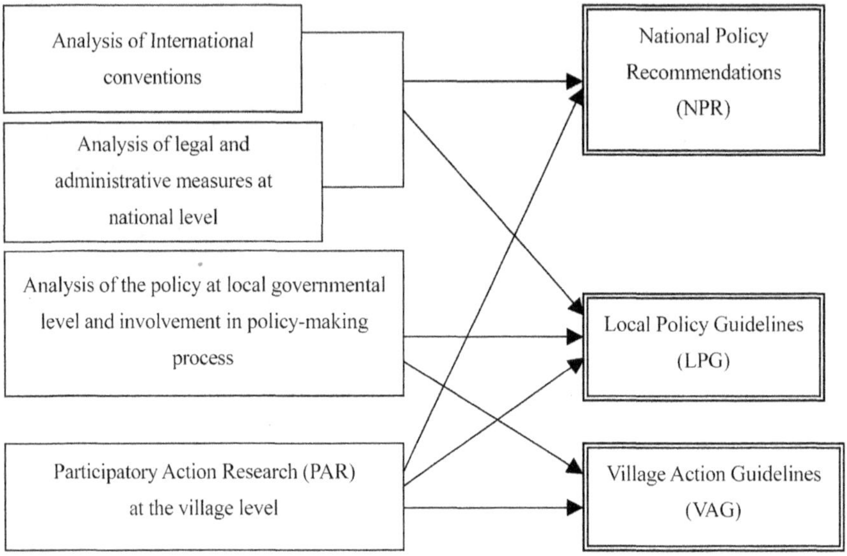

Figure 1: Overall process to develop three outputs of the Forest Conservation Project.

These two sets of guidelines will be presented at international conferences. This effort will provide opportunities for the local activities to be supported by the international community.

2.4.1.2 Influence at the National Level

The results of this project to date can have implications for national forest policy in two ways. First, the *local policy guidelines* can be used by local people and local governments in discussions with their national governments; we expect that the guidelines will be useful to them in order to understand international obligations that their national governments should fulfill. Interpretation of international and national policy to the local governments and the local people is important, because they typically have only limited access to the policy-related information on the international arena and are often not equipped to analyze the intricacies of international conventions that affect forests.

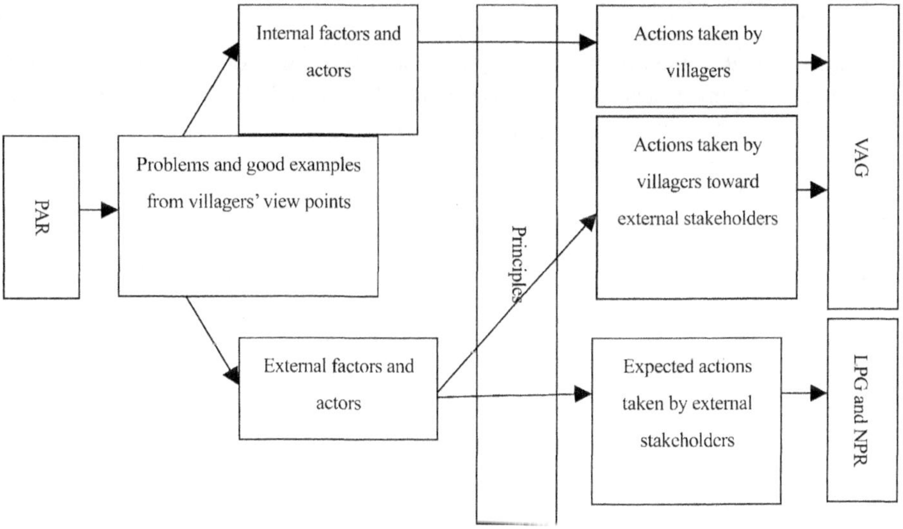

Figure 2: Concrete process to develop village action guidelines (VAG), local policy guidelines (LPG), and national policy recommendations (NPR) through the local approach.

Second, the *national policy recommendations*, derived from international treaties, can be applied in the process of policy reformation. Such top-down influences should be ensured through the involvement of high government officials in the process of developing recommendations.

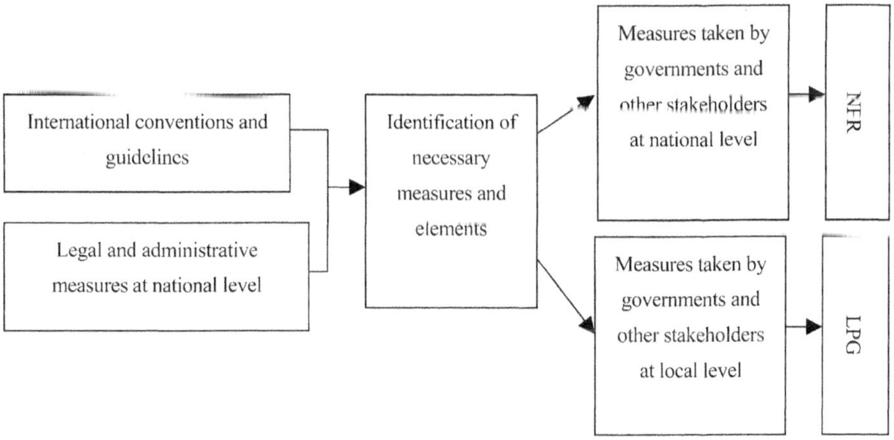

Figure 3: Concrete process to develop national policy recommendations (NPR) and local policy guidelines (LPG) through the international approach.

2.4.2 Collaborative Organizations

In order to realize the policy impacts mentioned above, we collaborated closely with the relevant organizations in each country, after concluding memorandums of understanding to strengthen inter-organizational relations.

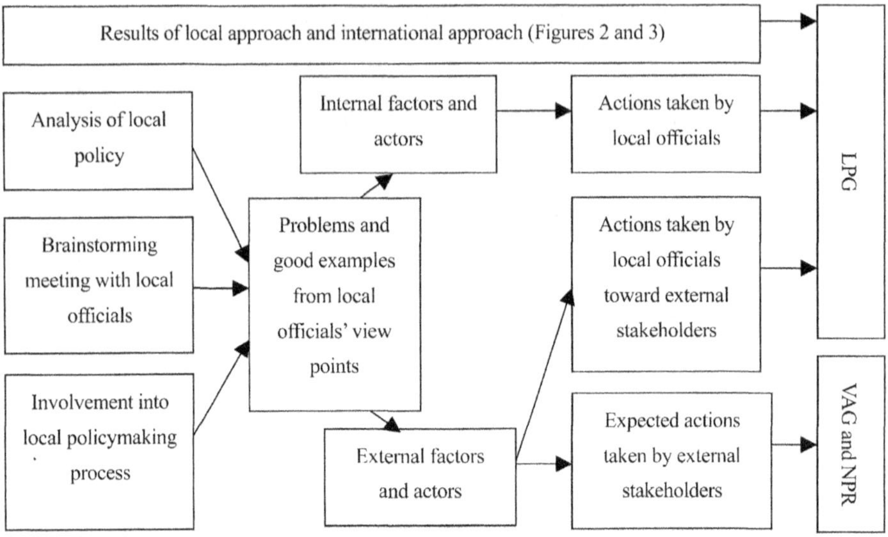

Figure 4: Integrated process to develop local policy guidelines (LPG).

3. IMPORTANCE OF PARTICIPATORY ACTION RESEARCH (PAR)

Here we focus on the local approach applied in the second phase of the project (and still on-going).

3.1 Paradigm Shifts in the Field of Rural Development

3.1.1 The First Paradigm Shift: The Emergence of Rapid Rural Appraisal (RRA)

Traditional methodologies to understand target areas as a basis for development planning consist of conventional short-term **appraisals** by development specialists, and orthodox **research** by academics, including broad or superficial surveys based on questionnaires, and detailed or in-depth fieldwork based on participant observation.

Both methodologies have been found to have certain shortcomings, as described below (Chambers 1983).

1. Conventional **appraisal** is regarded as a "quick-and-dirty" method; it takes only a short time to reach a conclusion. Cost-effectiveness is quite low, however, because it is unable to reveal the actual condition of the area, on account of the following six biases:

    - Bias towards place—The tendency to focus on limited areas of outskirts and along paved roads.
    - Bias towards existing projects—The tendency to ignore other places.
    - Bias towards informants—The tendency to get in touch with the local élite (men and special users) and ignore the poor.
    - Bias towards the dry season—The tendency to ignore the conditions in the rainy season.
    - Bias generated from politeness and shyness—The tendency to refrain from frankly questioning informants.
    - Bias towards special fields—The tendency to neglect other fields.

2. Orthodox **research** is regarded as a "long-and-dirty" method. It takes a long time to reach a conclusion—participant observation itself requires the researcher to stay a long time at the research site for fieldwork. Moreover, a lot of data collected through a survey is often not utilized in the process of analysis. In such cases, it is considered to be "dirty," even though it reveals the actual condition of the areas.

In the late 1970s and the 1980s, the *rapid rural appraisal* (RRA) method was developed, and was considered to be a "fairly-quick-and-clean" method. In this case, "clean" means highly cost-effective. Outsiders or development specialists pay attention to the local and indigenous knowledge. There are two important principles of RRA compared to the conventional "top-down" and "teaching-to-the-people" approaches (Project PLA 2000).

1. Reversal of learning—Development specialists apply the approach of "learning from the people" in a fairly short period of time.
2. Tools for quick and step-by-step learning—Participation of the local people in the decision-making process, application of visualized tools, optimal ignorance, and triangulation are important.

The first paradigm shift was thus characterized by the slogan of "putting the last first" (Chambers 1983). The application of sociological and anthropological viewpoints (Cernea 1991) to understand the reality perceived, not by the specialist but by the people, seems to be placed in this context.

3.1.2 The Second Paradigm Shift: Evolution to Participatory Rural Appraisal and Participatory Learning and Action

Under the first paradigm shift, the local people were used simply as sources of information and were exploited by development specialists. As it became clear that attaching importance to local knowledge was not enough, the second paradigm shift emerged, characterized by the slogan of "putting the first last" (Chambers 1997). This applied the approach of "learning with the people."

In this approach, outsiders such as development specialists and academics are expected to function as facilitators or catalysts to assist the process of empowerment of the people. In the late 1980s and the 1990s this approach was mainly developed by NGOs as *participatory rural appraisal* (PRA). This is a people-centered revision of *rapid rural appraisal* and even uses the same tools.

But the term *participatory rural appraisal* (PRA) is likely to cause misunderstanding. This is because only an appraisal utilizing various tools was stressed in the process of PRA, although the principle of "putting the first last" is essential. Sometimes PRA was applied simply as a tool to collect information and data for academic research. It seems that the term *appraisal* may not be suitable for the current situation in the field of developmental assistance.

This is the reason why we use the term *participatory learning and action* (PLA) (Project PLA 2000). In the process of PLA, the role of the outsider is limited to facilitation of the process in order to create a new institution and sustainable action by the people or leading actors.

3.1.3 Participatory Action Research as the Other Side of Participatory Learning and Action

Participatory action research (PAR) is defined as follows:

- The central idea of *participatory action research* is that the people who are to take action must also be involved in the research process from the very beginning (Chein et al. 1948). The survey is then planned and carried out jointly by responsible people in the community and social scientists with experience in all the technical details of community surveys.
- Participatory action research is a process of inquiry through which people work together on issues they consider relevant in order to bring about an improvement (Emadi et al. 1992).

Participatory action research has been especially popular in Latin America, and is aimed at changing socio-economic conditions and empowering the people (Chambers 1997). It seems that participatory action research, including a process of planning, acting, observing, and reflecting with maximum participation of the people, is parallel to and the other side of *participatory learning and action*. When focusing on an aspect of the research activities, the term *participatory action research* is appropriate. The term *participatory learning and action* is appropriate when focusing on an aspect of development practice. The implementation of participatory action research will contribute to the larger context of village development.

Hence, a field researcher makes a real contribution to the people he or she visits. In many cases, the local people expect the researcher to contribute to the development of the village, not just to collect information. Such an expectation creates pressure on the field researcher as an outsider. Participatory action research can release the field researcher from that pressure in a positive way. In contrast, conventional fieldwork and surveys rarely made a direct contribution to the people.

It is clear that the involvement of the people in making guidelines should be considered part of the long process of participatory learning and action, and the guidelines must be useful to the people, as well as to the local government, because it is based on their own experience.

3.2 Framework: Participatory Action Research in the Process of Developing Village Action Guidelines

3.2.1 Conceptual Framework

The implementation of *participatory action research* contributes to the larger context of village development. It is clear that the involvement of local people in making guidelines will be considered as part of the long process of *participatory learning and action* (Project PLA 2000), in which the role of the outsider is limited to facilitation of the process in order to create a new institution and sustainable action, with the local people being regarded as leading actors.

The long process of participatory action research can be conceptualized in five phases: (1) the initiation phase, which includes preparation and investigation to understand the present situation and identify problems; (2) the planning phase, including negotiation and preparation for action; (3) the implementation phase, in which the local people take action; (4) the monitoring and evaluation phase, in which the actions related to forest management are examined; and (5) the observation and reflection phase, which is carried out in connection with all the other phases (Figure 5).

The local people and researchers implement most of the activities in each phase to establish and improve participatory forest management jointly at the local level. The researchers then identify points to pay attention to when elaborating the *village action guidelines* in cooperation with the local people.

3.2.2 Key Concepts

Key concepts of *participatory learning and action* (Project PLA 2000) are applied as with that of *participatory action research* as follows:

- Local knowledge and the reality perceived by the people

- Various stakeholders such as the local people, local government, NGOs, and academics
- Learning process where the stakeholders construct a relationship of mutual trust
- Flexible use and application of appropriate tools (see Box 1)
- Facilitation to create an opportunity to build and realize consensus
- Trust and partnership between the people and facilitators
- Cyclical and dialogical process of verification and reflection
- Optimal ignorance and multi-sectoral viewpoint
- Triangulation to ensure the diversity of viewpoints
- Visualization to allow more people participate in participatory action research more easily

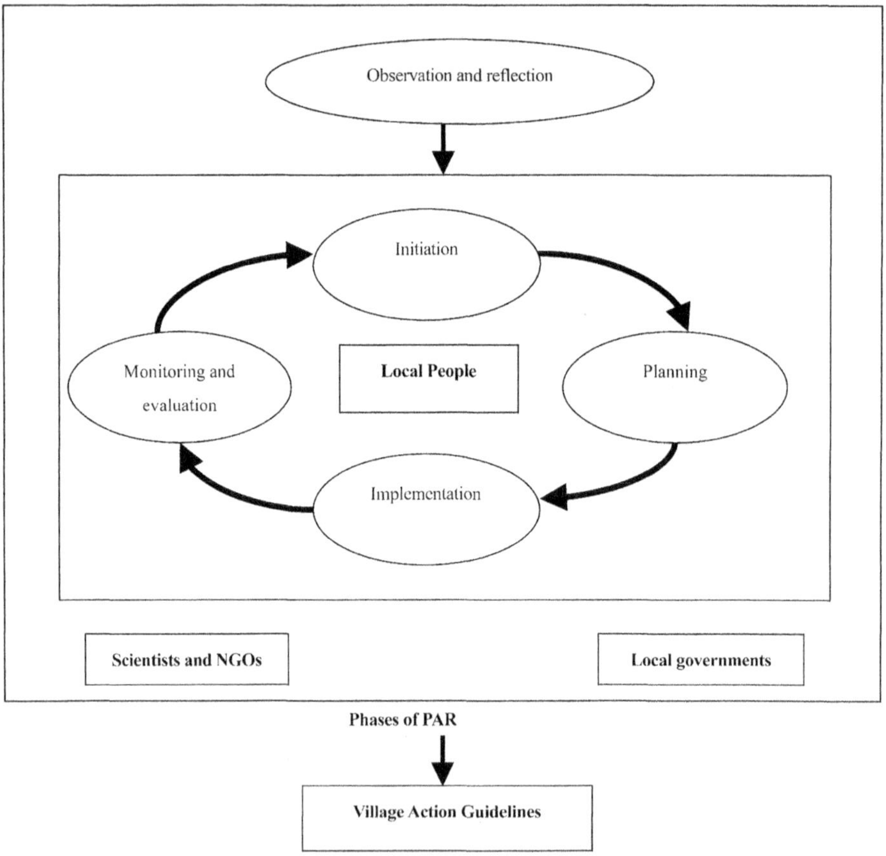

Figure 5: Conceptual framework of participatory action research (PAR) and guidelines.

Box 1: Tools for Participatory Learning and Action and Rapid Rural Appraisal.

Typical tools for PLA (or RRA) listed below can be applied as tools for PAR. The core team can use a flexible combination of the tools based on the principle of triangulation.

1. Tools related to dialogue

- Informal interview—Open-ended interviews such as those taken while standing and chatting or walking.
- Semi-structured interview—Interview with a checklist of key questions prepared in advance.
- Focus group interview—Including focus group discussion.
- Key informant interview—Interviewing the key informant about specific issues.
- Direct observation.
- Structured interview—Interview with a questionnaire. (Be careful: structured interviews often neglect the reality of the people.)

2. Tools related to space

- Mapping—Drawing maps of village overview, natural resources, land utilization, etc., perceived sensuously and mentally by the people.
- Three-dimensional model—Making a three-dimensional model of the village territory. (This may be difficult for our project because of the lack of technical capacity.)
- Transect—Walking around the village territory to observe, discuss, and record various issues.

3. Tools related to time

- Chronological table—Making the table through discussions.
- Daily routine table—Making the table of an informant whose daily routine is regarded as typical within a certain group. The pattern of the routine depends on the season, gender, and social group.
- Seasonal calendar—Making the calendar through group interviews.
- Vision drawing—Drawing a picture of a vision in the near future and discussing it with each other.

4. Tools related to social structure

- Venn diagram—Listing all the organizations that exist in the village, drawing diagrams to show inter-relationships, and discussing the function and the importance of the organization. Perception may differ among social groups.
- (Role-playing and flow charts are not so important for our project.)

5. Tools related to order

- Pair-wise ranking—Ranking made collectively.
- Matrix ranking—Ranking through aggregation of each villager.
- Wealth ranking—Identifying the criteria of the wealth and grouping the villagers by using cards.

6. Tools to establish rapport

- Joint working—Helping the villagers with farming, festivals, etc.
- Recreation—Relaxing with villagers.
- Eating together—Eating together is important in the case where the core team can not stay at a villager's house.

7. Utilization of secondary data

- Aerial/satellite pictures—Very useful.
- Statistics and publications—Collaborative organizations are requested to compile the publication.

3.2.3 Activities of Each Phase

3.2.3.1 Initiation Phase

First, the research sites should be selected. It is expected that two categories of research sites will be proposed: (1) villages in which the local NGOs have already started certain activities to support the people, and (2) villages where the people have the will to take action towards participatory forest management, but there has been no support from outside until now. The selection of the research sites is an important task for collaborative organizations that are working in cooperation with IGES.

Second, a small group of volunteers from the villages who are aware of the situation and want to improve it should be recruited from each village. They are candidates to form a "core team" with researchers (Opatpatanakit 1998). The core team of each village consists of a few villagers and a few researchers.

Third, the researchers can make full use of the information obtained in the process of the four phases for their research by mutual consent with the people.

Fourth, the core team facilitates the sharing of an understanding between the villagers and researchers of the following items:

- history of the village
- changes in production system
- social structure and social group
- village resources
- customary forest management system
- local marketing system
- development project and activities coming from outside
- facing problems
- causes of problems
- expected solutions of problems
- need for guidelines

It is very important for the core team to create an atmosphere for the people to express their opinions freely, without self-censorship from being sensitive to the government.

Fifth, the core team identifies the differing perceptions of various actors about participatory forest management and the concept of sustainability by means of in-depth interviews. The problems perceived by the local government (or local forest service) and the need for guidelines should also be identified.

3.2.3.2 Planning Phase

First, the core team facilitates the participants in taking the following actions:

- demarcation and mapping of the target forest area

- consensus-building of the demarcation among the villagers and between the villagers and other stakeholders
- definition of the people having rights to manage the forest
- formation of a management body
- making the rules to manage the forest, including monitoring
- discussion of the possibility to apply and improve the existing local practices
- clarification of the rights and duties of each villager
- examination of the need for external support
- planning action

Second, the core team assists the management body in submitting documents to the local government.

Third, the core team facilitates to provide an opportunity for negotiation and collaborative action between the villagers and other stakeholders (Emadi et al. 1992).

Fourth, the core team monitors the process of inspection conducted by the local government.

3.2.3.3 Implementation Phase

The core team assists the villagers with the necessary arrangements for tools and equipment, and in tackling the difficulties that arise during the implementation process.

3.2.3.4 Evaluation Phase

The core team facilitates for the villagers to evaluate their actions for forest management.

3.2.3.5 Observation and Reflection Phase

The core team facilitates for the villagers to observe, reflect, search out solutions, conduct necessary investigations, and revise the plan to improve the situation.

4. PRIORITIZED POINTS OF VIEW FOR LOCAL PARTICIPATION

Here, we would like to propose the following prioritized points of view that are important for the promotion of participatory forest management.

4.1 Degree and Type of Participation

4.1.1 Review of Existing Typologies of Participation

Several typologies of participation have been proposed in the past, as shown in Table 1, where the focus of discussion has been the degree of transferal or devolution of power to the local people, and this was adopted as a criterion to classify participation.

First, we should review the "ladder of participation," as elaborated by Arnstein (1969). There are eight steps in the ladder. This typology is logical and definite because the criterion of classification was clearly stated: it is "the extent of citizens' power in determining the end product." But "manipulation" and "therapy" should be excluded from the typology of participation, because Arnstein himself regarded them as non-participation. Taking account of the present situation in the tropics, "citizen control" in forest management is beyond our scope for the time being, although it is very important, even in developed countries.

The United Nations Development Programme (UNDP 2001) adopted "levels/degrees of participation." In this typology, "manipulation" should be excluded. It seems that further classification between "consensus-building" and "partnership" is not necessary, because the difference between "decision-making" and "risk-sharing" is very subtle.

Wilcox (2001) suggested "levels/stances of participation" to outsiders for offering increasing degrees of control to the others involved. In this typology, it is hardly possible for us to consider whether the stance of "acting together" is a higher-grade participation than the stance of "deciding together." In some cases "deciding together" will be higher-grade participation; in other cases it will be "acting together."

Pretty (1994) proposed a "typology of participation" in which "participation for material incentives" is problematic. It should not be given the position of an independent type of participation, because the dimension of the material incentives is different from that of other types, and the material incentives are also accompanied with other types of participation, such as passive participation, participation by consultation, and functional participation.

Typology by Hobley (1996) corresponds to that of Pretty (1994). But the participation in information-giving of Pretty is deleted, and manipulative participation is listed, which is not in Pretty's types. This typology has the same problem as with Pretty's types in terms of "participation for material incentives." And "manipulative participation" should be excluded from the typology, because it is not regarded by him as participation.

Inoue (2000) categorized various types of participation into three levels in terms of the relationship between local people and external agents in the decision-making process. This is the simplest typology of participation. But it may be too simple for field staff to use as a guide to promote local participation. In such cases, the three levels of participation will be appropriate to be utilized as a frame of the typology.

Table 1: Degree, type, and level of participation in the field of development studies.

Ladder of participation (Arnstein 1969)		**Level/ degree of participation (UNDP 2001)**	**Level/ Stance of participation (Wilcox 2001)**	**A typology of participation (Pretty 1994)**	**Types of participation (Hobley 1996)**	**Level of participation (Inoue 2000)**
Manipulation	Non-participation	Manipulation			Manipulative participation	
Therapy						
Information	Degree of tokenism	Information	Information	Passive participation	Passive participation	Participatory top-down approach
				Participation in information giving		
Consultation		Consultation	Consultation	Participation by consultation	Participation by consultation	
				Participaton for material incentives	Participation for material incentives	
Conciliation		Consensus-building		Functional participation	Functional participation	Professional-guided participatory approach
		Decision-making	Deciding together			
		Risk-sharing	Acting together			
Partnership	Degree of citizen power	Partnership	Supporting independent community initiatives	Interactive participation	Interactive participation	Endogenous bottom up approach
Delegated power		Self-Management		Self-mobilization	Self-mobilization	
Citizen control						

4.1.2 Proposed New Typology

A basis of the new typology we propose is the "ladder of participation" defined by Arnstein (1969), because it is logically most consistent. Based on a review of this research, the lowest steps of the ladder, such as manipulation and therapy, should be deleted, as should the highest step of the ladder, citizen control. The proposed typology in our project is shown below.

1. **Informing**—The local people are simply told what has been decided and unilateral announcements are made by external agents such as administrators

and development specialists. It is a one-way communication in which the information flows to the local people with neither a channel for feedback nor power for negotiation.

2. **Information gathering**—The local people participate by answering questions posed by outsiders, such as researchers and development specialists, using questionnaire surveys or similar approaches. It is also a one-way communication in which the information flows from the local people. But the people do not have the opportunity to influence proceedings, as the findings of the research are neither shared nor checked for accuracy.
3. **Consultation**—This level entails two-way communication, and the local people are consulted, but analysis and decisions are made by external agents. The most frequent approaches to consultation are chaired meetings, where the local people do not contribute to the agenda, public hearings, and surveys. In such cases, it is just a window-dressing ritual.
4. **Conciliation**—The local people may be involved in decision-making, but this tends to be only after major decisions have been made by external agents. They may be simply placated. This is regarded as "functional" or "ceremonial" participation.
5. **Partnership**—The local people participate in joint activities and decision-making in all processes, such as appraisal and investigation, development of action plans, formation or strengthening of local institutions, implementation, and evaluation. Participation is a right, not an obligation to achieve a goal. Various responsibilities are often shared, e.g., through joint committees. It is also called "interactive participation."
6. **Self-mobilization**—Independent initiatives by the local people are realized while advised and supported by external agents. They retain control over decisions and resource use; external agents facilitate them.

Informing, information gathering, and consultation are labeled the *participatory top-down approach*, as defined by Inoue (2000); conciliation, as the *professional-guided participatory approach;* and partnership and self-mobilization, as the *endogenous bottom-up approach* (Table 2). Partnership and self-mobilization can be categorized as *collaboration*, which refers to a wider spectrum of cooperation among concerned parties than does the term *partnership*. In any case, the participatory top-down approach should be avoided, because it is usually considered by local people to be nominal and fake.

A matrix can be used to indicate relations between the typology of participation and the phases of action taken by the local people (Table 3). At first, the local people will clarify and evaluate actual conditions in terms of typology of participation in each phase and discuss desirable types of participation. For example, the local people will try to raise the degree at least to conciliation, when they can only take part in the initiation phase through information gathering. In the planning phase, the people are expected to identify the forest users, organize the management body, and decide on the decision-making process with the assistance of facilitators. In this case, an ideal degree of partnership will be achieved.

Table 2: Proposed typology of participation.

Degrees/types of participation	Principle	Frames of approach
1. Informing 2. Information gathering 3. Consultation	Tokenism	Participatory top-down approach
4. Conciliation	Functionalism	Professional-guided participatory approach
5. Partnership 6. Self-mobilization	Collaboration	Endogenous bottom-up approach

Source: Drafted by Inoue after discussion with core staff of the IGES Forest Conservation Project.

Table 3: Matrix indicating the state of participation.

Typology phase	Informing	Info. gathering	Consulta-tion	Concilia-tion	Partner-ship	Self-Mobiliza-tion
Initiation		● ——	———	—▶?		
Planning			● ——	———	—▶?	
Implementation						
Monitoring & evaluation						

Note: Arrows do not indicate actual conditions.

Generalizations cannot be made, however, because conditions vary from place to place, and the expected degree of promotion might depend on your will, as well as the policy framework in each country.

4.2 State of Human Networks and Connectedness, Membership in Social Organizations, and Social Norms and Values

Social resources, such as human networks and connectedness, membership in social organizations, and social norms that people draw upon in pursuit of their livelihood objectives, are called "social capital" (DFID 1999), even if it is not important to use the terminology in the field of development studies.

Human networks and connectedness increase people's trust and ability to work together and expand their access to wider institutions such as political and civic bodies. Membership in more formalized organizations and groups often entails adherence to mutually agreed upon or commonly accepted rules, norms, and sanctions. Social norms and values usually function to sustain human networks and connectedness and membership in social organizations; human networks and the membership strengthen and reproduce the norms and values.

It is helpful to classify these social resources into two categories. The first is "bonding social resources," which strengthen the cohesion within the community, and the second is "bridging social resources" to connect the villagers with external stakeholders. Figure 6 shows target stakeholders. It is desirable that the local people construct good relationships among themselves (even between the élites and other villagers) and with external stakeholders, such as other communities, local government, local NGOs, scientists, local forest-related industries, and even with forest concessionaires with whom they have often been in conflict. The thick arrows in the figure show the most important human networks (or bridging social resources) to be constructed.

It is expected that the behavior of local governmental officials and forest concessionaires will be changed and improved through the construction of human networks and connectedness with the local people, although governmental officials tend to maximize the authority and budget and the forest concessionaires aim to maximize profit in general.

It is, however, an important fact that social resources have a downside. For example, social and peer pressure created by the networks and connectedness might prevent innovative people from initiating new participatory forest management systems. It is desirable that the local people reduce the use of such negative social resources.

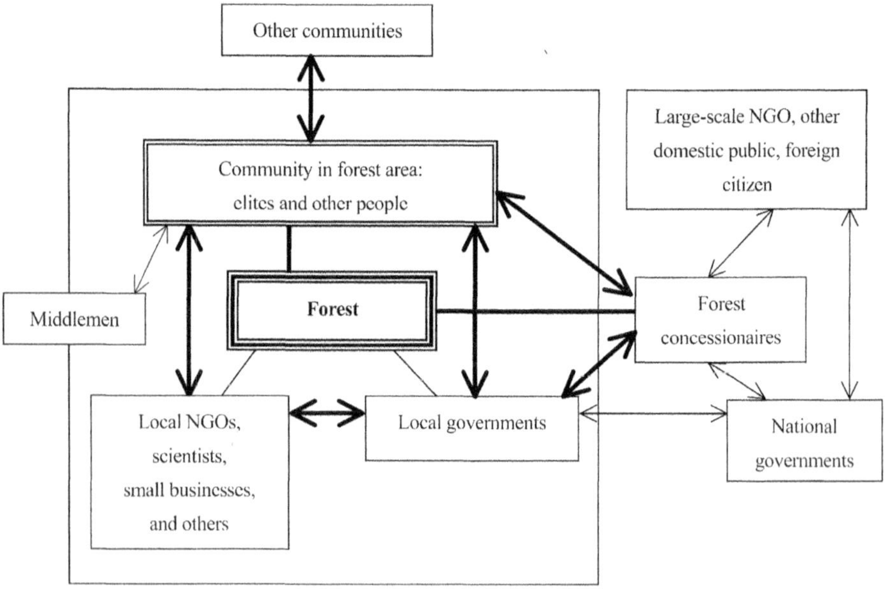

Figure 6: Main stakeholders with an interest in the forest.

4.3 Local Practices Evaluated in Terms of Sustainability

Local practices consist of established sustainable and unsustainable practices or customs. The gap between the "emic" (inner) and "etic" (outer) viewpoints stimulated us to conceptualize three categories of sustainable utilization (Inoue 1997; Inoue 2000) as shown below.

1. *Haphazard sustainable use* is defined as the mode of utilization where unconscious action brings about sustainable use of the resources.
2. *Incidental sustainable use* is defined as the mode of utilization where conscious actions for other main purposes achieve sustainable use.
3. *Intended sustainable use* is defined as where sustainable resource management is consciously carried out. Then some regulations are incorporated into customary law. For example, there were regulations on tools, season, and species in utilizing "Iriai-rin," or communal forests, in Japan.

Intended sustainable use can be applied as if it is a forest management system. Incidental sustainable use can be applied after devising countermeasures to ensure the existence of main purposes. Haphazard sustainable use should be applied after being recognized by the people that the use could be sustainable.

5. REMAINING ISSUES

We are now in the process of developing the *village action guidelines*, *national policy recommendations*, and *local policy guidelines*. In midst of this process, we have come to understand the importance of *incentive mechanisms* to materialize and actualize the participation of the local people in forest management—mechanisms such as the Clean Development Mechanism (CDM), labeling schemes, multi-purpose forest use (including eco-tourism, production of non-timber forest products [NTFP], hunting, etc.), and revitalization of forest related industries.

At the same time, in the process of *participatory action research,* the local people themselves recognized the importance of collaboration with other important stakeholders, such as the local governments, well-intentioned forest enterprises, NGOs, and academics. *Collaborative forest management systems* involving the local people, local government, timber concessionaires, and foreign companies, NGOs, and academics can be expected to help alleviate poverty of the local people as well as to prevent illegal logging and forest fires.

The potential for application of these guidelines to other villages and countries should be carefully examined. Any attempts at generalization should be done with discretion, however, because the local reality varies from village to village and country to country.

REFERENCES

Arnstein, Sherry. 1969. A ladder of citizen participation. *Journal of American Institute of Planners* 35:216–224.

Cernea, Michael, ed. 1991. *Putting people first: Sociological variables in rural development.* Second edition. Washington, D.C.: The World Bank Group.

Chambers, Robert. 1983. *Rural development: Putting the last first.* Harlow: Longman Scientific & Technical.

———. 1997. *Whose reality counts?* London: Intermediate Technology Publications.

Chein, Isidor, Stuart W. Cook and John Harding. 1948. The field of action research. *American Psychologist* 3:43–50.

Department for International Development (DFID). 1999. *Sustainable livelihoods guidance sheets*. London: DFID.

Emadi, Mohammad H., R. J. Fisher and Robert Woog. 1992. Action research as an approach to research and development among Iranian nomads. Paper prepared for the international conference "Nomadism and Development: Survival Strategies & Development Policies," 1–6 September, at Isfahan, Iran.

Hobley, M. 1996. Participatory forestry: The process of change in India and Nepal. In *Rural development study guide 3.* London: Rural Development Forestry Network.

Inoue, Makoto. 1997. Examination of tropical forests as commons: Based on positive survey in Kalimantan, Indonesia (in Japanese with English summary). *Journal of Environmental Sociology* 3:15–32.

———. 2000. Participatory forest management. In *Rainforest ecosystems of East Kalimantan: El Nino, drought, fire, and human impacts,* ed. Edi Guharidja et al., 299–307. Tokyo: Springer-Verlag.

Opatpatanakit, Avorn. 1998. Collaborative action research: An alternative methodology to create sustainable rural development. Paper presented for the CSF workshop on "Participatory Action Research," 4–9 May, at Center for Social Forestry, Mulawarman University, Samarinda, East Kalimantan, Indonesia.

Pretty, J. N. 1994. Alternative systems of inquiry for sustainable agriculture. *IDS Bulletin* 25(2): 37–48.

Project PLA. 2000. Introduction to social development: Participatory learning and action (in Japanese). *International Development Journal* 2000.

United Nations Development Programme (UNDP). 2001. Empowering people: A guide to participation. <http://www.undp.org/csopp/paguide.htm> (December 2001).

Wilcox, David. 2001. The guide to effective participation. <http://www.partnerships.org.uk/guide/main1.html> (December 2001).

AFTERWORD AND ACKNOWLEDGEMENTS

The first phase of research of the Forest Conservation Project (Hiroji Isozaki, Project Leader) of the Institute for Global Environmental Strategies (IGES) focused on analyzing various topics critical to forest conservation, including the underlying causes of deforestation, timber trade policy, participatory forest management, and legal and administrative measures. Most of the countries in the Asia-Pacific region were targeted for research; this is why we were able to comprehensively analyze the state of forest-related issues, and concluded that the participation of the people living in proximity to forests is a key component for sustainable forest management. The exclusion of local people from the management of forests, which has long been the basis of forest policy in many countries, has exacerbated the situation.

In the second research phase of the Forest Conservation Project (Makoto Inoue, Project Leader), we are concentrating our research on the *participation of the local people* whose livelihood depends to a large extent on forest resources, with a special focus on three target countries: Indonesia, Laos, and Russia (more specifically, the Far East region). The outputs of the Project's second phase (see Conclusion chapter of this book) will be applicable to other regions, as well, for the promotion of participation of local people in forest management. This local participation can be expected to help achieve poverty alleviation and sustainable forest management at the same time, when one recognizes that timber and non-timber forest products (NTFP) have historically been important income sources for the local people.

In the process of developing these guidelines, we came to better understand the importance of incentive mechanisms for materializing and actualizing the participation of local people in forest management. At the same time, in the process of participatory action research (PAR), the local people themselves recognized the importance of collaboration with private companies and support from the local governments, NGOs, and academics. Ideally, the financial, physical, and human capital held by private companies should be used in a way that is benefit to both the local people and the management of forests.

More follow-up is needed on these concepts. Subject to the approval of the IGES board of directors and trustees, the research plan for the Forest Conservation Project that I drafted with Project staff will likely delve further into these concepts during the third phase of research (fiscal years 2004–2006).

All papers in this book were peer-reviewed, and then revised based on the reviewers' comments. We believe that this process helped to ensure the quality of each chapter, although we regret that some papers had to be dropped in the process. We would like to express our sincere gratitude to three reviewers in particular: Professor Dinah Shelton, of Notre Dame Law School, University of Notre Dame; Dr. Hwan Ok Ma, Project Manager at the International Tropical Timber Organization (ITTO); and Dr. Mafa Chipeta, Chief of Policy Coordination, Technical Assistance Department, Food and Agricultural Organization (FAO).

We thank Professor Akio Morishima, President and Chairman of the board of directors of IGES, Mr. Shinichi Arai, former Secretary General of IGES, Ms. Saho

Moriguchi, former secretary of the project, Ms. Soko Fujimori, present secretary of the project, and other staff of the secretariat for their kind assistance of our research activities. We extend our heartfelt appreciation to Dr. Paul Roos and Betty van Herk of Kluwer Academic Publishers for their generous guidance and support for this publication. We are also deeply grateful to Randal Helten, for coordinating the preparations for publication, and his assistants: Greg Helten, for proofreading, and Susan Juby, for layout. It would not have been possible to publish this book without the marvelous cooperation and help of all the individuals and institutions concerned with our project, and we would like to express our appreciation to all of them.

We hope that this book will be a valuable contribution for those involved in forest issues, especially academics, policymakers, and of course, the local people who depend on forests.

July 2003

Makoto Inoue, Ph.D.
Associate Professor of Forest Policy, The University of Tokyo
Project Leader of the Forest Conservation Project's Second Phase
Institute for Global Environmental Strategies

INDEX

9 781402 016844

GPSR Compliance

The European Union's (EU) General Product Safety Regulation (GPSR) is a set of rules that requires consumer products to be safe and our obligations to ensure this.

If you have any concerns about our products, you can contact us on ProductSafety@springernature.com

In case Publisher is established outside the EU, the EU authorized representative is:

Springer Nature Customer Service Center GmbH
Europaplatz 3
69115 Heidelberg, Germany

Batch number: 10809439

Printed by Printforce, the Netherlands